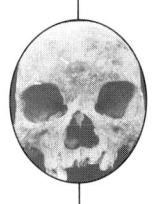

Fossil Primates

Reconstructing the paleobiology of fossil non-human primates, this book is intended as an exposition of non-human primate evolution that includes information about evolutionary theory and processes, paleobiology, paleoenvironment, how fossils are formed, how fossils illustrate evolutionary processes, the reconstruction of life from fossils, the formation of the primate fossil record, functional anatomy, and the genetic bases of anatomy. Throughout, the emphasis of the book is on the biology of fossil primates, not their taxonomic classification or systematics, or formal species descriptions. The author draws detailed pictures of the paleoenvironment of fossil primates, including contemporary animals and plants, and ancient primate communities, emphasizing our ability to reconstruct lifeways from fragmentary bones and teeth, using functional anatomy, stable isotopes from enamel and collagen, and high-resolution CT scans of the cranium.

Fossil Primates will be essential reading for advanced undergraduates and graduate students in evolutionary anthropology, primatology and vertebrate paleobiology.

Susan Cachel is Professor of Physical Anthropology at Rutgers University. She has been on the Executive Committee of the Rutgers Center for Human Evolutionary Studies (CHES) since 2010 and a member of the graduate interdisciplinary Quaternary Studies Program at Rutgers since 2000. She has taught and performed research at the Koobi Fora Field School in northern Kenya, and she is currently a research associate of the Kenya National Museums (Nairobi). She was recently elected a Fellow of the American Association for the Advancement of Science for "incisive contributions to hominization theory, the role of nutritional fat in human occupation of high latitudes, and primate evolution." Her previous title, *Primate and Human Evolution*, was published by Cambridge University Press in 2006.

Cambridge Studies in Biological and Evolutionary Anthropology

Consulting editors

C. G. Nicholas Mascie-Taylor, *University of Cambridge*
Robert A. Foley, *University of Cambridge*

Series editors

Agustín Fuentes, *University of Notre Dame*
Sir Peter Gluckman, *The Liggins Institute, The University of Auckland*
Nina G. Jablonski, *Penn State University*
Clark Spencer Larsen, *The Ohio State University*
Michael P. Muehlenbein, *Indiana University, Bloomington*
Dennis H. O'Rourke, *The University of Utah*
Karen B. Strier, *University of Wisconsin*
David P. Watts, *Yale University*

Also available in the series

53. *Technique and Application in Dental Anthropology* Joel D. Irish & Greg C. Nelson (editors) 978 0 521 87061 0
54. *Western Diseases: An Evolutionary Perspective* Tessa M. Pollard 978 0 521 61737 6
55. *Spider Monkeys: The Biology, Behavior and Ecology of the Genus Ateles* Christina J. Campbell 978 0 521 86750 4
56. *Between Biology and Culture* Holger Schutkowski (editor) 978 0 521 85936 3
57. *Primate Parasite Ecology: The Dynamics and Study of Host–Parasite Relationships* Michael A. Huffman & Colin A. Chapman (editors) 978 0 521 87246 1
58. *The Evolutionary Biology of Human Body Fatness: Thrift and Control* Jonathan C. K. Wells 978 0 521 88420 4
59. *Reproduction and Adaptation: Topics in Human Reproductive Ecology* C. G. Nicholas Mascie-Taylor & Lyliane Rosetta (editors) 978 0 521 50963 3
60. *Monkeys on the Edge: Ecology and Management of Long-Tailed Macaques and their Interface with Humans* Michael D. Gumert, Agustín Fuentes, & Lisa Jones-Engel (editors) 978 0 521 76433 9
61. *The Monkeys of Stormy Mountain: 60 Years of Primatological Research on the Japanese Macaques of Arashiyama* Jean-Baptiste Leca, Michael A. Huffman, & Paul L. Vasey (editors) 978 0 521 76185 7
62. *African Genesis: Perspectives on Hominin Evolution* Sally C. Reynolds & Andrew Gallagher (eds.) 978 1 107 01995 9
63. *Consanguinity in Context* Alan H. Bittles 978 0 521 78186 2
64. *Evolving Human Nutrition: Implications for Public Health* Stanley Ulijaszek, Neil Mann & Sarah Elton (eds.) 978 0 521 86916 4
65. *Evolutionary Biology and Conservation of Titis, Sakis & Uacaris* Liza M. Veiga, Adrian A. Barnett, Stephen F. Ferrari & Marilyn A. Norconk (eds.) 978 0 521 88158 6
66. *Anthropological Perspectives on Tooth Morphology: Genetics, Evolution, Variation* G. Richard Scott & Joel D. Irish (eds.) 978 1 107 01145 8

67. *Bioarchaeological and Forensic Perspectives on Violence: How Violent Death is Interpreted from Skeletal Remains* Debra L. Martin & Cheryl P. Anderson (eds.) 978 1 107 04544 6
68. *The Foragers of Point Hope: The Biology and Archaeology of Humans on the Edge of the Alaskan Arctic* C. E. Hilton, B. M. Auerbach & L. W. Cowgill (eds.) 978 1 107 02250 8
69. *Bioarchaeology: Interpreting Behavior from the Human Skeleton, 2nd Ed.* Clark Spencer Larsen 978 0 521 83869 6 & 978 0 521 54748 2.

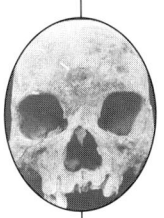

Fossil Primates

SUSAN CACHEL
Rutgers University, NJ, USA

CAMBRIDGE
UNIVERSITY PRESS

University Printing House, Cambridge CB2 8BS, United Kingdom

Cambridge University Press is part of the University of Cambridge.

It furthers the University's mission by disseminating knowledge in the pursuit of education, learning and research at the highest international levels of excellence.

www.cambridge.org
Information on this title: www.cambridge.org/9781107005303

© Susan Cachel 2015

This publication is in copyright. Subject to statutory exception and to the provisions of relevant collective licensing agreements, no reproduction of any part may take place without the written permission of Cambridge University Press.

First published 2015

Printed in the United Kingdom by TJ International Ltd. Padstow, Cornwall.

A catalogue record for this publication is available from the British Library

Library of Congress Cataloguing in Publication data
Cachel, Susan, 1949–
Fossil primates / Susan Cachel, Rutgers University, NJ, USA.
    pages   cm. – (Cambridge studies in biological and evolutionary anthropology)
Includes bibliographical references and index.
ISBN 978-1-107-00530-3 (Hardback) – ISBN 978-0-521-18302-4 (Paperback)
1. Primates, Fossil. I. Title.
QE882.P7C33 2014
569'.8–dc23   2014031792

ISBN 978-1-107-00530-3 Hardback
ISBN 978-0-521-18302-4 Paperback

Cambridge University Press has no responsibility for the persistence or accuracy of URLs for external or third-party internet websites referred to in this publication, and does not guarantee that any content on such websites is, or will remain, accurate or appropriate.

To my family:
Najpierw pobijemy a potem policzemy

Acknowledgments		*page* xii
Preface		xiii

1 Introduction: primates in evolutionary time — 1
Primates among the mammals — 9
Stratigraphy and rock units — 11
Geological time — 12

2 Primate taxonomy — 14
What is taxonomy? — 14
The Linnaean hierarchy — 15
Species definitions — 18
Variation — 25
Taxonomic methods — 27
Primate taxonomy — 31
Taxonomic categories — 34

3 Fossils and fossilization — 36
The origin of fossils — 36
Types of fossil materials — 36
Trace fossils — 43
Lagerstätten — 49
Taphonomy and taphonomic processes — 51
Can one trust the fossil record? — 54

4 The world of the past — 59
The origin of continents and oceans — 59
Climates of the past — 60
Habitat reconstruction — 62

5 The lifeways of extinct animals — 66
Introduction — 66
Body size — 67
Diet — 74
Inferring behavior from morphology — 81
   Locomotion — 81
   Ranging behavior — 84
   Temporal patterning of behavior — 86
Additional behavioral inferences — 88

6 Evolutionary processes and the pattern of primate evolution — 90
What drives evolution? Physical environment versus biological factors — 90
Natural selection and adaptation demonstrated — 93
Systematics, evolutionary trees, and homoplasy — 96
Evolution and development — 98
Comparative genomics — 101

7 Primate origins — 103
The Cretaceous world — 103
The Cretaceous/Tertiary mass extinction — 105
Defining primates — 106
Primate origins — 110
   Arboreal locomotion in small branches — 114
Comparative studies that shed light on primate origins — 116

8 The Paleocene primate radiation — 120
The plesiadapoid primates — 120
The adaptive zone of plesiadapoid primates — 127
The Paleocene/Eocene Thermal Maximum — 133

9 The Eocene primate radiation — 136
Primates in the High Arctic — 136
The first euprimates — 142
Adapoids and omomyids — 143
Prosimian descendants of the Eocene radiation — 152

10 The Malagasy primate radiation — 155
Lost Lemuria — 155
Malagasy natural history — 157
Colonizing Madagascar — 158
Taxonomic inflation in the living Malagasy lemurs — 162
The subfossil lemurs — 163
A peculiarity of Malagasy existence – extreme seasonality — 169
The human colonization of Madagascar — 170
Lemur extinctions and loss of disparity — 172

11 The Oligocene bottleneck — 174

12 Rise of the anthropoids — 179
Convergent origin of the anthropoids? — 182
The Fayum primates — 186
Position of the Parapithecoidea — 190
Other Fayum taxa — 192
Later catarrhine divergence — 193

13 The platyrrhine radiation — 195
Colonizing South America — 197
Platyrrhine diversity — 202
Inferring locomotion without postcrania in fossil platyrrhines — 205
Centers of platyrrhine diversity — 207
Fossil mammals from Patagonia — 208

14 The Miocene hominoid radiation — 214
Introduction — 214
Postcranial anatomy of the crown hominoids — 217
Discovery of the hominoid radiation — 220
Inferring lifeways in fossil hominoids — 223
    Niche structure in sympatric Early Miocene East African hominoids — 225
Ancestors for the living apes? — 228
Late Miocene extinctions — 229
    Oreopithecus bambolii — 233
Last survivor: *Gigantopithecus* — 237
The origin of hominids — 240
Hominoid locomotor adaptations — 241

15 The cercopithecoid radiation — 247
Plio-Pleistocene diversity and sympatry — 253

16 Late Cenozoic climate changes — 255

17 Conclusions — 259

References — 264
Index — 293

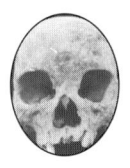

ACKNOWLEDGMENTS

The Rutgers Center for Human Evolutionary Studies (CHES) provided funding for original illustrations. This original artwork was cheerfully and eagerly done by two artists: Ms. Angela J. Tritz of Pittsville, Wisconsin and the University of Wisconsin, Stevens Point; and Ms. Irene V. Hort of Rutgers University, New Brunswick. Ms. Devin Ward of Rutgers University, New Brunswick, assisted with scans and other preparation of the artwork. Dr. Christopher Scotese, of the University of Texas, Austin, sent me high-resolution copies of plate tectonic maps of the ancient earth generated by the Paleomap Project.

Professor Dimitri Metaxas, Director of the Computational Biomedicine, Imaging and Modeling Center (CBIM) of Rutgers University, provided the space and equipment for the three-dimensional analysis of early human locomotion carried out by my advisee, Ms. Melanie Crisfield, and cited in Chapter 5. Professor Julia Lee-Thorp, Chair of the School of Archaeology, Oxford University, and Head of the Research Laboratory for Archaeology and the History of Art (RLAHA), provided an internship for my advisee, Mr. Renè Studer-Halbach, to study stable isotope analysis. His initial work on diet and inferring niche structure in sympatric Old World monkeys at the Pliocene site of Laetoli, Tanzania, is cited in Chapter 5.

My Rutgers colleague Professor Craig Feibel (Departments of Anthropology and Earth and Planetary Sciences) discussed questions about stratigraphy and site formation processes, and provided information about floating vegetation mats on lakes in northern and central Kenya.

The following Rutgers University graduate students helped me to elucidate explanations and examples: Susan Coiner-Collier, Melanie Crisfield, Stephanie Green, Sarah Hlubik, James Lister, Jay Reti, Lauren Saville, Darshana Shapiro, and Renè Studer-Halbach. The following Rutgers University undergraduates suggested improvements to parts of the text used in class: Nicolette Bronisevsky, Ralph Cretella, Lily Flast, Jennifer Giannini, Samantha Harrison, Morgan Hill, Katherine Kearney, Michael Kennedy, Anna Latka, Kara Lipinski, Marissa Lugo, Lawrence Lyons, Caitlin McCabe, Lindsay Modugno, Valerie Park, John Peters, Stephanie Ricciardi, Daniel Saldana, Kruti Shah, Jillianne Tiongko, Anthony Tricarico, Victoria Versprille, Hilary Veth, Nicolette Waksmundzki, Gandhi Yetish, David Zaitz, and Amadeusz Zajac.

PREFACE

While reading websites, blogs, newspapers, or popular magazines, one frequently encounters a statement like this: "The discovery of new human fossil X completely rewrites the textbooks!" Many editors would set this entire sentence in bold capital letters. Or, "New fossil primate is the first monkey ..." or, "New higher primate is the first human ancestor." Such hysteria has become a normal part of press hyperbole. One expects that virtually every new primate or human fossil will completely rewrite the textbooks. But is it true? Dinosaur paleontology also receives a great deal of attention from both the public and the press. Do new dinosaur fossils mandate a complete rewriting of the textbooks?

A study has been conducted on both Old World higher primates (catarrhines) and dinosaurs, testing to see whether new fossils result in a complete re-vamping of evolutionary history—that is, do new fossil finds repeatedly rewrite the evolutionary history of a group? Tarver et al. (2010) discover that this is not true for catarrhine primates over the last 200 years of study. The basic outline of catarrhine evolution has remained the same since the early twentieth century. New dinosaur fossils, on the other hand, do continually and radically shift our understanding of dinosaur evolutionary history. Many new lineages have been discovered, and new fossils expand our understanding of the geographic expansion of dinosaurs. Our understanding of dinosaur evolution changes rapidly and wildly. Yet, fossils of new catarrhine primates result in virtually no change in the understanding of their fossil record and evolutionary history. Clearly, the mass media is unduly fixated on catarrhine primates. The principal reason for this is that humans are catarrhine primates, and the merest scrap of a new human fossil generates hysteria in the popular press. This also reflects a funding bias. Funding agencies are more apt to focus on primate (including human) paleontology, than paleontological work on other animal groups. Dinosaurs are clearly an exception—major dinosaur research programs have been funded by private donations alone.[1] This is why a test of whether new catarrhine primate or dinosaur fossils truly do rewrite evolutionary history is important. As a physical anthropologist, I am irreverent in pointing this out: dinosaur discoveries trump those of primates in terms of the advance of knowledge. Why study primates at all? Is this just stubborn single-mindedness, or a simple exercise in human vanity?

Testing whether new fossils necessarily rewrite evolutionary history (Tarver et al., 2010) is important in a general sense. Our understanding of evolutionary patterns is not entirely dependent on the discovery of new fossils. This may be true for dinosaur history, which still has unknown dimensions. Primate history, on the

---

[1] For example, the research of Dr. John R. (Jack) Horner, Museum of the Rockies, Bozeman, Montana, on dinosaur paleobiology has been abundantly funded by private donations.

other hand, can be discerned from the fossils that we already know. Why should one write another book on primate fossils? A major reason is to establish that primates do, indeed, conform to the evolutionary processes that can be observed in other mammals. Traditionally, primate evolution is viewed as an inevitable progression from the lowest and least to the best of all. As T. H. Huxley first and famously phrased it, "Perhaps no order of mammals presents us with so extraordinary a series of gradations as this—leading us insensibly from the crown and summit of the animal creation down to creatures, from which there is but a step, as it seems, to the lowest, smallest, and least intelligent of the placental Mammalia" (Huxley, 1863:124–125). Huxley was mired in a deep debate about whether organic evolution had occurred at all, and can be excused some rhetorical flourishes. Since that time, however, many anthropologists and primatologists take the special status of the primate order as a given. The human-like or anthropoid primates, particularly the great apes, are especially revered, and debates now occur over whether they should be accorded the same legal rights as human beings. Yet, what does the fossil record show? Are primates subject to the same forces generating new species or determining species extinctions as other mammals? Have higher primates arisen independently in the Old and New Worlds? An overview of the primate fossil record immediately shows that major extinctions have occurred, including a recent major ape extinction. Some major researchers respond to this—I think indefensibly—by arguing that primate species have been continually expanding in number since the beginning of the order. Clearly, the prospects of primate extinction are emotionally disturbing. Other interesting questions arise: how many species of primates should one expect to see? How fast do primates evolve? Are climatic triggers important in primate evolution? What is primate niche structure like? What happens to primates isolated on islands? Do primates experience resource competition from their fellow primates? What place do primates have in community structure?

Another reason to examine fossil primates is that primates, along with birds, are often the only creatures still studied as whole animals in university curricula—veterinary schools excepted. The remainder of the animal world is now often reduced to the study of molecules, cells, DNA, or genes. Furthermore, as whole animals, primates are embedded in tropical ecosystems. Anyone concerned about the fate of these ecosystems and their preservation will be concerned about primates. Living primate species (the sifaka, the muriqui, the orangutan) often stand as heraldic figures that animate the worldwide fight for conservation of other endangered species or habitats.

An additional reason to study fossil primates is that they remain one of the last groups where traditional vertebrate paleontology can be taught and practiced. Many university geology departments have abandoned the teaching and study of vertebrate fossils altogether, except for teaching an introductory course on dinosaurs. Invertebrate paleontology has economic significance in stratigraphic analysis, dating, climatic research, and petroleum exploration—this will never be the case for vertebrate fossils, because they are too rare. Universities and museums are disbanding their vertebrate collections. Vertebrate curators are no longer needed. Many researchers fear that vertebrate paleontology will itself become extinct as a science. "I believe that the fate of the paleontologist is in jeopardy. Where are the specialists who will focus on the new fossils still to be recovered, and where will they be trained? Perhaps more importantly, where will they be employed?" (Reed, 2011:77).

One might think that primate history, anatomy, and morphology are immune to this trend, because of the medical importance of studying human anatomy and functional morphology. Unfortunately, many medical schools in the USA are abandoning the traditional study of gross anatomy through dissection. Instead, they increasingly rely on computer software programs that teach anatomy through simulated dissections. Thus, the fate of the human anatomist may also be in jeopardy. Should one applaud this? Can medical personnel be adequately trained without reference to cadavers? Can they appreciate the vast array of variation and variability that living humans encompass if they only study computer simulations of ideal human anatomy? In addition, physicians still receive virtually no training in evolution, and, for over 30 years, have roundly rejected the call for introducing it into basic medical science (Ewald, 1980). A later call for a "dawn of Darwinian medicine" never saw the sun rise on this endeavor (Williams & Nesse, 1991). Yet, the concept of evolutionary medicine, based on natural selection, adaptation, and population-level differences in humans, would revolutionize medicine by altering merely descriptive or mechanistic approaches to disease. Evolution is capable of offering powerful new explanations for alterations in human life-history variables such as growth, reproduction, and lifespan, as well as the onset of ageing, and disease.

In summary, although the focus of this book is primate evolution, I also intend it to be a resource for those interested in both exploring evolutionary processes, as well as the broad shape and pattern of mammal evolution. Along the way, I will emphasize the possibility and utility of studying function and behavior from the remains of fossil animals. I will reinforce the general importance of studying evolution and functional anatomy in biology, as well as in other disciplines, such as conservation biology and medicine.

Finally, I introduce a novel perspective on primate and mammal evolution by interjecting new research into the narrative on the genetic bases for anatomical shape and form. This new research, called evolutionary development, promises to unveil mysteries about the appearance of new anatomical structures. Without knowledge of evolutionary development, paleontologists could merely describe novel structures and compare them to similar structures in living animals. Paleontology was description and phylogeny, the attempt to reconstruct ancestor-descendant relationships between fossil organisms. The origin of anatomical innovation remained a mystery. But the genetic bases for new anatomical structures are now becoming clear. Evolutionary development is therefore becoming the linchpin between evolutionary processes affecting variation within populations and the grand procession of life as revealed by classic paleontology. Thus, there is continuity between evolution at the level of individuals in populations (microevolution), and evolution at the level of the origin of new families, orders, and classes of animals (macroevolution). This firm linkage between microevolution and macroevolution represents a break from paleontology during the 1980s and 1990s, when some major figures (e.g. S. J. Gould, N. Eldredge) argued that microevolution was caused by relatively weak processes like natural selection, and was decoupled from macroevolution, which was largely directed by unpredictable accidents and catastrophes.

The future importance of evolutionary development to human and non-human primate evolution is indicated by a major symposium mounted at the 2012 annual meeting of the American Association of Physical Anthropologists: "Finding our inner animal: Understanding human evolutionary variation via experimental model systems" (Young & Devlin, 2012). Papers presented at this symposium illustrate how morphology and adaptation can be studied through experimental comparative anatomy and comparative genomics. The way in which natural selection affects variation in the body size, teeth, skulls, and limbs of humans and non-human primates is examined through animal experimentation. The results are used to generate and test hypotheses about evolution. Thus, adaptation and evolution can sometimes be studied by experimentation. This counters the arguments of skeptics who state that the study of adaptation and functional morphology in fossils or in living organisms is nothing more than a series of ad hoc stories (Gould & Lewontin, 1979).

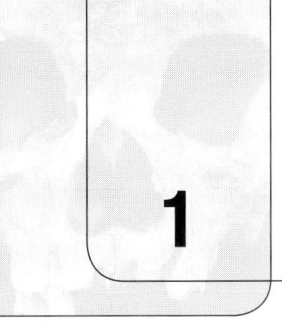

1 Introduction: primates in evolutionary time

A major goal of this book is to show that fossil primates (including fossil humans) fit within evolutionary patterns seen among other mammals. What is a mammal? Mammals are a class of animals. That is, they are technically arranged in the zoological Class Mammalia, which is formally characterized by a number of distinctive traits. With the exception of birds, most of the animals with which we are most familiar are mammals. There are about 5,000 living mammal species, although they comprise only about 5 percent of all known animal species. The term "crown species" is often used to refer to living species, in contrast to fossil species, because the living animals appear at the top or crown of an evolutionary tree. Mammals have a very ancient and complicated evolutionary history. Figure 1.1 presents a simplified version of this evolutionary past. A complete and continuous fossil record documents the transition from first reptiles to first mammals. Although a number of "mammal-like reptile" or proto-mammal groups independently evolved mammalian traits, the first creatures identified as true mammals emerge about 200 mya (million years ago). Living mammals occur in three main groups: the ancient monotreme mammals of Australasia, and the closely related placental and marsupial mammals, whose origins are much more recent.

Mammals are animals that have backbones or spinal (vertebral) columns, composed of individual vertebrae or vertebral bones. They are thus members of a major division of animals called vertebrates (Subphylum Vertebrata). Most mammals, the placental and marsupial mammals, give birth to live young. The exceptions are the monotreme mammals, the duck-billed platypus and the echidnas, or spiny anteaters. These monotreme mammals lay eggs. Mammals nurse their young, or lactate with milk produced from mammary glands, which are modified sweat glands. In monotremes, milk oozes through to the surface of a mother's fur, where it is licked up by the young. In other mammals, the young actively suckle milk, as they latch onto a discrete nipple or teat. Mammal hearts are four-chambered, and completely divided vertically into right and left halves. Mammals are endothermic, using physiology to maintain a stable internal body temperature, despite the changing ambient temperature. They possess hair, usually of different types (a fine undercoat and a longer outer coat), and specialized sensory hairs or whiskers (vibrissae) on the head. *Megaconus*, a mammal-like fossil from 165–164 mya, preserves a halo of guard hairs and denser undercoat hairs in fine sediments immediately surrounding the skeleton, although the middle ear and ankle region resemble those of mammal-like reptiles (Zhou *et al.*, 2013).

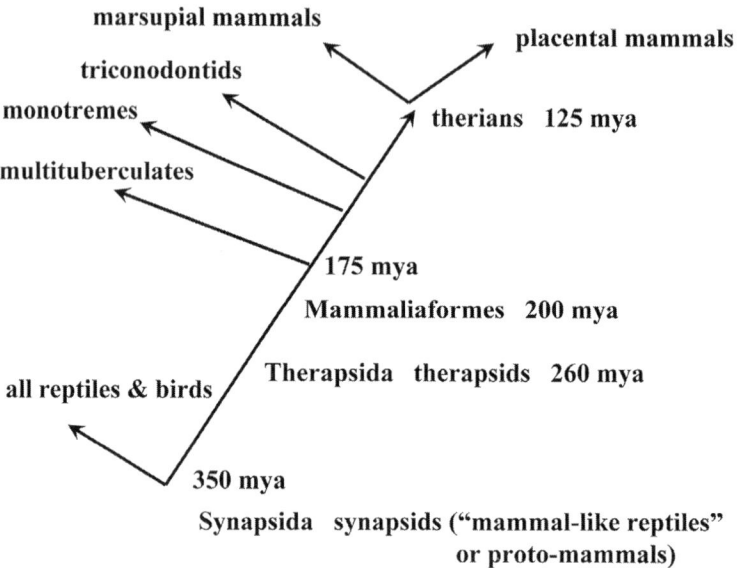

Figure 1.1. Mammal evolution. This figure illustrates the evolution of mammals from synapsids, and the origin of major mammal groups.

A furry pelt thus originated long before the true mammals appear. Unlike fish or reptiles, mammals have a distinct and restricted period of growth. They do not continue growing as long as they live.

Mammal teeth have a core of bone-like dentine encased in a fabulously hard coat of enamel. Enamel, which is largely calcium phosphate, is one of the toughest of organic materials. When the root cavity of a tooth seals up from the bottom, the tooth then stops growing. Some mammals, such as rodents, have teeth that grow throughout life. Mammals develop only two sets of teeth—a milk or deciduous set that is lost during juvenile life, and an adult or permanent set. Unlike the teeth of fish or reptiles, the teeth of mammals are heterodont. They have four different shapes, depending on their position within the jaws. From the front of the jaw to the back, these teeth are respectively called incisors, canines, premolars, and molars. Molar teeth are never replaced—they do not have deciduous versions. Milk or deciduous molars are, in fact, premolars. These different types of teeth perform different functions. Every mammal species has a characteristic number of teeth of each type. These numbers are summarized by a dental formula, which lists the number of teeth of each type in one half of the head. The basic placental dental formula is

$$\frac{I3.C1.P4.M3}{I3.C1.P4.M3}$$

where I stands for incisor, C for canine, P for premolar, and M for molar. Each tooth is numbered from front to back. A deciduous incisor tooth is indicated by the following convention: dI. Isolated teeth are identified by superscripts or subscripts. Thus, P^4 refers to the last upper premolar; P_4 refers to the last lower premolar.

The upper and lower molar teeth of mammals meet together or occlude. As a result, these molars can tear, crush, or chew or masticate food, depending on the shape of the molars.[1] As seen in Figure 1.1, marsupial and placental mammals are identified as therians. The ancestral therian possessed a distinctive form of molar tooth: a tribosphenic molar in which both upper and lower molars have three principal cusps, and the lower molar has a distinctive posterior talonid basin which occludes with the principal cusp, the protocone, of the upper molar. This molar shape allows therians both to crush food inside the talonid basin, and to shear food along the anterior vertical face of the basin. Figure 1.2 illustrates the basic therian molar pattern, from which the molar teeth of all marsupial and placental mammals are derived.

Mammals develop a secondary palate, a plate of bone that grows in from the edges of the upper jaws, and envelopes the original synapsid palate in a cylinder. The internal air passage opens at the back of the throat. Mammals can therefore eat and breathe at the same time. This is important, because mammals have a high metabolic rate—they need to process food quickly, in order to survive. They cannot hold their breath for a long time, while trying to ingest struggling, intractable prey, or a large bolus of food. The mammalian lower jaw is formed by a separate dentary bone in each half of the lower jaw, joined together by a symphysis in the middle. Alternatively, the symphysis may be fused, and the lower jaw is then a single, complete bone, called a mandible. The jaw joint is formed by an articulation between the dentary bone in the mandible and the squamosal bone in the cranium. There is a cheek bone (zygomatic arch), and ancient ribs in the neck region (cervical ribs) are lost or reduced. The ball joint (occipital condyle) that attaches the cranium to the first neck vertebra becomes a double ball joint in mammals, allowing the head to be more flexible and mobile.

Mammalian limb structure is entirely reorganized. The shoulder or pectoral girdle becomes stronger and more flexible, as does the pelvic girdle, which has more vertebrae fused to the hip bones. In therian mammals (marsupials and placentals), the separate coracoid bone of the shoulder girdle becomes fused to the shoulder blade to form the coracoid process. In addition, the ankle joint of therians is formed only between the tibia bone and the astragulus bone. There is no contact between the calcaneus ankle bone and the tibia. The limbs of mammals are held vertically under the body; they do not sprawl out to either side. Each digit of the extremities (hands and feet) has a characteristic number of phalangeal bones or phalanges. Some mammals walk with their feet flat on the ground, but most walk up on their toes or digits. In any case, the digits are shortened, and do not splay out sideways. Ribs are lost in mammals, until only an anterior rib cage

---

[1] There is a website on mammalian teeth that contains a large database of both living and extinct mammal species. It is accessible via the downloadable Morphobrowser web interface. This allows for 3D visualization, rotation, and scaling of tooth images, as well as allowing for the import and export of dental data. This website can be accessed via the following URL: http://www.biocenter.helsinki.fi/bi/evodevo/morphobrowser/.

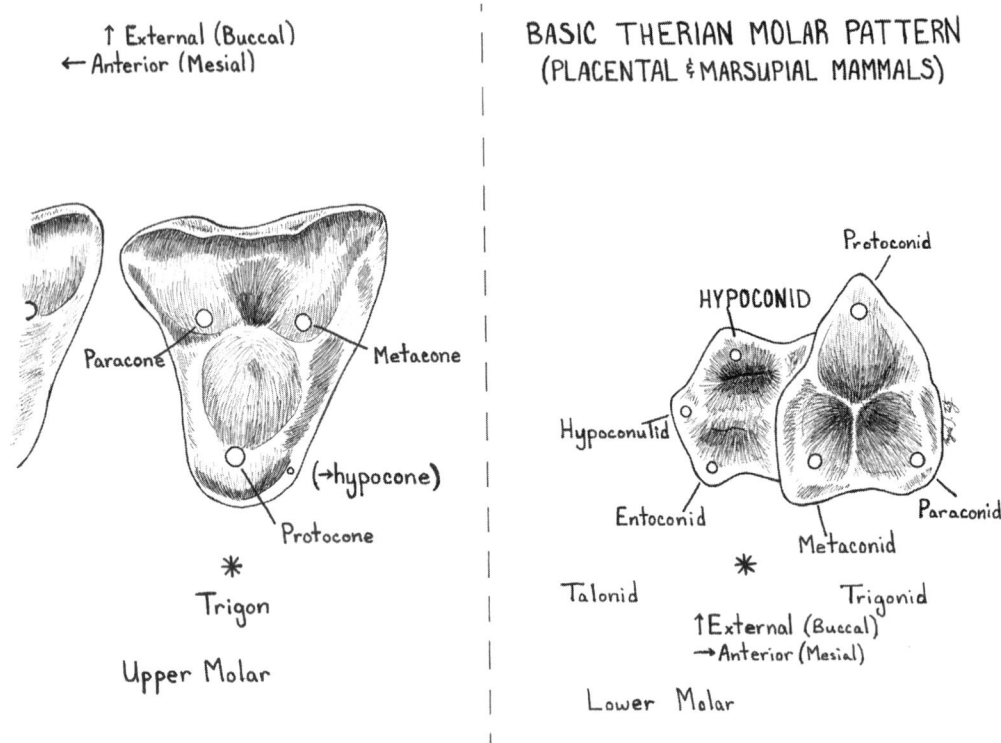

Figure 1.2. The basic therian molar tooth pattern. Note that the upper and lower molars are reversed. Photocopy this figure. Then fold it along the center line, hold it up to the light, and approximate the two stars in each half of the figure. The figure will then illustrate occlusion between the upper and lower molar teeth. Note that the principal cusp of the upper molar (the protocone) fits into the posterior basin (the talonid basin) of the lower molar. Illustration by Angela J. Tritz.

remains. The remaining ribs are locked together, forming a rather rigid thorax. Unlike the ribs of reptiles, mammalian ribs do not markedly expand and contract. Mammals therefore breathe in an entirely different fashion. They use a muscular wall (the diaphragm) that separates the lung and abdominal cavities to pump air in and out of the lungs through muscular effort. Figure 1.3 shows a familiar living mammal.

Mammal brain evolution has been studied in detail.[2] Because the first mammals were nocturnal, both the sense of touch and the sense of smell were accentuated, rather than vision. Each of the sensory vibrissae on a mammal's snout can move

---

[2] A website devoted to mammalian brain anatomy and evolution contains photos, atlases, and descriptions of the brains of more than 175 species of living mammals. These include the brains of living non-human primates. It can be found at the following URL: http://www.brainmuseum.org/.

(a)

Figure 1.3. This domesticated dog illustrates several traits that are characteristic of mammals. (a) Besides the normal hair of the pelt, vibrissae on the snout function for the sense of touch or the tactile sense. The importance of olfaction is shown by the long snout and the moist, naked nose or rhinarium. The teeth are heterodont. (b) The legs are vertical, and are placed directly under the body. As is true of many mammals, the dog is standing up on its toes or digits, illustrating a form of posture or locomotion called digitigrady. Digitigrady increases stride length, and therefore minimizes the cost of locomotion.

independently of the others, and, when an individual whisker touches an object, a mammal's brain records the location of the object and the pressure that it exerts on the whisker. The brain then integrates information from the array of whiskers to unveil the shape of the unknown object (Towal *et al.*, 2011). When the first mammals emerge 200 mya, their brains already contain relatively large olfactory lobes—areas of the brain devoted to the processing of olfactory information. Improved smell and tactile abilities occur concomitantly with superior coordination, and are followed by a further enhancement of olfaction (Rowe *et al.*, 2011). The ancestors of living mammals experience a third pulse in olfactory ability, along with an expansion of epithelial tissue lining delicate, scrolled turbinal bones in the snout. Reflecting the primacy of the sense of smell, olfactory receptor genes in living mammals comprise a major portion of the genome. In fact, the general augmentation of the mammal brain relative to body size is a gift of the olfactory sense (Rowe *et al.*, 2011:Fig. 3).

(b)

Figure 1.3. (*cont.*)

Mammals evolve from synapsids. In synapsids, the lower jaw is formed by multiple small bones. Two of these bones (the articular and quadrate) ultimately end up in the middle ear region of mammals, becoming the incus and malleus, respectively. This transformation is shown not only in fossils, but also during embryonic development. Thus, these cartilaginous embryonic jaw bones become bones of the middle ear in living mammals. The evidence of mammal evolution is seen not only in fossils, but also in embryology.

Multiple lineages or clades of synapsids independently acquired characteristic mammal traits. They acquired these traits in a mosaic fashion, not as a complete suite of traits that arrive all at once. Thus, during a particular point in geological time, a number of different synapsid lineages were experimenting with a mammal-like way of life. The living monotreme mammals seem to be the descendants of a separate lineage of animals that crossed over the threshold into mammal status. This is shown by paleontology, which demonstrates the different way in which bones of the monotreme middle ear are derived from bones of the synapsid lower jaw (Rich *et al.*, 2005). Thus, a key trait for mammals evolves independently between monotremes and other mammals. This trait is the transformation of bones in the synapsid jaw into the incus and malleus bones of the middle ear. It is likely that this separation from the lower jaw of bones that now function in mammalian hearing occurred independently in a number of other extinct lineages. A 120 mya fossil triconodont from China has a middle ear that is still not completely free from the lower jaw. This discovery implies that the definitive mammalian middle ear, fully separated from the lower jaw, evolved independently in three mammal lineages—the monotremes, the extinct multituberculates, and the therian mammals (Meng et al., 2011:Fig. 4).

The live-bearing, therian mammals (marsupials and placentals) are clearly distinct from the egg-laying monotreme mammals. The extinct multituberculates, the longest lived of any mammal group (Figure 1.1, Chapter 8), also appear to have given birth to live young, in contrast to the monotremes. Multituberculates had epipubic bones that supported a pouch, as in living marsupials (Jenkins & Krause, 1983:Fig. 2). Genome analysis further illustrates the distinct nature of monotreme mammals. The genome of the platypus, a monotreme, has its own unique traits. It has major differences from the genomes of both placental and marsupial mammals, and also shares some genetic features with reptiles and birds (Warren et al., 2008). At the genetic level, the platypus is as much a composite as its body is. Yet, genetic analysis reveals what fossils cannot: endothermy and lactation were acquired before other mammalian traits.

As mammals evolve, they demonstrate patterns in time and space that relate to origin, extinction, competition, and dispersion. Mammals respond to alterations in geography, global sea level, and climatic fluctuations. Through time, continents show a succession of faunas and floras. Some of these are distinctive and virtually unique; others are admixtures formed by the dispersal of species from separate areas of origin. Fossil primates are no different from other mammals. They conform to patterns of origin and extinction that characterize other mammals.

There are over 20 orders or major, formal zoological groups of living mammals. Among the orders of mammals is the Order Primates: the First or Top Animals, when this word is translated from Latin. This name reflects an old idea that humans are not only the natural end point of Nature, but they are the very reason why Nature exists. This idea argued that Nature is subordinate to humans, and only exists to fulfill human needs. Humans are outside, and above, Nature. Living and fossil humans are members of the Order Primates. Other members of this order are ape-like or monkey-like animals, or animals less well defined in the popular mind, such as the lemurs, lorises, and tarsiers. If humans are the nearest approach to perfection in the natural world, then animals resembling humans in either anatomy or behavior must also be of the utmost importance. This is one approach to the non-human primates—to consider them an inherently interesting and charming group of creatures whose study may illuminate questions about the origins of human anatomy, behavior, intelligence, and morality. The civilizations of India, China, and Japan have traditionally taken this approach. This is enhanced by the Hindu and Buddhist religions, which minimize the differences between human and animal. Thus, the macaque monkeys native to Japan (*Macaca fuscata*) appear in countless depictions that illustrate the Buddhist moral teaching to "See no evil, hear no evil, speak no evil."

Figure 1.4. Stela at the entrance of the Museum of Natural History, Oxford, memorializing the famous June 30, 1860 debate that took place there between Thomas Henry Huxley, Bishop Wilberforce and others on Darwin's *Origin of Species*.

There is an alternative viewpoint. In Western Civilization, primates were often considered to be tainted or ruined creatures precisely because their resemblance to humans appeared to mock the human condition. This view is ancient in the West, where the Greek word *pithekos* or "trickster" was applied to Barbary macaques and baboons, called "apes" by the old Greek scholars. Because the human mind and body were considered the zenith of the natural world—the gods themselves being merely exaggerated versions of mortal humans—human-like animals were especially revolting. This view of primates as ruined animals that mocked the human condition was continued in the Bestiaries written by Medieval Christian scholars, as well as in the traditional folklore of European nations. If non-human primates were peculiarly degraded animals, one might then regard them in the same way that the eighteenth-century English novelist Jonathan Swift used his invented creatures, the Yahoos, to depict the unique corruptions of humans. This view of non-human primates ultimately hindered the easy acceptance of the idea of human evolution for many people. They experienced a visceral disgust at the notion of being related to apes and monkeys. Thus, when a great public debate on evolution took place in Oxford, England, in June 1860, Bishop Samuel Wilberforce mockingly inquired of Thomas Henry Huxley whether he had apes on his grandfather's or his grandmother's side of the family. Huxley replied that he would rather have "a miserable ape for a grandfather," rather than an intelligent and influential man who subverted scientific discussion through the use of ridicule (Desmond, 1997:279). This retort (and Huxley had many others like it) effectively silenced the bishop and many other critics of evolution (Figure 1.4).

Primates among the mammals

Living primates are subject to intense scrutiny and interest, not only from scholars, but also from members of the general public. At the present time, much of this examination concerns conservation and the fate of endangered species. A continually updated database of all the living primates is maintained at the following website: http://alltheworldsprimates.org. Although the focus of the website is conservation, the website includes descriptions, photographs, and aerial satellite and normal road maps showing the ranges of all known primate species and subspecies identified by formal taxonomic and common names, as well as video and audio clips for about 16 percent of them. The database is relational, which allows for different modes of searching for information. However, users need to join Primate Conservation, Inc., and pay a steep fee in order to log on to the website, although institutional memberships exist.[3]

As the nineteenth century drew to a close, it became clear to most scholars that evolution was real, that human evolution had occurred, and that humans were certainly a member of the Order Primates. Yet, what is the relationship of this order to other orders of mammals? The traits that define primates, both living and fossil, will be discussed in Chapter 7. This definition is problematic, because primates appear to possess no unique keystone or signature features that ineluctably allow

---

[3] As an example of the kind of data that can be retrieved from this primate database, detailed information about time spent feeding, time spent feeding on different food types, and the nutritional

scholars to identify them. This is true even when living species are examined. Thus, within the twentieth century, tree-shrews (Order Scandentia), elephant-shrews (Order Macroscelidea), colugos (Order Dermoptera), and fruit bats (a subgroup of Order Chiroptera) have been identified as primates by eminent scholars. Yet, how can gliding colugos, flying bats, and tree-climbing primates be grouped together? The answer lies in convergent evolution, the independent origin of traits among lineages that are distant in ancestry. Thus, as discussed above, the middle ear of monotremes is convergent to that of placental and marsupial mammals. The close evolutionary relationship between tree-shrews, colugos, and primates is often now recognized by a taxonomic category called Euarchonta. A Grandorder Archonta may also include bats. Figures 1.5 and 1.6 illustrate two distinguished attempts to define the relationships of primates and other placental mammals. Figure 1.5 is based on an analysis of morphology (Novacek *et al.*, 1988). Figure 1.6 is based on an analysis of molecular evidence (Murphy *et al.*, 2001). Note in Figure 1.6 that elephant-shrews and bats are no longer considered to be closely affiliated to primates. Also, note that DNA and molecular evidence currently indicate that colugos are the closest living relatives of primates (Janečka *et al.*, 2007; Perelman *et al.*, 2011). I will not use subordinal distinctions between primates of modern aspect (euprimates) and the extinct plesiadapiform primates for two reasons: we cannot retrieve ancient DNA or molecular evidence from the plesiadapiform primates; and accumulating fossil evidence (Chapter 8) indicates that some plesiadapiform species had acquired traits characteristic of euprimates.

The controversy surrounding primate definition becomes even more intense when fossils are studied, because no soft-tissue traits are present, and bones and teeth may be fragmentary. Another factor is that the earliest primates were apparently adapted for ecological niches now occupied by rodents, and they therefore possess many traits not seen in living primates (Chapter 8). Major changes in niche structure therefore affect primate traits through evolutionary time.

It is nevertheless clear that primates, far from being the First Animals—a literal translation of the Latin word *Primates*—are not specialized. They are generalized in dentition and body form. One need only consider kangaroos, elephants, moles, sloths, and whales to realize that this statement is true. All of these animals have radically transformed their dentition and body plan to reflect highly specialized lifeways. On the contrary, primates have very generalized teeth and skeletons, and they even retain bones like the collarbone or clavicle that are otherwise lost in most other mammals. Primates in the distant geological past (Deep Time) are also closer

---

breakdown of these food types is available for some species. This allows a researcher to examine and compare dietary quality in these species. It is regrettable that access to this database is not free. This is not true for similar databases for fossil mammals (e.g. NOW—the Neogene Mammal Database [http://www.helsinki.fi/science/now/]) or mammalian anatomy (e.g. Comparative Mammalian Brain Collections [http://www.brainmuseum.org]). The database for All the World's Primates justifies its fees by arguing that the dire problems of primate conservation make it necessary to charge fees. However, fund-raising for conservation never seems to end, and worthwhile research is thwarted by high fees.

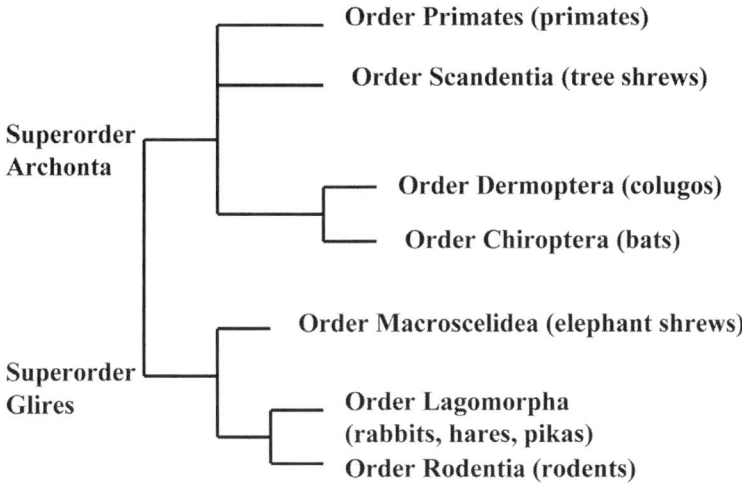

Figure 1.5. The relationship of the Order Primates to other placental mammal orders, based on morphology (Novacek *et al.*, 1988).

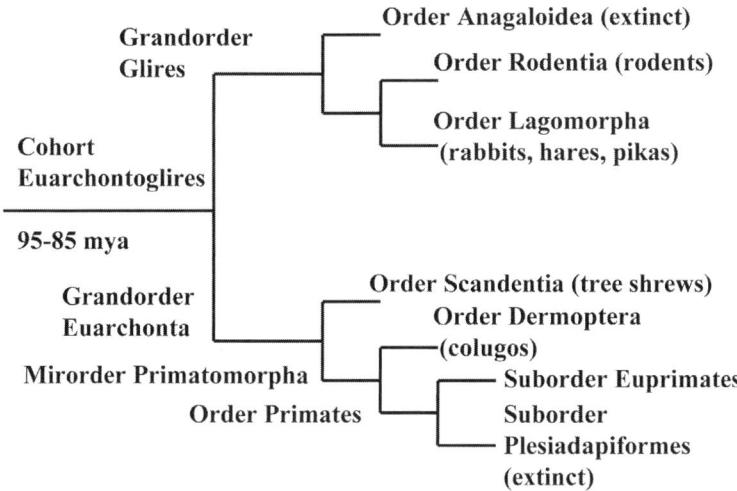

Figure 1.6. The relationship of the Order Primates to other placental mammal orders, based on DNA and molecular evidence (Murphy *et al.*, 2001).

to the origin of other mammalian orders. Because these orders are also differentiating, dental and skeletal traits all tend to be more generalized. Identifying primates from other mammals becomes even more complicated in these distant time frames.

Stratigraphy and rock units

Scientists now understand that the earth is very ancient. In fact, it is so old that it is difficult for humans to comprehend the span of time during which the earth and life on earth has existed. This unfathomable span of time is often referred to by the term "Deep Time." Studying primates and other mammals in Deep Time needs a vocabulary of terms originally devised by geologists studying earth history. Basic geological processes such as sedimentation, plate tectonics, and global climate and sea level changes will be discussed throughout this book. Processes affecting ancient geography and climate will be detailed in Chapter 4. However, terms related to the physical characteristics of rocks and strata in particular areas must be introduced here. A reference scale for the geological time over which primates evolve will also be introduced.

Sediments are laid down in horizons called "strata" (singular, stratum). The study of strata is so complicated that a sub-discipline of geology has emerged from it: stratigraphy. Fossils within strata are used in biostratigraphy. Fossil taxa can give an idea about relative time across broad geographic expanses, if they have several features. They must have hard parts that fossilize, be abundant within strata, disperse or migrate quickly, and have abrupt origins and extinctions. They can then serve as good "index fossils" for stratigraphic correlation.

Field geologists working in a certain area make maps of the local rocks and strata. These mappable rock units are technically termed "formations." The geologists define a formation using characteristics normally observable under field conditions. These characteristics typically include such things as the gross lithology (e.g. sandstone, mudstone, limestone), color, and fossil content. A general relative time frame within the formation is given by recognizing the physical position of material. It can be early (lower in the formation), middle (intermediate in position), or late (in the upper portion of the formation). Each formation must have a formally designated type section. Modern geologists rarely have the opportunity to describe new formations, given the long history of field geology and stratigraphy. Stratigraphy is correlated from one formation to another, until there is some general understanding about the geological history of an area. Datable material within strata allows dates to be extended to correlated formations. A formation name is often used as a time signature for fossil material. Thus, the Legetet Formation of Western Kenya contains the Early Miocene primate *Micropithecus clarki*. As geologists continue to map, several formations that are sequential in time or lateral to each other in space may be designated as a "group." Subdivisions of a formation that are easily recognizable in the field are termed "members."

Geological time

Table 1.1 shows the time framework over which primates evolve. An International Commission on Stratigraphy (ICS) is responsible for establishing a standard reference scale for geological time. This scale is periodically refined and updated. The latest revision took place in 2004 (Gradstein *et al.*, 2004). Here standard units of geological time that are recognized worldwide are associated with chronometric dates based on the rates of radioactive decay of argon and uranium isotopes. For the Cenozoic Era, direct chronometric dating is supplemented with paleomagnetic data and information on sea-floor spreading. In addition, the last 23 million years

Table 1.1. Geological time and primate evolution.[4]

Epoch	Age (mya)
Holocene through the Paleocene = The Cenozoic Era	
Holocene	0.0115 (11,500 yrs B.P.)
Pleistocene	1.81
Pleistocene + Holocene = The Quaternary Period	
Paleocene through the Pliocene = The Tertiary Period	
Pliocene	5.33
Miocene	23.03
Oligocene	33.90
Eocene	55.80
Paleocene	65.50
The Cretaceous Period	145.50–65.50

have been tuned with 40,000 yr oscillations associated with changes in the earth's orbit. Dates recognized by the ICS are given in Table 1.1. In May 2009, the ICS recommended that the beginning of the Pleistocene epoch be moved back from 1.8 mya to 2.6 mya. This recommendation was made on the basis of the onset of cold global climate. Paleoanthropologists and other researchers in Quaternary time have resisted this recommendation for two reasons: they were not consulted about this modification, and, more importantly, no significant biotic changes occur at this time. Traditionally, biotic changes were the principal indicators of a new time division (Van Couvering *et al.*, 2009). Thus, Table 1.1 continues to use the 2004 reference scale for geological time.

---

[4] The dates for these time divisions are given by the latest revision established by the International Commission on Stratigraphy (Gradstein *et al.*, 2004). These dates are universally recognized by geologists, until the next, updated version is published.

2 Primate taxonomy

What is taxonomy?

Taxonomy is the study, identification, and sorting of organisms. It is sometimes satirized as mere "beetle-collecting," but both Darwin and Wallace began as ardent hunters of beetles and emerged as co-discoverers of natural selection (Berry & Browne, 2008). At its simplest level, taxonomy is a highly practical exercise. Taxonomists identify things. For example, exterminators examining insects that are infesting a house are doing taxonomy when they identify the culprit species; fish biologists are doing taxonomy when they identify fish species in a freshwater lake. A more complex level of taxonomy is classification. This is the formal, scientific identification and categorizing or sorting of organisms. A scientifically accepted group of organisms of any rank is then called a taxon (plural, taxa). The dataset of organisms being studied is provisionally called an operational taxonomic unit (OTU), until a decision is made about the final sorting.

Members of the public often consider taxonomy to be merely mindless and obsessive collecting, saving, and cataloguing, but it is much more than this. For example, the correct identification of pests or vermin afflicting domesticated plants and animals is necessary for the preservation of an adequate food supply; the correct identification of pathogens responsible for human disease is necessary for the maintenance of public health. Conservation biologists working within nature reserves must be able to identify hundreds of species, assess biodiversity, and predict the likelihood of maintaining biodiversity within a reserve. Some recent taxonomic revisions have major effects on both human health and mammal conservation. *Anopheles* mosquitoes with no discernible physical distinctions have been newly separated into multiple species based on DNA differences. Some of these species spread malaria to humans, while others do not (Paskewitz, 2011). The giraffe has recently been separated into six or more species, and the African elephant has been separated into two distinct species. The African lion exists in two separate subspecies, and one of these has only 500 individuals dispersed across eight West and Central African nations. Instead of being widespread and abundant, these large, well-known, and beloved mammals may therefore be highly endangered taxa (Conniff, 2010; Patterson, 2013).

For animals, the International Commission on Zoological Nomenclature (ICZN), created in 1895, maintains standards for scientific names. Based at the Natural History Museum in London, it establishes norms and criteria by constructing and revising the International Code of Zoological Nomenclature. The fourth edition of the code is now available online (http://www.ICZN.org/ICZN/index), and appears in both English and French. The ICZN preserves the stability of scientific names and publishes rules for the creation of valid new species names. It is concerned with the safe archiving of taxonomic information. The code is highly legalistic, and unfortunately reads like the tax codes of Western nations. Like the tax codes, it is just as unforgiving, and specialists knowledgeable about the rules for creating new species names must often be consulted. The ICZN rules consider January 1, 1758 to be the origin of the modern taxonomic system. This is because 1758 was the publication date of the 10th edition of a monograph entitled *Systema Naturae*, written by a Swedish researcher whose Latinized name was Carolus Linnaeus. Working during the age of major European exploration, Linnaeus devised a method of organizing knowledge about the vast numbers of new species being discovered by travelers in new lands.

The Linnaean hierarchy

There are two practical aspects to Linnaeus's work. First, organisms are given a universally recognized name. No biological work can proceed unless there is agreement about what to call something. Before there can be communication, a common language must exist. As Linnaeus himself observed: "each object ought to be clearly grasped and clearly named, for if one neglects this, the great amount of things will necessarily overwhelm us and, lacking a common language, all exchange of knowledge will be in vain" (Hopper, 2007:1097). Today, this stable naming system allows access to a huge and growing realm of information, much of it online. Second, the hierarchical nature of Linnaean taxonomy means that the classification has implicit predictions. For example, when taxa are compared at the same rank (e.g. genus to genus), contrasts are examined at a lower taxonomic significance than when taxa are compared at a higher rank (e.g. family to family). Originally, these hierarchical levels were thought to reflect merely phenotypic traits in taxa that were unchangeable and eternal. However, with the advent of evolutionary thinking, it became clear that greater or lesser phenotypic similarity reflected greater or lesser evolutionary distance. In general, the more recent the common ancestor is, the more similar the phenotype is.

The concept of species is today intertwined with the idea of evolution, and debates about the existence of evolution and evolutionary processes. However, the word "species" is very ancient, and comes from the Latin verb *specere* ("to behold," "to regard"). In a literal sense, species are what we behold or observe when we examine organisms in public museum dioramas, in the sacrosanct confines of museum research collections, or during a stroll through the fields on a summer afternoon. The term "species" has therefore been used for hundreds of years, although it acquired a standardized meaning with Linnaeus, and a radically different meaning with the advent of evolutionary thinking.

One of the novelties of the Linnaean system is that species are named with a combination of two words, or a binomen. This naming system is therefore formally termed binomial nomenclature. These two names are the genus and species, and the double combination is unique to the species. The generic name, always capitalized, appears first; the specific name, always in lower case, appears second. Both these names are italicized in print, to indicate their peculiar status. Before Linnaeus, species names were long Latin descriptions, sometimes more than 60 words in length. The wood anemone flower is now called *Anemone nemorosa*, but botanists once called it *Anemone seminibus acutis foliolis incises caule uniflora*. That is, an anemone with pointed seeds, incised leaves, and stalk with one flower (http://www.linnaeus.uu.se/online/animal/1_13.html). The species name *Anemone nemorosa* is indivisible—it can be shortened to *A. nemorosa*, but the second half of the name cannot appear by itself. Because the name is an abstract collection of every member of the species, living or dead, the name cannot be used with definite or indefinite articles.

Among other things, the International Commission on Zoological Nomenclature (ICZN) exerts final authority over the proper description of new species. There are iron-clad rules relating to the valid description of a new species. A type specimen must be designated. This is the actual physical specimen used to describe a new species. Traits must then be listed that distinguish the new species from similar species. The type specimen serves as the ultimate reference when arguments arise over species identification. New zoological names are then published either in print or on a CD, and copies must be deposited in at least five major public libraries. Botanists now allow new species to be described online only. Zoologists are lagging behind, but they may be allowed to publish new species descriptions online in the near future (Anonymous, 2011b). A fossil primate sits at the heart of this debate. The discovery of *Darwinius masillae*, a remarkably complete fossil primate whose describers argued was an ancestor of higher primates, was announced in the online journal *PLoS One*. In addition to debate about the evolutionary significance of this species (Chapter 9), its initial publication in an electronic journal generated controversy—was this a valid new species description? The International Code of Zoological Nomenclature does not currently allow for electronic description of a new species.

Although Linnaeus classified living species, fossil species must also be positioned within the Linnaean hierarchy. Thus, the hierarchy functions for both the living and the extinct. This is an astounding concept, because it is obvious that information about the phenotype of extinct organisms is very limited. In fact, sample sizes may be represented only by a single specimen, the type specimen, and this may be fragmentary or distorted. Nevertheless, the goal of paleontologists is to arrange fossil material in its correct location within the Linnaean hierarchy, regardless of a paucity of data. Sometimes this is impossible. At this point, the Latin term *incertae sedis* is used to indicate that the taxonomic affiliation of the material is currently mysterious. Usually, this means that the proper family, superfamily, or order designation is unknown.

Furthermore, there are groups of organisms that routinely defy taxonomic sorting. These are referred to as "Problematica" (Jenner & Littlewood, 2008). Problematica have equivocal traits that defy any attempt to relate them to other taxa. There are three factors that often cause ambiguity in classifying fossils. First, the process of fossilization itself can obscure or distort anatomy, and inferences about behavior may be forever impossible. Second, the rarity of fossils makes it difficult to assess variability or geographic differences. Lastly, interpretation of fossil anatomy depends on information about living species. The fossil record of 600–500 mya is rich in Problematica, and this seems to be because early animal body shape and form is diversifying at this time. Many animals from that distant time have blueprints for shape and form that are not represented in the modern world. Yet, many Problematica are also found in living organisms, especially in cases where much morphological evolution has occurred, and DNA or molecular evidence is ambiguous. One important conclusion from the fossil record is that lineages tend to undergo convergent evolution with increasing evolutionary age (Jenner & Littlewood, 2008). This causes similar phenotypic traits to evolve in organisms that are very different in ancestry. The technical term for this is homoplasy (Chapter 6).

In addition to the traits studied by Linnaeus and his successors, realms of science unknown to Linnaeus, such as genetics and protein structure, are now used as traits in his classification system. Genome sequencing or comparative genomics is used to discern evolutionary trees, or the relationships between different groups of organisms. These trees may be difficult to generate from anatomy alone, if organisms have simple structures. In this case, gene sequencing, gene position, or unusual genetic events can reveal hidden relationships (Whitfield, 2007). However, it is becoming abundantly clear that the use of mitochondrial DNA (mtDNA) alone to infer evolutionary trees is problematic. When evidence from many nuclear DNA genes is compared to mtDNA genes, the results are jarringly discordant (Miller *et al.*, 2012). This is because nuclear DNA genes reflect events at a population level, as well as events in a species history. Evolutionary analysis of species differences using nuclear DNA creates true species trees, but mtDNA creates gene trees. Changes in mtDNA are caused by positive selection on mtDNA genes, as well as chance demographic events, population bottlenecks, and large-scale migration. This is not an abstract consideration, because it affects questions about primate and human evolution. For about 20 years, an evolutionary tree using only mtDNA genes appeared to support the idea that anatomically modern humans had emerged in a small, founding population in sub-Saharan Africa, and had quickly swept through the Old World, driving local human populations into extinction. Analyses of nuclear DNA from Neanderthal humans (Green *et al.*, 2010), as well as nuclear DNA from fossils in Denisova Cave, Siberia, clearly demonstrate that this accepted and well-received idea was incorrect.

In May 2007, on the 300th anniversary of the birth of Linnaeus, an electronic database was initiated to catalog the 1.8 million known species of living organism. This database is informally known as the "Encyclopedia of Life." The digitalization of data has altered the slow, fine-grained analysis of organisms traditionally performed by taxonomists—taxonomic data in cyberspace is in a state of perpetual flux (Hine, 2008). Besides commemoration of Linnaeus's birth, there is a major practical reason for starting an electronic database now. Many living species face impending extinctions. Without a database of all modern life, the loss of species diversity could not be grasped. Whether the Encyclopedia of Life succeeds or not, much taxonomic work is now published on websites. The obscurity of many print journals has inhibited taxonomic work. As a result, specialized websites have been created to publicize new species names published elsewhere. Plans are under way to streamline species identification in the tropics by submitting molecular data and DNA barcodes from individual specimens directly to the Web. In fact, some researchers argue that future taxonomists will use only globally accessible electronic databases, and not traditional, paper-based modes of publication (Godfray, 2007; Knapp et al., 2007). Type specimens published on the Web ("cybertypes") would include images and molecular data. They would be collectively accessible from the Web—researchers would not need to travel from collection to collection, performing comparative studies. Furthermore, the type specimen would remain pristine, preserved in cyberspace, and safe from decay, or the inevitable wear and tear caused by normal physical investigation. Given that many type specimens were originally described by Linnaeus, one can appreciate how 250 years of handling and measurement can alter or ruin specimens. However, philosophical problems inherent in the idea of a type specimen—that it represents all of the variation in a species through evolutionary time by a single specimen (even a single pelt, a single tooth)—would still remain.

Practical human limits to the quick identification and description of species are now being overcome by the use of automated computer systems that classify or categorize digital images within seconds using object recognition software. An example of this is the Digital Automated Identification System (DAISY) used by personnel at the Natural History Museum in London (MacLeod et al., 2010). This system is 100 percent accurate in identifying 15 species of parasitic wasp from digital images of wing patterns. Three-dimensional images can also be examined using ultraviolet and infrared light. Calls and vocalizations, including calls outside the normal range of human hearing (e.g. bat calls), can be classified using these automated techniques.

Lastly, I must note that a project called the PhyloCode, based on a taxonomic method called cladistics (see "Taxonomic methods"), proposes to abandon the Linnaean hierarchy in favor of lineages on a Great Tree of Life (Yoon, 2009). The hierarchical nesting of families within orders within classes, etc., would be jettisoned by the PhyloCode project. This project has generated such intense controversy that the Linnaean hierarchy seems safe and secure.

Species definitions

The International Code of Zoological Nomenclature implies that species definitions are legalistic and absolute. A properly defined species is forever immortalized in the taxonomic literature. Yet, the ugly truth is that the species concept and its definition is a very contentious issue in biology. Most biologists would agree that species truly exist. That is, they are not abstract concepts formed in the mind as the human brain imposes order on the external world. Species would exist regardless of whether humans were around to study them. Furthermore, most biologists would agree that species are the fundamental unit of evolution. The transformation of species is responsible for the procession of life through geological time. Evolutionary events responsible for the creation of new species are therefore the subject of intense research. Taxonomic arguments about species and subspecies also have a practical impact on the politics of conservation—for example, is a threatened or endangered population truly a distinct species or subspecies that exists nowhere else? Taxonomic distinctiveness or the function of a species as a keystone within a multilevel ecosystem also factors into conservation questions of rescue or triage (Marris, 2007b). Species definition may also have a fundamental bearing on human health and disease. For example, identifying the true insect vector of malaria and other parasitic diseases depends upon sorting out mosquito species complexes using DNA analysis (Paskewitz, 2011).

A great number of species definitions exist (Winker, 2010). The number of definitions partly reflects philosophical issues about the permeability of species boundaries, as well as different views about the relative importance of factors that create new species. Geographic boundaries that prevent breeding are generally thought to be a major factor in speciation. This type of speciation is called allopatric speciation. There is good evidence that allopatric speciation occurs in living primates. New World monkey subspecies are often separated by river boundaries. The formation of the River Congo 2 mya led to the origin of the bonobo species (*Pan paniscus*) when animals became separated from other chimpanzees by a northward-veering loop of the river. This separation was rapid and enduring, because common chimpanzees on the other side of the River Congo are no more related to bonobos than chimpanzees in West Africa (Prüfer *et al.*, 2012). Speciation can occur even if there is no complete geographic isolation. For example, polar bears evolved from brown bears without geographic isolation (Miller *et al.*, 2012). Arguments also occur about other modes of speciation—for example, whether new species can form from within a single population, or from hybridization. Hybridization may be especially important during periods when many closely related species are evolving (Seehausen, 2004). These periods are termed "adaptive radiations." The course of primate evolution is distinguished by several adaptive radiations. And species boundaries can be permeable. Polar bears and brown bears form hybrids today, and there is a widespread transfer of genes between them—the spread of mtDNA is especially broad and rapid between them (Miller *et al.*, 2012). Ultimately, evolution itself is responsible for the fact that researchers may disagree about the allocation of specimens to distinct species, or may argue about species definitions. The origin of species occurs along a continuum through evolutionary time, and the exact point of divergence may be indistinct.

I will present five major species concepts here—the phenetic species, the biological species, the ecological species, the recognition species concept, and the phylogenetic species. However, the phenetic species concept is the principal one used by paleontologists, given the limitations of the fossil record.

The phenetic species is based on regularly observable differences of anatomy (including microscopic anatomy), physiology, and behavior in normal individuals. The phenetic species is the oldest of species concepts. It dates back to Aristotle, who studied dead specimens in much the same way as modern researchers study museum collections today. However, the use of instruments to detect and measure microscopic and biochemical differences became routine during the late nineteenth and early twentieth centuries. Paleontologists study the anatomy of fossil organisms almost exclusively. Traces of behavior (e.g. tracks, nests) are very rare, and it is impossible to extract ancient DNA from most fossils. This is because DNA degrades so quickly after death. There are some exceptions. Ancient DNA may be preserved in cadavers encased in amber, or may be extracted from fossil bones from cold environments in recent time ranges, or from mummified tissues in even more recent time ranges. The oldest DNA yet recovered is mitochondrial DNA from a cave bear and a contemporary fossil human dating over 300,000 years B.P. (Dabney *et al.*, 2013; Meyer *et al.*, 2013). This DNA was extracted from bone collagen in fossils at a deep, subterranean cave site called Sima de los Huesos in Spain, where the fossils were protected since their deposition, and where the ambient temperature and humidity have been constant. If genetic data are available, they can be amazingly informative. Genetic data can be used to analyze population variability, and to reconstruct detailed scenarios of population dispersion and contraction. For example, ancient mitochondrial DNA from Arctic foxes shows that this species did not migrate as its habitat shifted in response to changing climate after the end of the Ice Age (Dalén *et al.*, 2007). Instead, this species suffered local extinction as its habitat disappeared, and survived only in refuge areas. Arctic foxes then re-colonized areas from refugia as climate improved. This is a rare instance when genetic data can be used by paleontologists. It yields insights into how species respond to local habitat change—information about local extinctions occurring across a broad geographic distribution is usually invisible to paleontology.

Paleontologists principally study the anatomy of hard tissues, although casts and impressions of soft tissues can sometimes be used. Thus, paleontologists overwhelmingly use morphology when studying fossil species. They realize that other aspects of an organism, such as behavior, are fundamentally important, but will probably never be known. Hence, they often use the term "morphospecies" when referring to fossil taxa. In theory, paleontologists understand that they must acknowledge the possibility of age and sex differences, as well as differences between populations and geographic areas when classifying extinct organisms. In practice, however, a lack of information about variation caused by biological factors and landscape distribution causes paleontologists to rely largely on the distinctive morphology of fossils and recognize morphospecies. But how different does morphology need to be before one suspects the presence of another fossil species? Among mammalian paleontologists, a general rule of thumb is that, if a trait being examined in fossils encompasses more than ± 2 standard deviations, then one might reasonably infer the presence of another species. This is because, in a normal distribution, 2 standard deviations above and below the mean of a trait in living organisms cover 95 percent of the variability. This is true whether one is studying head and tail length, body mass, or behavior, as long as the trait is normally distributed.

Some researchers despair about ever knowing the true status of fossils—whether they represent distinct species or not. They are convinced that the actual number of species will be underestimated, because of the absence of soft-tissue traits and behavioral data. For this reason, they adopt the term "similum" for any definable group of fossils (Henneberg & Brush, 1994; Henneberg, 1997). They believe that a species, as defined from living organisms, can never be recognized from fossils. Although most researchers agree about the unknown status of fossil species, most do not succumb to hopelessness—that is, arguing that fossil species are forever unknowable.

The biological species concept identifies a species as a group of interbreeding populations that do not reproduce with members of other species. Features exist to foil such interbreeding. Under natural circumstances, mating either does not occur, or else hybrids are completely sterile or relatively infertile. The result is that different species are reproductively isolated. The biological species concept dates back to the seventeenth century, but was strongly promoted in modern times by the systematist Ernst Mayr, who refined the idea from the 1940s through the 1960s (Mayr, 1942, 1963). It is now the definition most commonly used by zoologists working with living animals, but it has obvious drawbacks for paleontologists working with extinct animals. Reproductive behavior or fertility underlie the concept of biological species, but cannot be unequivocally determined from the fossil record. This can be demonstrated by arguments about hybridization among fossil humans—for example, whether the Lagar Velho specimen from Portugal represents hybridization between Neanderthals and modern humans. Only study of the morphology of hybrids between living primate species, or the evidence of ancient DNA (assuming this were available), could resolve such arguments. Two introduced species of marmoset (*Callithrix jacchus* and *Callithrix penicillata*) in Brazil show extensive hybridization, based on phenotypic characters and mtDNA (Malukiewicz et al., 2012). In the case of Neanderthals, nuclear DNA has been preserved. Recent publication of the Neanderthal genome now shows that Neanderthal DNA occurs in modern humans from non-African populations—that is, Neanderthals are not a separate species from modern humans (Green et al., 2010). An illustration of how morphology could be used to detect hybridization in fossils is given by recognizing characteristic morphological traits that separate baboon hybrids from their yellow baboon and olive baboon parents (Ackermann et al., 2006). When dental and cranial traits are examined, the hybrids are significantly larger and more variable than the parental species.

From material shown by Schillaci *et al.* (2005) and Ackermann (2007), I estimate that hybrid skulls are roughly 20 percent larger than those of their parents. Novel cranial phenotypes and sutural anomalies occur, and the dentition of the hybrids is especially variable. In theory, this type of morphology might allow hybrid individuals to be identified in the fossil record. Ackermann lists some possible examples of hybridization in the early human fossil record based on dental anomalies, and illustrates six possible examples of Neanderthal and modern human hybridization from fossil dental evidence (Ackermann, 2010:Fig. 4). Nevertheless, reproductive behavior in fossil species will always be mysterious. In summary, although reproductive barriers are considered by many researchers to be the most fundamental aspect of a valid species, because they maintain species boundaries, reproductive isolation is invisible to paleontologists, and cannot play a role in diagnosing fossil species.

The ecological species concept is based on the observation that every living species has a distinctive ecological niche. Even if several closely related species with similar lifeways occur in the same area, each species will be different enough in ecology to allow them to divide up the common resource space and coexist. Species boundaries are maintained by ecological adaptations to local environments and resources. Geographic variation within a species is caused by environmental variation across the range of its distribution. Genetic differences related to local adaptations or random genetic changes occurring in small populations trigger speciation. Subtle changes in behavior can theoretically alter an ecological niche through time, leading to the prediction that new species can arise through shifts in ecology. Biologists studying species under field conditions have established that slight differences in behavior can be important in separating related species. Historical records confirm that these differences can result in morphological and behavioral changes that are distinctive enough to represent incipient speciation. Because these changes can occur over a human lifespan—sometimes only over the course of several decades—they characterize what is now called "Contemporary Evolution." This is evolution that can be witnessed by living humans. It does not necessarily need long spans of geological time in order to occur. In a general sense, the importance of contemporary evolution is that it affirms the existence of evolution per se. In a particular sense, contemporary evolution establishes that subtle ecological changes can be crucial to the origin of new species. Contemporary evolution is not an esoteric subject, because cases of it are widespread and instantly comprehensible. For example, many lakes, rivers, and streams throughout the United States have posted limits on the numbers and sizes of fish species that can be legally taken. Fish below a certain size must be released back into the water; only fish above that length can be caught and taken home. These measures were instituted for reasons of conservation, to ensure that over-fishing would not occur. The species would flourish in perpetuity, rather than being exterminated from some bodies of water. However, these policies have resulted in easily visible and unexpected results. The average size of many game fish species is declining, and old national size records remain unbroken. Some size records have not been broken since the early twentieth century. Why is this? The effort to preserve smaller individuals by taking out only large fish has created selection pressure favoring smaller individuals. Hunting creates similar selection pressures. Thus, hunting and fishing records faithfully maintained by sportsmen yield examples of contemporary evolution occurring over the span of decades, not millennia.

The recognition species concept is based on the reproductive barriers that exist between species. Genetic differences cause speciation, because even geographically distant members of the same species can be sterile or relatively infertile when they are bred under captive conditions. Members of such distant populations would be unlikely to encounter each other in natural circumstances. However, behavioral differences related to mate recognition and sexual selection can also create new species. Swarms of closely related species such as the cichlid fish species of East African lakes or the guenon monkeys of West Africa rainforests demonstrate how this occurs. Because these species are closely related and are sympatric, or occur in the same area, there is always the possibility that members of different species may mate—hybrid offspring may even be born. Under these conditions, appropriate mates can be identified by complex courtship behaviors, visual signals, or vocalizations. Animals supplying the suitable signals are identified as potential mates by signal recipients. A correct suite of behavioral, visual, and vocal signals identifies the right mate. Inappropriate signals deter sexual interest or terminate mating attempts. This suite of characters may appear more fragile than gross morphological differences such as body size and shape. However, this suite is responsible for maintaining species boundaries, and is therefore a fundamental aspect of species identification. The recognition concept of species is based on the complex sexual interplay occurring during reproduction. Competition between members of the same sex for access to mates (sexual selection) can lead to changes in the mate recognition system. Different sexual strategies pursued by males and females can also lead to such changes.

Brilliantly colored patterns on the head and hindquarters, as well as species-specific vocalizations, define mate recognition in West African guenon monkeys. These monkeys are otherwise very similar in morphology and coat color. Sometimes mate recognition fails, and inappropriate mating occurs. Viable hybrid offspring may be born. The percentage of guenon hybrids born in the wild is probably negligible, although this would need to be analyzed with genetic testing. The proportion of hybrids that naturally occur in the wild is theoretically interesting (Arnold, 1997). Such hybrids imply that evolution may be very rapid. There are obvious implications for conservation. Hybridization does not threaten the reproductive isolation of the biological species concept, because this concept is inherently probabilistic in nature. Hybridization does threaten the phylogenetic species concept. About 10 percent of all non-marine bird species hybridize; 2 percent hybridize on a regular basis, and 3 percent do so occasionally (Grant & Grant, 1992). It is interesting to consider how high a proportion of hybrids can exist in an area without collapsing the integrity of the parent species. The vast array of cichlid fish species is declining in East African lakes. The cause of this decline is the growth of algae in the lakes, triggered by runoff of fertilizers used by local farmers. The presence of algae in the upper levels of the lakes creates a dimmer environment lower down—sunlight cannot penetrate past the algae barrier. The result is that brilliant colors forming the visual signals underlying mate recognition are muted. Unsuitable mating is frequent. The resulting hybrids are viable; they lead to loss of the parent species as species boundaries are conflated.

The original advocate of the recognition concept argued that it would be useful to paleontologists, because one would not need to know about reproductive boundaries and gene flow. One would only need to infer the presence of species-specific mate recognition systems (Paterson, 1981, 1982, 1986). Although biologists confirm the importance of mate recognition systems and the recognition species concept in modern field studies, this species definition is generally impractical for paleontologists. Evidence of species-specific mate recognition systems would need to be found in the fossil record. In living animals, mate recognition is usually based on behaviors like elaborate courtship displays, fur or plumage colors, or vocalizations. These traits obviously do not fossilize. In some cases, hard-tissue traits unique to one sex, such as bony horn cores distinctive to an antelope species, can be used to infer sexual selection, and, indirectly, mate recognition systems. Primates lack such species-specific traits, although body size and canine tooth size differences can be large between the sexes in living primates. Sexual selection has therefore sometimes been inferred in fossil primates. However, the recognition species concept remains as impalpable for fossil primates as it does for members of other mammalian orders.

The phylogenetic species concept defines a species as "an irreducible (basal) cluster of organisms, diagnosably distinct from other such organisms, and within which there is a parental pattern of ancestry and descent" (Cracraft, 1989:34–35). That is, a species is a unique entity. It has a fundamental existence that transcends that of individuals in populations. It is a group of organisms—not variable individuals in populations—that can be diagnosed by a unique combination of traits. A species exists if it is diagnosably different. Any group of organisms that shares at least one unique trait is a distinct species. Even a single soft-tissue or behavioral trait, such as pelage color or vocalization, can be used for this diagnosis. A good example of the application of this species concept was the 2007 announcement of the discovery of a new species of clouded leopard (*Neofelis diardi*) found on the islands of Borneo and Sumatra. Only coat color and pattern differences are used to define this new species, even though such differences often occur in the same species in different geographic locations (Meiri & Mace, 2007). Mainland Asia is sufficiently different in both physical geography and biology from Borneo and Sumatra to account for minor pelage differences. That is, species can be polytypic. Living humans are a good example. Thus, traditionally, different species used to be recognized only when major differences in morphology and genetics occurred between geographical variants. In one sense, the phylogenetic species concept is another version of the phenetic species concept, because reproductive boundaries are not of interest. Yet, phenetic boundaries are completely inflexible here. Using this concept, variation between populations of the same species might cause these populations to be classified as different species, even if they could interbreed. Furthermore, a species in this concept is an entity with a distinctive position on a phylogenetic or evolutionary tree. It can be identified in relation to the phylogenetic pattern shown by its ancestral, descendant, and related species. It has an exclusive position in the phylogenetic pattern—it exists after a founding species and before a descendant species.

It might be expected that paleontologists, who constantly think about evolutionary relationships, would approve of this species concept, especially because it is grounded on regularly observable differences in phenotype, including morphology. The proliferation of software programs that generate phylogenies has simplified this task, and has led practitioners to be confident of their results. There is another explanation for the spread of the phylogenetic species concept. Some researchers find it more philosophically satisfying than more traditional species concepts. They believe that a species is an entity. A species has actuality and authenticity. It is more real and true than the individual organisms that constitute it (Eldredge & Cracraft, 1980). Species are real and unvarying. Because evolution creates species, only the true evolutionary events, recorded in phylogenies, have any validity. Yet, are the phylogenies a true and faithful record of evolutionary events? Practitioners of the cladistic method of taxonomy or cladistics argue that they can produce phylogenies that are completely faithful renditions of evolutionary history. But, if this were so, why are so many conflicting phylogenies produced with the same methodology (see below and Chapter 6)?

Variation

One of the major problems affecting taxonomy is the dilemma of individuality. Clearly, every individual organism is distinct, unless it possesses an identical twin. Even identical twins may differ, because of separate environmental events that occur from intrauterine life onwards. These distinct organisms are members of a variable population, and a number of populations constitute a species. Yet, how distinct must individuals be before they are classified as members of a different species? A complicating factor is that the species definition is based on a single specimen (the type specimen), which itself varies from other members of the same population. Because data for fossil species are so limited, variability parameters for extinct populations may never be known. Much discussion about the novelty of fossil species therefore centers on the standard deviation of a dataset. Traditionally, because 2 standard deviations encompass 95 percent of the variation within a living population, fossils that exceed the boundaries of 2 standard deviations are considered to have a reasonably good probability of belonging to a different species. Nevertheless, it is obvious that datasets for fossil species will always be more limited than datasets for living species. The number of fossil specimens is constrained by the accidents of fossilization that winnow numbers

Figure 2.1 Head of the Late Cretaceous dinosaur *Triceratops*, which may be an immature form of the contemporary genus *Torosaurus*. This illustrates how ontogenetic or age-related changes can affect taxonomy. The genus *Triceratops* has historical priority, so material previously identified as *Torosaurus* would now be identified as *Triceratops*.

down (Chapter 3). A fossil species may be known by only a single fragmentary specimen. Furthermore, questions about whether fossil specimens differ because of age, sex, or separate population, and the limits of population variability in extinct species may never be answered. Mammals, at least, unlike many other organisms, do not undergo metamorphosis or moult an exoskeleton–species diagnosis is even more problematic when fossils of such animals are being considered.

A concrete example of how age-related or ontogenetic differences can affect taxonomy has been the recent recognition that the iconic dinosaur genus *Triceratops* is an immature form of what was previously regarded as another genus, *Torosaurus* (Figure 2.1). This recognition could only occur because of the density and fine dating of dinosaur fossils that are present in the Latest Cretacous of North America (Scannella & Horner, 2010). In practical terms, under the rules of the ICZN, because the two genera are synonymous, the genus with the earlier recognition date–*Triceratops*, recognized in 1889–takes precedence over *Torosaurus*, recognized in 1891. The end result is that fossil material hitherto identified as

Torosaurus is collapsed into *Triceratops*. The genus *Torosaurus* is eliminated. But this is a mere accident of the time of publication. If *Torosaurus* had prior publication, we would now be calling the famous dinosaur with a huge, three-horned head *Torosaurus*.

In summary, philosophical issues about defining species have plagued biology for a long time. Yet, practical issues of conservation have now made the species concept an area of desperate concern. This is because of the great expansion in species numbers that has occurred recently. This "taxonomic inflation" is caused by the elevation of subspecies to species level, especially in groups that are well studied (Marris, 2007a). This precipitous rise in species numbers has debased the meaning and value of species diversity in conservation research, in the same way that "bad money drives out good" when currency is inflated. Taxonomic inflation also afflicts the primate order. A major contributor to taxonomic inflation is the cladistic method of taxonomy (Chapter 6).

Fossils are not immune to taxonomic inflation. The charismatic dinosaurs are evidence of this. In fact, indefatigable dinosaur hunters appear to have been the most interested in splitting species (Benton, 2010). A similar trend appeared among human paleontologists in the first half of the twentieth century. This was not stopped until major figures interested in systematics intervened (Mayr, 1950; Le Gros Clark, 1964). Fossil non-human primates, however, do not seem to suffer from taxonomic inflation. Nevertheless, taxonomic work on newly discovered non-human primate fossils does not seem to rewrite or revolutionize their evolutionary history. In spite of the fanfare often associated with the discovery of new catarrhine fossils, these fossils appear to have long-term pedestrian effects (Tarver *et al.*, 2010).

Taxonomic methods

Different approaches have been made to the classification of organisms. Through time, these approaches have been formalized into distinct schools of taxonomy. Beginning in the 1960s, there have been virulent interactions between practitioners of these schools. Interactions between numerical taxonomists and practitioners of cladistics grew so heated that they became the subject of a scholarly work on the history of science that used these taxonomy wars to exemplify how science works (Hull, 1988).

Phenetics is the most ancient approach to the classification of organisms. It sorts and catalogs organisms according to their degree of phenotypic resemblance. At its simplest, the number of traits that separate taxa are counted up, and a taxonomy is devised to show the degree of similarity or difference between taxa. Before the advent of evolutionary theory, there were no implications that traits could transform. A strict phenetic approach today is also not concerned with whether traits are primitive or derived, and simply assesses their degree of similarity to traits in other organisms. Shared traits are added up, as are divergent traits. The number of shared traits determines overall similarity between organisms. Traits are not weighted for relative importance—all are judged equally significant. During the 1960s, two biometricians re-vamped phenetics using newly available personal computers, and invented the school called numerical taxonomy (Sokal & Sneath, 1963). The advent of personal computers was absolutely necessary for this, because numerical taxonomy requires large datasets and multivariate statistics that measure distance and assess the existence of clusters. Stratophenetics merges together phenetics and stratigraphy—the study of the layers or strata within a geological sequence. Stratophenetics can only be used when the fossil record is dense, and the relative position of fossils in a geological sequence is unambiguous. Stratophenetics has been applied to abundant primate fossils in the American West that come from well-studied localities with a continuous fossil sequence (Gingerich, 1979). Stratophenetics can demonstrate transformation within separate fossil lineages, and the type and degree of differentiation between contemporary lineages. Thus, it can be very useful to paleontologists, given the conditions of a dense fossil record and a very complete stratigraphic sequence. These conditions rarely occur on land, but they are the norm in deep ocean sediments.

Traditional taxonomic methods relied on overall similarity of organisms, as does phenetics. With the advent of evolutionary thinking, it became clear that the resemblance between organisms was caused by descent from a common ancestor. However, distantly related organisms adapting to similar environmental circumstances could evolve strikingly similar traits through natural selection. The existence of these convergent adaptations creates phenotypic similarity in organisms that are not closely related. Researchers who recognize this problem and weight traits accordingly are termed evolutionary taxonomists. They utilize both phenetic and evolutionary methods. Furthermore, when a group of organisms diverges strikingly from its ancestral group, these taxonomists acknowledge the existence of a major novelty by sorting the group into a different rank in the Linnaean hierarchy. Thus, birds, by virtue of their specializations for powered flight, become members of a new class of vertebrates (Aves), even though they are descendants of archosaur reptiles. This school of taxonomy was the principal approach from the mid nineteenth century until the last quarter of the twentieth century.

Systematics is the philosophy of taxonomy; it searches for guiding principles underlying classification. During World War II, Willi Hennig, a German entomologist working independently of the outside academic world, devised a taxonomic method that he considered both objective and reflective of the actual course of evolutionary events. He termed the method phylogenetic systematics. It received only modest attention until Hennig's book was translated into English (Hennig, 1966). This method is now usually called cladistics. It has been so ardently embraced by some researchers that a novel species definition—the phylogenetic species concept—was based on this method (Eldredge & Cracraft, 1980).

Cladistic methodology makes certain questionable assumptions about speciation and the origin of different taxonomic ranks (Chapter 6). The cladistic methodology promised to yield objective classifications based on evolutionary events as they truly occurred. True phylogenies were supposed to be revealed. However, it soon became clear that the methodology produced a multitude of possible or alternative phylogenies, even when researchers examined the same data. Elaborate statistical manipulations became necessary to search out the most likely phylogeny. A major complication is that, as the number of taxa included in the analysis increases, the number of convergent character traits (homoplasies) also increases. This occurs whether phenotypic traits or molecular traits are used. Homoplasies are vexing to cladists, because they obscure the "true" phylogenetic signal. This phenomenon is so well known to molecular systematists that it is termed the "node density effect" (Jenner & Littlewood, 2008).

Another taxonomic method ultimately developed from cladistics. Transformed cladistics results when cladists begin to doubt that true phylogenies will be revealed. Transformed cladistics rejects any evolutionary implications, or consideration of time, and simply uses the cladistic methodology to do practical taxonomy (Scott-Ram, 1990).

Taxonomists traditionally were not explicit about the rules that they used to sort material into different species, genera, etc. The late Joseph Camin, a systematist at the University of Kansas, therefore invented a series of artificial organisms to study taxonomic decision-making. Some of these living and fossil caminalcules (Camin + animalcules [Latin for little animals]) are shown in Figure 2.2. There are 29 "living" and 48 "fossil" caminalcules distributed over a series of time periods, and they are known only by their morphology, because Camin did not invent a behavioral component to the phenotype. Vertebrate specialists sorted caminalcules into categories that emphasized such things as locomotor differences, based on inferences about the presumed functional morphology of the limbs. Invertebrate specialists emphasized surface features on the caminalcule body, arguing that these represented respiratory pores, rather than trivial color variations (Sokal, 1974). After Camin's death, images of all 77 of the caminalcules were published, along with their true evolutionary tree, which Camin had created using known (although implicit) evolutionary processes (Sokal, 1983a, 1983b, 1983c, 1983d). Sokal argued that phenetic taxonomy using the largest number of available traits gave the best approximation to the true evolutionary tree. However, because Sokal was one of the creators of the numerical taxonomy school, which re-vamped phenetics through the use of computers and multivariate statistics, there was some skepticism about this conclusion.

Specialists in fossil human and non-human primates are now usually explicit about the traits that they use in classification, and the protocols that they use to evaluate or weight these traits, if the traits are weighted. However, even though the same fossils and methodological procedures (e.g. cladistics) are used, different phylogenetic trees can be generated. The end result is that some specialists now argue that primate fossil data cannot be used to reconstruct an evolutionary tree (Wood & Collard, 1999; Collard & Wood, 2000). This is clearly a theoretical quagmire that is brought about by strict adherence to the rules of cladistics that do not recognize biological events brought about by evolution itself (Hawks,

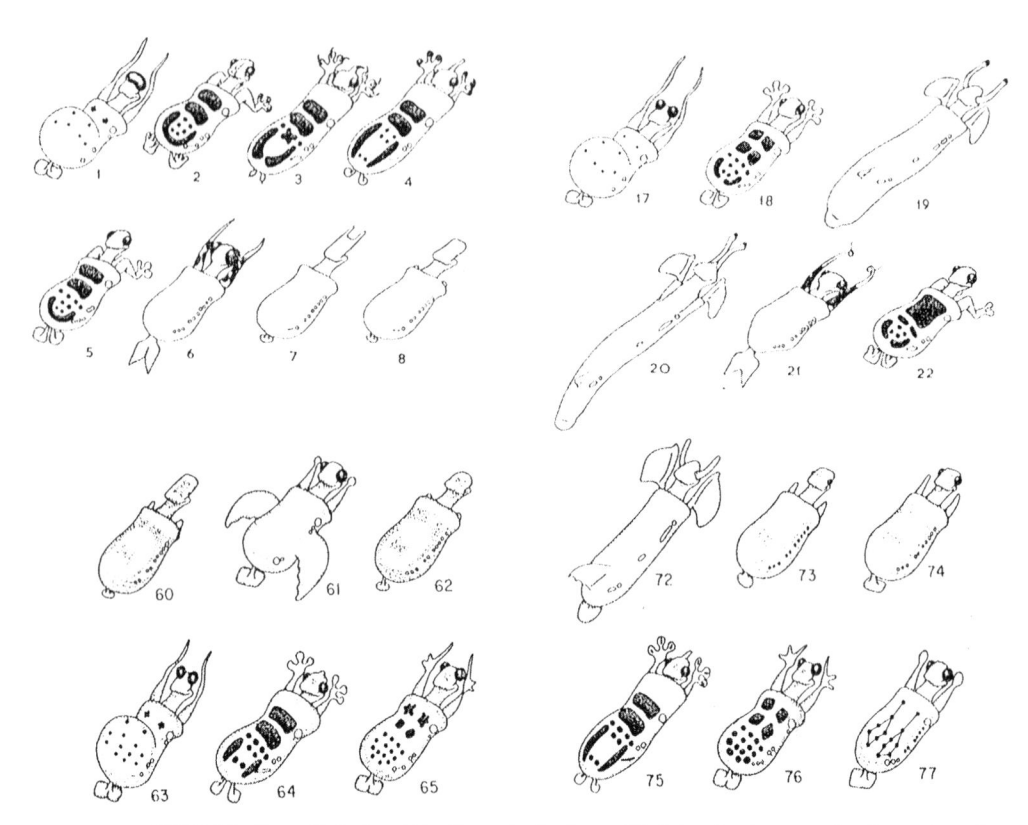

Figure 2.2. Some living and fossil caminalcules, artificial animals created to investigate how taxonomists make decisions when they classify organisms. Top two rows: living caminalcules. Bottom two rows: fossil caminalcules. From Sokal (1983a:Figs 1 and 2).

2004). Such events, for example, would be speciation events that do not adhere to strict dichotomous bifurcations, the persistence of an ancestral species so that it is sympatric with a descendant species, and the existence of homoplasy.

A recent suggestion for standardizing species recognition in birds compares phenotypic traits (morphology, plumage color and pattern, vocalizations, and behavior) between closely related sympatric or parapatric pairs of species that do not interbreed. The novelty of this work is that species thresholds are pragmatically or empirically derived from the data. The thresholds between species are quantified using 58 species pairs distributed across 29 bird families (Tobias *et al.*, 2010). Researchers studying other organisms could utilize phenotypic characteristics not found in birds (e.g. teeth or genitals), and estimate different species thresholds for the groups at hand. The practical value of this approach is that it can be performed on phenotypic traits that are easily studied in the field or in museum collections, and it results in objective taxonomic decisions. The approach does not rely on DNA sequence data that are not available to most taxonomists, or which are patchily available (Brooks & Helgen, 2010). Furthermore, there is disagreement about how DNA data apply to species boundaries.

Primate taxonomy

Primates are one of the most well-studied mammalian orders. As a result, they have received an extraordinary amount of taxonomic attention—some might argue, too much attention, if this results in the multiplication of species. Groves (2001:39–49) summarizes the history of primate taxonomy from Greco-Roman antiquity to the late twentieth century, including interesting vignettes of some of the major figures who studied primates. He refers to the span of time from the mid nineteenth century to the mid twentieth century as "the age of prolixity," because researchers with no appreciation of natural variation in the wild (e.g. age or sex variation) endlessly multiplied the number of primate species as they collected new specimens in the wild, or encountered new pelts or craniodental material preserved in major natural history museums. It should be noted that the number of living primate species that Groves (2001) recognizes is over 90 more than the number generally acknowledged. He arrives at a figure of 356 living species, whereas Nowak (1999), utilizing a widely accepted and more conservative taxonomy, recognizes 263 living species of primates. This is because Groves uses his ability to diagnose a distinctive new taxon (i.e. the criterion of diagnosability) when naming new taxa. I will explain how this criterion is applied below. A conservation-driven website devoted to living primates (http://alltheworldsprimates.org) has further expanded the number of living primate species and subspecies to 612. This is a notable example of "taxonomic inflation," because not all of the new taxa recognized on the website will endure future scrutiny.

Groves is not unique in promoting what has been termed "taxonomic inflation." Similar procedures applied to mammals in general have multiplied the total number of recognized mammal species from 4,659 in 1993 to 5,418 in 2005 (Meiri & Mace, 2007). Note that most of this inflation is not caused by exploration of unknown areas in the wild. Instead, it is caused by taxonomists either raising subspecies to species status, or recognizing novel species with no morphological differences among a species hypodigm. These are so-called "cryptic species," which are morphologically indistinguishable from each other, but are recognized by genetic differences. Sometimes these genetic differences are chromosomal, and actually hinder the production of viable hybrid offspring. A primate example is found in the New World owl monkey genus *Aotus*, where chromosomal differences have established as many as 10 species from animals that were originally considered to be a single species (*Aotus trivirgatus*). These chromosomal differences explained the mystery of owl monkeys in biomedical breeding colonies that mated, but failed to reproduce. Sometimes relatively minor genetic differences are used to establish new species, with no indication of any reproductive boundary between them. An example is mtDNA sequences that have established two new species of lemur (*Avahi mooreorum* and *Lepilemur scottorum*) at Masoala National Park in Madagascar (Lei et al., 2008). It is not generally known how different DNA must be in order to establish a new mammal species. If one applied the same genetic difference criterion to living humans that is applied to living lemurs, modern humans would probably be separated into different subspecies, and possibly species. For example, distinctive structural differences in chromosomes can be identified between major human geographic populations in Europe, China, and West Africa. And, when phylogenetic trees of human genetic differences are produced, the Khoisan people of South Africa are radically separated from all other living humans. Are they a different species? Yet, many biologists applaud species multiplication, because they believe that it can fuel extraordinary conservation efforts (Chaitra *et al.*, 2004).

Groves (2001) published what is currently the most widely used handbook of primate taxonomy. He uses different species concepts for living and fossil primates: the phylogenetic species concept in living animals, and the phylogenetic species concept applied with conveniently recognized species boundaries for evolving lineages in fossil animals. Because speciation events are almost always invisible in the fossil record, new species can be recognized only when they are distinguishable using the available fossil evidence. Genuine fossil species indistinguishable by preserved morphology could never be recognized. This is the criterion of diagnosability, which Groves uses also for living species. A species has fixed, inherited differences that allow it to be diagnosed or differentiated from other species. The adjectives "fixed" and "inherited" mean that individual variations caused by disease or accident would not be used in a species diagnosis, even though they result in morphological differences.

Taxonomy can be distinguished by three levels of analysis. The fundamental description of species, subspecies, and genera is known as alpha taxonomy. This is a straightforward description of the material being examined. The emphasis is on explaining why the material represents a hitherto unknown taxon. This is done by differentiating it from known taxa. Field biologists also practice alpha taxonomy when they compile lists of species that exist within a given area, such as a reserve or national park. Beta taxonomy occurs when researchers attempt to understand the relationships between higher taxa, such as how different families and orders are connected. This often necessitates a complete overhaul of existing classifications. The entire Class Mammalia has been the subject of several such revisions, through the practice of beta taxonomy. Gamma taxonomy occurs when relatively abstract evolutionary or ecological studies are performed. For example, if comparisons are made about rates of evolutionary change in different orders, or if questions are asked about factors that affect species abundance in an area, gamma taxonomy is taking place.

A "protocol for alpha taxonomy" is described by Groves (2001:51–53). He lists 10 steps, beginning with deciding which group of organisms to study, and ending with giving a new name to a novel species. This protocol clearly marks out the route necessary for basic taxonomic description. The only step missing, however, is one in which the researcher is sternly admonished to subdue all feelings of doubt and horror about the overwhelming magnitude of the biological variation that occurs in the material. This has always been my response to museum collections. However, a good taxonomist must cultivate optimism and good cheer. Because Groves is a primate taxonomist, he leaves us with insight into the stages he uses in recognizing and describing living primate taxa. Yet, primate paleontologists do not have the luxury of consulting enormous collections such as those available for living organisms. These are scattered through the natural history museums of the world. A new fossil species may often be based on very small samples. Sometimes the sample size is one! This, of course, is caused by the vagaries of fossil preservation. The limits of age, sex, and geographic differences can be observed in living organisms. They can only be inferred for fossil organisms. It is often assumed that age, sex, and geographic differences are the same in living and fossil organisms. This is a convenient assumption, but it is not necessarily true. Because it is not necessarily true, some human paleontologists argue that any fossil that is distinguishable from known fossils should be recognized as a new species, as long as possible age and sex differences are taken into account. This is the position taken by Tattersall (1986), but he must then argue that living primate subspecies—caused by population-level differences invisible to paleontology—are trivial and unimportant in evolutionary terms. This is a position that paleontologists working with other mammalian orders would find unsupportable. Using this rule of diagnosability, mammal groups with rich evolutionary records (e.g. rodents, antelopes, horses) would need a mountain of new species names. Furthermore, the idea that evolutionary processes observable in the modern world should be ignored by paleontologists would also be considered untenable by mammalian paleontologists.

By convention, the names of extinct taxa are often printed with a dagger (†) preceding the name. I will not do this here. The fossil status of these taxa will be obvious. However, the dagger symbol often appears when a broad outline of primate taxonomy is presented. It is important to note that extinct taxa must appear alongside living taxa in the same classification system, and the dagger instantly allows one to identify the living and the dead. The inclusion of both living and dead taxa in the same classification creates a remarkable problem. In spite of a paucity of data, paleontologists must nevertheless assess the meaning of variation and the relative rank of extinct taxa, and sort them along with living taxa. It is obviously difficult to do this. For this reason, adjustments in the classification of extinct taxa are frequently made. If data are very scarce—for example, if a fossil taxon is known only by a single tooth or a handful of battered cranial bones—taxonomic adjustments may be almost continual. This accounts for some (but not all) of the acrimony attending the study of fossil primates. Finally, in spite of great efforts to analyze and interpret the fossils, some material remains mysterious. In these circumstances, the term *incertae sedis* is used. It has remarkable utility. It indicates that the family, superfamily, or order of a fossil is unknown.[1]

---

[1] Even higher taxonomic ranks can be uncertain. For example, the extinct species *Tullimonstrum gregarium*, the state fossil of Illinois, is known to be an invertebrate. Beyond that, even its phylum is unknown. It is found only in the famous locality called Mazon Creek, and nowhere else on earth. Its preservation here is caused by a remarkable series of accidents that preserve many fossils with remarkable completeness—even soft tissues are preserved. The technical name for such fossil localities is *Lagerstätten* (singular *Lagerstätte*, Chapter 3). Despite the preservation of the soft tissues of *T. gregarium* at Mazon Creek, its anatomy remains so mysterious that the Latin term *incertae sedis* is used to indicate that its higher-order affinities to other invertebrates are unknown.

Taxonomic categories

Seven major divisions exist in the Linnaean hierarchy. These divisions or taxa are the following: Kingdom, Phylum, Class, Order, Family, Genus, and Species. A fabled mnemonic has helped generations of biology students to memorize this sequence: King Philip Came Over From Genoa Spain. The sequence begins with Kingdom, and moves downwards in succession. Thus, it is a hierarchy, because each division incorporates the divisions below it. Additional categories are sometimes used for primates. Living primates are often differentiated by subspecies. This is almost never done for fossil mammals, including fossil primates.

Table 2.1. Taxonomic categories in primates.

First example
Kingdom Animalia (animals)
Phylum Chordata (chordates)
Subphylum Vertebrata (vertebrates)
Class Mammalia (mammals)
Order Primates (primates)
Suborder Anthropoidea (anthropoids)
Infraorder Catarrhini (catarrhines)
Superfamily Cercopithecoidea (cercopithecoids; the superfamily name suffix is -oidea, -oid)
Family Cercopithecidae (cercopithecids; the family name suffix is -idae, -id)
Subfamily Cercopithecinae (cercopithecines; the subfamily name suffix is -inae, -ine)
Tribe Papionini (papionins; the tribe name suffix is -ini, -in)
Genus *Papio*
Species *Papio anubis*
(the living olive baboon)
Second example
Kingdom Animalia
Phylum Chordata
Subphylum Vertebrata
Class Mammalia
Order Primates
Suborder Anthropoidea
Infraorder Catarrhini
Superfamily Cercopithecoidea
Family Cercopithecidae
Subfamily Colobinae
Genus *Mesopithecus*
Species *Mesopithecus pentelici*
(an extinct species of Old World leaf-eating monkey)

This is because subspecies are often based on soft-tissue differences that are not discernible in fossils.

I give two examples of how formal Linnaean categories are used in primate taxonomy (Table 2.1). The examples are here printed formatted with successive indents to illustrate the inclusive nature of the taxonomic hierarchy. In actual practice, researchers would not use such formatting. The first example gives the classification of a living baboon species, the olive baboon; the second example gives the classification of a Late Miocene fossil leaf-eating monkey from Eurasia.

In his textbook on primate evolution, Conroy (1990) uses Cenozoic time divisions as a framework for his chapters. He begins discussion of fossil taxa in each chapter with a useful table that includes genera and species, their geographic origin, and their known time ranges. I will begin the chapters on fossil primates (Chapters 7–15) with a general discussion of global climate and significant changes in mammal evolution. Chapters that deal with older time ranges will also begin with a global paleogeographic map. This is because plate tectonic movements that may account for only centimeters/year can yield significant alterations in the relationships of continents and oceans over tens of millions of years. The distribution of fossil primates, as well as factors like global climate and seasonality, can be affected by the presence of land bridges and the latitudinal position of land masses.

3 Fossils and fossilization

The origin of fossils

After an animal dies, its behavior immediately stops. Of course! Thus, behavior is the first component of the phenotype to be lost. After this, DNA and soft tissues are also rapidly lost. Large and small animals may eat or scavenge the carcass, dismembering the body, stripping away flesh, and breaking open bones that are rich in marrow. Bacteria and fungi alter soft tissues as decay takes place. Nevertheless, soft-bodied organisms may be preserved as flattened carbon films, preserved as calcium phosphate, or altered by early mineralization. For example, in the 425 mya Eramosa Formation of Canada, animal tissues containing melanin were altered by sulfur early after death; this caused resistance to bacterial decay (von Bitter et al., 2007). Exceptionally well-preserved material from this formation disproved an idea that shallow marine fossils after the Cambrian would be unlikely to fossilize in great detail. It had previously been thought that an increase in burrowing organisms after the Cambrian would irretrievably alter sediments. Specimens from the Eramosa Formation show that this is not necessarily the case.

However, it must be understood that the likelihood of any single ancient organism being preserved is miniscule. It is only the multitude of organisms living over vast reaches of geological time that allows these faint probabilities to emerge as recognizable fossils. A special sub-discipline of paleontology has been created to study all of the processes that affect an organism immediately after death until its discovery as a fossil. This is taphonomy. Techniques used by paleontologists to study taphonomy and taphonomic processes affecting fossils are also used by archaeologists when analyzing archaeological materials and sites. They are also used by forensic anthropologists when analyzing material from a crime scene, especially when considering events around the time of death and after death (Klepinger, 2006).

Types of fossil materials

Fossils occur in two forms: the actual remains of organisms, which are generally incomplete, including three-dimensional molds of their external body, and the remains of an animal's activity, such as trackways. The remains of an animal's activity are called trace fossils. Trace fossils will be dealt with in the next section.

The fossil record is always biased in favor of hard parts. Enamel, which covers the crowns of teeth, is the hardest biological substance known. Enamel consists of about 98 percent apatite, which is a mineral. Enamel is essentially rock-like, even during the life of an animal. Thus, the fossil record of many groups (including primates) is over-represented by teeth. Bones of the skeleton are also hard. Living vertebrate bone is typically formed from a hard, brittle material called hydroxyapatite or calcium phosphate ($CaPO_4$) and a soft, flexible material called collagen. Collagen tends to be lost quickly after death, although it may survive longer in cold climates. In some Pleistocene European sites, collagen has been retrieved from the bones of both fossil animals and humans. Stable isotopes from both collagen and enamel have been used to reconstruct ancient food webs (Chapter 5). Bone may be hard, but it is alive while an animal is alive. Living bone is highly vascularized. Bone cells exist in an organic matrix imbued with inorganic minerals. Bone is continually created and destroyed or remodeled during life, and its structure therefore reflects the movements and activities of an animal (Chapter 5). After death, the skeleton may become disarticulated, fragmented, and abraded by both biological activities like predation or scavenging, and by physical activities like water transport that knock rocks or gravel against bone. Shells and pollen are other hard tissues that survive in the fossil record. They can also suffer fragmentation and abrasion. Sometimes shells and bones are eroded by organisms like insects, worms, and fungi after death. Figures 3.1–3.3 illustrate the search for and discovery of ancient fossil mollusks dating to 450 mya. Note that delicate morphology and color patterns may be preserved despite the age of the fossils.

A body starts to decay immediately after death. This process is caused by bacteria or fungi. The rate of decay is affected by external temperature, the pH of surrounding materials, and the supply of oxygen. Where oxygen is present, bacteria convert organic carbon into carbon dioxide and water. Bacterial decay can also occur without oxygen, but other materials (e.g. iron oxide) must be present before decay occurs. The higher the ambient temperature is, the higher is the rate of decay. Most sediments have a neutral pH, which promotes decay. Acidic environments such as swamps slow down or even halt microbial decay. During the Bronze and Iron Ages of northern Europe, many people were ritually murdered and interred in peat bogs (Glob, 1971). These bodies are startlingly intact. They retain their hair and skin, and internal organs such as the brain and gut, which otherwise quickly decay. The soft tissues in bog bodies are actually tanned, although their bones are decalcified and soft.

Usually, fossilization begins when an organism is covered by sediments (soil, dust, sand, mud, ash). There are rare circumstances when a dead organism is covered by other materials. Small organisms such as insects can be quickly and completely encased in resin, which becomes fossilized as amber. Decay stops at this point, and there is even some potential for ancient DNA to be retrieved from animal and plant inclusions in amber. Tar is another unusual material that can quickly cover a carcass. Under these circumstances, soft tissues are not preserved,

Figure 3.1. Hunting invertebrate fossils in the limestone of the Lone Star Quarry, near LaSalle, Illinois.

although complete individual skeletons are present. The Rancho la Brea Tar Pits, located in what is now downtown Los Angeles, dating to 20,000 yrs B.P., are an example of tar preservation. The quantity of fossil material reflects the fact that this is a catastrophic site. The seeping tar was covered by pools of rainwater. Animals coming to drink at these pools were mired in the tar. Predators and scavengers preying on struggling animals were also mired. The nature of the site and the density of fossils allow paleontologists to study topics that would be impossible at most other sites. For example, they can reconstruct an ancient ecosystem, examine niche separation in sympatric animals, and infer behaviors like intra-specific competition. However, under normal circumstances, decay continues either until the entire carcass disappears, or until the carcass becomes mineralized. Scavenging insects like beetles and carnivorous animals remove dead flesh. Usually, all of the soft tissues are removed before hard tissues become mineralized.

Fossilization takes place when the tissues of an organism are replaced by minerals. Sometimes replicas of soft tissue are preserved by bacteria that leave a mineral coat of phosphate or pyrite over the surface of the tissue. Alternatively, fine-grained sediments may sometimes preserve a hardened external cast of an organism. In addition to hard parts (pollen, shell, bone, teeth) that are

Figure 3.2. An Ordovician ammonite fossil from the Lone Star Quarry, near LaSalle, Illinois, dating to about 450 mya.

more apt to fossilize, other materials, such as wood, may also sometimes be fossilized. Fossil wood may be so abundant in some areas that partial reconstruction of an ancient forest may be possible. Sometimes the fossil wood is not mineralized. The cellulose has decayed, but the lignan may still be present. A partial fossil forest of the swamp cypress (*Taxodium*) has been preserved as lignan stumps in 8 mya deposits in Hungary. The trees had been quickly buried in sands that partly preserved the plant material up to the point of burial. Similar swamp forests from the Late Miocene of Hungary once housed *Rudapithecus hungaricus*, one of the last survivors of the once diverse European fossil ape radiation. Plants preserved as coal deposits may be so intact that they allow fine reconstruction of plant anatomy. In fact, coal deposits may contain such detailed information about fossil plants, and may be so extensive, that portions of ancient forests can be mapped. This is the case with the Herrin Coal deposits in Illinois, which preserve over 1,000 ha of an ancient swamp forest dated to 300 mya (DiMichele *et al.*, 2007). Morphospecies can be mapped, and ecological gradients can be identified in the Herrin Coal, in the same way that botanists can identify and map tree species in modern forests, allowing analysis of species abundance, diversity, community structure, and species turnover through time. The exceptional preservation of plants in the Herrin

Figure 3.3. Ordovician mollusks from the Leaf River Formation, Ogle County, Illinois, dating to about 450 mya. Note the survival of delicate shell morphology, as well as the preservation of intricate color patterns.

Coal is attributed to the abrupt collapse and flooding of this region when one or more earthquakes shifted part of an ancient forest several meters down into an ancient estuary. Casts or molds of plant parts can also allow fine reconstruction.

Sometimes the outline of soft tissues (skin, fur, feathers) or casts of mummified soft tissues are preserved in fine sediments. Very rarely, tissues themselves may be fossilized. A recent discovery, once trumpeted as being a putative dinosaur heart, is now considered to be dubious. Another example is the preservation of ligaments, tendons, and skin of a hadrosaur (duck-billed) dinosaur from the latest Cretaceous of North Dakota. Patterned bands of scales on the skin allow researchers to infer that the living animal had striped bands of color. Although decomposition was occurring, local chemical conditions caused soft tissues to be replaced by minerals faster than the tissues were decaying. True ancient tissue may rarely be preserved within permafrost or permanently frozen soil, salt, or amber. There is even one report of actual ancient tissue being preserved in normal sediments. Blood vessels and red blood cells have been retrieved from the marrow cavity of a *Tyrannosaurus rex* femur dating to 66 mya in Montana

(Schweitzer et al., 2005).[1] It is also possible that two fossil hominids of the species *Australopithecus sediba* from the site of Malapa, South Africa, have preserved skin tissue dating to 1.95–1.78 mya. An international, open-access collaborative project dedicated to the study of this material (The Malapa Soft tissue Project) can be accessed at this URL: http://johnhawks.net/malapa.

Endocasts or fossils that preserve the outer surface of ancient brains from the interior of the neurocranium offer insight into relative brain size and proportions of various parts of the brain. This is important in reconstructing the life of ancient animals. Brain size relative to body size allows researchers to assess whether the brain was larger than expected for any given body size. If so, one might infer that the extra neural tissue was involved in higher-order behaviors like cognition, rather than simple body maintenance. Because brain function is localized, the proportions of different parts of the brain yield information about specialized senses or activities. The neocortex of living brains has a complicated external structure, with folds of neural tissue (gyri) separated by furrows (sulci). In living mammals, sulci often form the boundaries of neocortical areas known to receive or process information from the senses or parts of the body. Thus, a paleontologist examining an endocast might infer that tactile information from the hands was important to an extinct animal, given enlargement of a certain neural area.[2]

Endocasts are created when very fine-grained sediments enter the neurocranium as brain tissue decays. These sediments then harden into a stony matrix that preserves the internal configuration of the interior of the neurocranium, and thus reveals details of the outer surface of the living brain. Fossil bones of the neurocranium may be lost, leaving only the endocast of the long-decayed brain. I once discovered a pretty little bovid endocast weathering out on the land surface at Koobi Fora, in northern Kenya. Dating to 1.5 mya, it had belonged to a little, even-toed, hoofed animal that was in the same taxonomic family containing antelopes, gazelles, and cattle (Family Bovidae). The animal had been only a little larger than the living dik-dik, which weighs between 3 and 7 kgs. The endocast perfectly preserved external details of the brain, and the large olfactory bulbs protruded from below the cerebral hemispheres on both sides. Nothing else remained of this fossil animal.[3]

There are a number of excellent natural endocasts from fossil primates. For example, the type specimen of *Australopithecus africanus* from the South African site of Taung, which was the first fossil australopithecine ever discovered,

---

[1] Specialists on mummification were assembled at a February 26, 2011 symposium on "The Anatomy of a Mummy" at the University of Pennsylvania Museum of Archaeology and Anthropology. During the open discussion, I asked a question about the conditions that might have preserved the tyrannosaur blood vessels and red blood cells. The consensus of the experts was that a heavy metal in the surrounding sediments, possibly uranium, could have been responsible for this exceptional preservation.

[2] Dr. Harry J. Jerison, University of California at Los Angeles, has prepared a discussion on endocasts and paleoneurology on the website of the Comparative Mammalian Brain Collections: http://www.brainmuseum.org/Evolution/paleo/index.html.

[3] How did I know that it was a bovid? It was an educated guess. I based this identification on the local abundance of bovids, its size, and the fact that it did not resemble the brain of a carnivore or primate.

contains an endocast, as does the type specimen of *Rooneyia viejaensis*, an Early Oligocene prosimian from West Texas. CT scans of intact fossil crania are also used to examine the interior volume of parts of an intact neurocranium (Conroy & Vannier, 1984). Through the use of CT scans, a fossil neurocranium is not destroyed or altered in order to create an artificial internal cast even when stony matrix fills the interior. Computer imaging systems recreate the internal dimensions of the neurocranium or other volumes.

Stomach contents can be preserved. These literally allow the last meal to be reconstructed, but also give detailed insight into an animal's diet and behavior. For example, a young *Psittacosaurus* dinosaur from the Cretaceous Jehol biota was found in the stomach of a contemporary mammal—a badger-sized animal that ruled as a giant among its fellow mammals. This was *Repenomamus*, dated to 128 mya. Fossilized feces or coprolites are occasionally discovered, and these contribute another line of evidence about ancient diet.

Dinosaur eggs have been found at over 200 sites. Intact embryos may sometimes be preserved within these eggs, and comparative development can be examined. Nests allow inferences about egg-laying and nest construction, and multiple nests allow one to infer the possibility of communal nesting or rookery areas. The fossils of young animals found near nests allow one to analyze rates of growth and development. The discovery of juvenile *Protoceratops* dinosaurs in Late Cretaceous Mongolia was first made in the early twentieth century, and recent expeditions to the same area have unearthed further juvenile material. In addition, inferences about parental care have been made, using discoveries of multiple hatchlings within and around a nest. At the 80-my-old Mongolian site of Ukhaa Tolgod, a nesting adult *Oviraptor* was found sitting on a nest of at least 20 unhatched eggs (Norell et al., 1995). Other nesting oviraptorid dinosaurs have also been found, and intact embryos have been retrieved from eggs, allowing detailed study of fetal growth. Ukhaa Tolgod was an oasis where precious water occurred in a Cretaceous desert—equivalent to the density of animals living in stabilized sand dunes surrounding a modern waterhole. The unexpected preservation of dinosaurs, mammals, and lizards at this site may have been caused either by animals perishing in ancient sandstorms, or by being engulfed in flash floods and mud slides after brief torrential desert rains (Dashzeveg et al., 1995; Loope et al., 1998).

Catastrophes that preserve many animals of the same species together have been used to infer the existence of social behavior in these animals. Herding behavior in fossil hoofed mammals such as horses or antelopes has been inferred by the discovery of multiple carcasses of the same species at the same site. The earliest example of social behavior in mammals is found in a catastrophic die-off of an extinct species of marsupial found in Bolivia just after 65 mya. Strong size differences between the sexes, as well as the gregarious nature of the species, are used to infer that males competed with each other for access to mates, and that individual males mated with multiple females (Ladevèze et al., 2011). Inferences about behavior can be made from other lines of evidence. For example, strontium isotopes in ancient tooth enamel preserve a history of ranging behavior (see Chapter 5).

Trace fossils

Trace fossils are literally that—they are not fossils per se, but mere traces of the activity and behavior of fossil animals. Nevertheless, they may sometimes include a great deal of information about the life of a long-extinct animal. These traces are impressed into the sedimentary record itself. Trace fossils are technically referred to as ichnofossils. Characteristic types of ichnofossils may be given genus or species names, and are called ichnogenera or ichnospecies. However, ichnofossils are not biological entities. In most cases, the actual makers of the trace fossils are unknown, and they will probably remain forever unknown. One of the leading experts in trace fossils, Martin Lockley (2007), argues that invertebrate and vertebrate ichnofossils have very different prospects for yielding information about the past. Invertebrate ichnofossils are essentially a record of behavior. But vertebrate ichnofossils reveal the morphology of the maker—behavior is generally revealed only when trackways, or a sequence of footprints, are uncovered. However, both invertebrate and vertebrate ichnology can uncover specific habitat information, such as the nature of the sediments, sediments distinctive for a particular local paleoenvironment (facies), or paleoecology (Lockley, 2007).

Invertebrate paleontologists can examine the burrows, feeding traces, or trackways of mollusks or worms that dwelt in the oceans or shallow ocean margins. These traces may be so linked to certain sedimentary environments that assemblages of these traces can be used to reconstruct fine details of the environment itself. For example, characteristic deep-sea assemblages of trace fossils appear when animals first colonize disturbed areas of the continental shelf; after sediments stabilize with time, other trace fossil assemblages appear. Invertebrate paleontologists have discovered practical, economic benefits in the study of trace fossils. Because petroleum exploration involves extensive geological survey, many sediment cores are produced to amplify samples being surveyed. The abundance and diversity of trace fossils can allow oil industry geologists to reconstruct fine details of sedimentation from very narrow-caliber cores. Furthermore, burrowing and sediment mixing caused by ancient animals can either decrease or enhance oil permeability through sediments, and affect the costs or benefits of oil extraction.

Invertebrates leave behind a density of trace fossils; vertebrates leave much poorer traces. Nevertheless, vertebrate paleontologists have detected dens, burrows, nests, and footprints.[4] In 2007, the vertebrate paleontologist Spencer G. Lucas announced the discovery of a fine-grained slab of rock from eastern

---

[4] My retired Rutgers colleague, Professor Robert Blumenschine, who is an archaeologist, began to use the term "trace fossils" for modifications made to animal bones by either carnivores or hominids. That is, when carnivores gnaw or break bones or hominids open bones for marrow, dismember a carcass, or slice meat off bones using stone tools, he refers to these modifications as "trace fossils." This use of the term is foreign to paleontology, and would ultimately be confusing, because paleontologists expect that trace fossils would appear in sediments. Archaeologists traditionally refer to these markings on animal bones as "modifications."

Pennsylvania with the full body imprints of three fossil amphibians. The find dates to 330 mya, in the Mississippian, or Lower Carboniferous Period. The skin preserved in the imprints is smooth. Four webbed toes can be seen on each extremity, and body proportions can be measured. One might infer some type of aggregation behavior or social behavior from the fact that the imprints of three animals lie so close together. Why should individuals cluster so close together? Are males competing for access to mates?

A multitude of dinosaur footprints have been discovered. Besides analysis of individual prints, many of these prints have been studied in detail by dinosaur specialists to infer gait from sequences of fossil footprints, or trackways. Robert McNeill Alexander, an expert in animal biomechanics, even devised an equation to determine dinosaur speed from stride length and the height of the hip joint above the ground (Alexander, 1989). In fact, once body size is taken into account by measuring hip height, there is a constant relationship between stride length and speed in animals, regardless of whether the animal is a biped or a quadruped. Experiments with living animals examine how body weight and sediment consistency affect the formation of tracks (Milàn & Bromley, 2007). Modern research also studies surface deformation as a foot is impressed into sediment, and the subsequent loss of three-dimensional track structure with increasing depth inside the sediment (Milàn & Bromley, 2006). The nature of the sediment itself can affect the morphology of prints and their preservation. Thus, salts and clays within a sediment can cause variation in prints created by a single species, and can alter the distribution of preserved prints (Scott et al., 2010)

Mammal footprints are much rarer than dinosaur prints, but mammal paleontologists, including primate paleontologists, have also studied footprints. The record of mammal prints and tracks becomes fairly dense around 30 mya, and the prints and tracks of hoofed mammals or ungulates are particularly abundant. Details of gait (e.g. walking, trotting, gallop, or pacing locomotion), preferences for straight-line or curvilinear movement, stamping patterns used for conspecific communication, and herding behavior can be inferred from the trackways left by ancient ungulates (Lockley, 1999).

There are two famous trackways, or sequences of fossil footprints, made by fossil humans. One of these, and a newly discovered site with hominid prints, also includes the footprints of contemporary non-human primates. Here one can see another virtue of trackways: they preserve a record of contemporary animals, and thus give a narrow enough window of time to identify members of ancient ecosystems. The first trackway is from Laetoli, in Tanzania. Dated to 3.6 mya, it preserves the footprints of three fossil hominids, almost certainly members of the species *Australopithecus afarensis*. Fossils of this species are found at Laetoli. These footprints were impressed in a soft, plastic layer of volcanic ash that had been ejected from a nearby volcano, and was falling in a gentle rain shower. The volcanic ash allows a chronometric date to be generated using potassium/argon (K/Ar) dating (Chapter 4). Detailed analyses of the prints indicate that the hominids had a fully bipedal gait. Many other animal prints occur in the same horizon, representing about 20 different species. These include baboons, *Hipparion* horses, shovel-tusked elephants, chalicotheres, a large lion-size cat (possibly a saber-tooth), hyaenas, two types of giraffe, several antelope species, rhinoceroses, pigs, and many birds. At 1.6–1.5 mya in paleontological collecting area 103 in the Koobi Fora region of northern Kenya, the footprints of a single *Homo erectus* individual awkwardly wading through shallow water were impressed into a muddy substrate that included hippopotamus prints (Behrensmeyer & LaPorte, 1981). Although an attempt was made to conserve the original footprints by re-burying them, they have since been virtually destroyed by erosion, as documented by their re-excavation during the summer of 2008. However, latex molds of these prints are preserved in the National Museums of Kenya, Nairobi.

The site of FwJj 14 East in paleontological collecting area 1A in the Ileret region of Koobi Fora was discovered in 2006 (Figure 3.4). The Northern and Lower Ileret tuffs date the site to between 1.54 and 1.52 mya. Several hominid footprints, not laid-down trackways, occur with the prints of a baboon and a smaller monkey species (Figures 3.5 and 3.6). Other contemporary animals can be identified by

Figure 3.4. Mammal and bird footprints preserved at FwJj 14 East, Ileret, northern Kenya, dating to 1.5 mya. Multiple overlapping animal prints are preserved from some type of ancient locality resembling a modern waterhole. Courtesy of Ms. Melanie Crisfield, M.A.

Figure 3.5. A single hominid footprint from FwJj 14 East, Ileret, northern Kenya. Courtesy of Ms. Melanie Crisfield, M.A.

their prints (Table 3.1). A multitude of other, unidentifiable prints also occur here. Continuing work during the summer of 2007 brought multiple levels of prints to light. A laser scanner was used on exposed, excavated surfaces in July 2007, in order to collect digital data on the prints. Hominid prints that appear modern in shape were thus revealed in three dimensions (Bennett et al., 2009). In addition, there is a small hominid print that shows a slight separation between the big toe and the remaining digits. Some hominid prints are well defined; others are not. As of the summer of 2010, five different trackway levels were exposed. Three of these contain hominid footprints. At least three hominid species were sympatric in the Koobi Fora region during this time range. Experiments have been performed with habitually unshod human subjects, the local Dassanetch people, who produced footprints in sediments excavated from the same geological stratum that preserves the ancient Ileret footprints. Contrary to Bennett et al. (2009), who only measured the shape of the prints, and did not investigate the functional anatomy of the ancient feet producing the prints, significant multiple differences in footprint depth were found between the ancient and the modern prints (Hatala et al., 2012). Either functional or anatomical differences existed between the feet of the ancient and modern humans. This illustrates the value of interjecting

Figure 3.6. A single baboon footprint from FwJj 14 East, Ileret, northern Kenya. Courtesy of Ms. Melanie Crisfield, M.A.

anatomical or physiological studies of living organisms into the study of fossil species. The 1.5-my-old prints were made by either *Homo erectus* or *Australopithecus boisei*. That is, they were made either by an australopithecine or by an early member of genus *Homo*.

The FwJj 14 East prints were laid down in a soft muddy stratum, traversed over and over again by many animals. The mud is therefore heavily churned, or bioturbated. Fresh water probably occurred nearby. The modern analog would be the soft ground around a waterhole impressed with many prints. The co-occurrence of these fossil prints indicates that their makers not only existed within the same ecosystem, but lived near enough to each other to leave their tracks behind in a stratum that preserves behavior from a single season—even perhaps as short a time frame as several weeks. Hoofed mammals are abundant, which is expected, given what we know about modern food webs. But carnivore species are also abundant: four families are represented. This might reflect carnivore hunting and scavenging behavior around a waterhole. Pigs, which are omnipresent in the East African fossil record, including the record from this area, are rare. Against this palimpsest of ancient life is recorded the presence of three types of primate: hominids, baboons, and a smaller monkey species.

Table 3.1. Identifiable animal tracks from FwJj 14 East, paleontological collecting area 1A, Ileret, Koobi Fora, 1.54–1.52 mya.[5]

Class Mammalia	Common name and details
Mammal Family	
Hominidae	Five well-defined individual prints (three prints from one individual [*Homo erectus*?]; no trackways)
Cercopithecidae	Baboon; a smaller primate species (vervet?)
Felidae	Lion
Hyaenidae	Hyaena
Canidae	Jackal
Herpestidae	Mongoose
Bovidae	Various antelope species, based on small to large size prints
Equidae	Zebra
Chalicotheriidae	Chalicothere (extinct, large odd-toed herbivore related to horses and rhinos)
Suidae	Pig (one single print)
Class Aves	
Bird species	Heron; various other bird species, based on small to large size prints

In general, fossil sites in East Africa that contain primates also contain pigs. The virtual absence of pig footprints at FwJj 14 East signifies subtly different ranging and foraging behavior between primates and pigs that is not recorded in fossils, but is visible in this trackway evidence. A difference in water requirements might account for this. Note also that the primates traverse a landscape in which diverse mammalian carnivores roam. Predation was a real risk, but primates desperate for fresh water took this risk, and were alert enough in the presence of many carnivores so that there was a probability of emerging unscathed after drinking at the water source. There is circumstantial evidence that the hominid prints that appear modern were produced by members of the species *Homo erectus*. Several fossil arm bones from an individual hominid (either *Homo erectus* or *Australopithecus boisei*) were excavated from FwJj 14 East. The species that the arm bones belong to cannot be identified. This is because the arm bones do not occur with teeth, jaws, or cranial fossils, and these are the kind of fossils that establish ancient taxa. Type specimens that establish ancient species are overwhelmingly represented by teeth, jaws, or cranial remains (Chapter 2). This is a general problem in vertebrate paleontology. Archaeological evidence indicates that *Homo erectus* was capable of intricate foraging and ranging behavior, as well as complex interactions with contemporary carnivores. The smaller print with a more distinct big toe may be attributed either to *Australopithecus boisei* or to *Homo habilis*. Alternatively, this print may represent a juvenile *Homo erectus* individual.

---

[5] Based on two seasons of excavation during 2006–2007 (Fitzgerald 2007; Bennett *et al.*, n.d.).

The presence of multiple hominid prints with those of large carnivores indicates an intimate knowledge of predator behavior and ecology, so that terrestrial hominids were able to traverse the landscape safely.

Lagerstätten

Even though the likelihood of fossils or trace fossils being preserved is very low, there are sometimes rare sites that exhibit exceptionally well-preserved fossils. These fossils may even include superb, detailed evidence of soft tissues, like the flowers, fruits, and leaves of plants, or the skin, muscle, fur, stomach contents, and gut microbes of animals. These soft tissues are otherwise irretrievably lost soon after death. Liaoning Province in north-eastern China contains a number of sites that preserve the Jehol Fauna from the Early Cretaceous, dating to 130–110 mya. These Jehol deposits contain plants, insects, and vertebrate remains—the latter represented in such detail that they have established not only the dinosaur origin of birds, but the existence of feathered dinosaurs through intricate details of a variety of feather types, plumage patterns, and individual feather morphology (Norell & Ellison, 2005). Another example of a site that contains both exceptionally well-preserved plant and animal remains is the Eocene site of Messel, near Frankfurt, Germany (Schaal & Ziegler, 1992). Primates are rare components of the Messel fauna, even though fossils at the site are found in lake sediments surrounded by dense tropical rainforest, where primates should be expected to flourish.

Exceptional sites like Messel are known by the German word *Lagerstätten* (singular *Lagerstätte*). In English, this translates to "Motherlode," and was originally used in the mining industry to designate outstandingly rich ore or coal deposits. Now the term refers to exceptionally rich fossil deposits that contain information unlikely to occur anywhere else. Table 3.2 lists the *Lagerstätten* referred to in this book.

In spite of the wonderful preservation of fossils in *Lagerstätten*, organisms that were originally three-dimensional structures are often flattened into two-dimensional surfaces. Even if they are preserved in three dimensions, there are problems of retrieving information about the fossils from the rocky matrix in which they are entombed. In fact, this particular problem about retrieving fossil information is widespread, and not confined to *Lagerstätten*. Beginning in the early twentieth century, paleontologists addressed this problem by laboriously sawing serially though the rocks containing fossils. They then reconstructed the original fossil structure by drawing or photographing the sequence of sections. Many small fossils, such as those in the Burgess Shale *Lagerstätten*, are excavated from the surrounding rocky matrix under a microscope, using modified dental drills (Conway Morris, 1998). Photographs record anatomy, in case part of the fossil is destroyed as drilling continues. Currently, datasets of a series of sectional images of a fossil are often compiled through X-ray CT scans. Fortunately, massive radiation is not a problem for long-dead organisms! Other scanning and medical imaging technologies (e.g. MRI) can also be used. Computers can then manipulate the digital datasets to yield virtual fossils (Sutton, 2008).

Table 3.2. Lagerstätten referred to in this book. Sites marked with an asterisk (*) contain primate fossils.

Precambrian		
Doushantuo Formation	600 mya	Guizhou Province, China
Ediacara Hills	565 mya	Southern Australia
Cambrian		
The Burgess Shale	505 mya	British Columbia, Canada
Silurian		
Eramosa Formation	425 mya	Ontario, Canada
Carboniferous		
Mazon Creek	300 mya	Illinois, USA
Herrin Coal deposits	300 mya	Illinois, USA
Jurassic		
Solnhofen limestone	149 mya	Bavaria, Germany
Cretaceous		
Jehol Biota	128–110 mya	Liaoning, China
Yixian Formation	125 mya	Liaoning, China
Ukhaa Tolgod	80 mya	Mongolia
Paleocene		
West Bijou Valley	64 mya	Colorado, USA
Eocene		
*Green River Formation	50 mya	Colorado, Utah, Wyoming, USA
*Messel Oil Shale	49 mya	Hessen, Germany
*The London Clay	54–48 mya	Southern England
Miocene		
Pebas Formation amber	15 mya	Peru
*Rusinga Island sites	14 mya	Kenya
Ashfall Fossil Beds	10 mya	Nebraska, USA
Pleistocene		
Rancho La Brea Tar Pits	20,000 ya	Los Angeles, California, USA

Sometimes the chemical composition of sediments that enclose fossils is responsible for their exceptional preservation. Fossil ape bones and molds of insects and plants occur at 14-my-old sites at Rusinga Island in Kenya. Ancient volcanoes whose ashfalls had a rare mineral composition were erupting as the fossil sites were forming. The ash from these volcanoes (carbonatite ash) contained more than 50 percent carbonate minerals and almost no silica (Fischer *et al.*, 2009). This exotic ash was responsible for the exceptional preservation of fossils at these sites. However, this ash distorts the reconstruction of ancient climate, because it creates a false impression of environments that are hyper-arid. Thus, these 14-my-old Kenyan sites were initially assessed as forming during periods of reduced rainfall and declining forest vegetation.

Taphonomy and taphonomic processes

Taphonomy is the study of all of the processes that occur after an organism dies and is eventually discovered as a fossil. This discovery sometimes occurs long ages of time after the death of the organism (Figure 3.7). Thus, it is obvious why taphonomy is the study of how information about the past is subtracted, altered, or skewed. In fact, taphonomy is typically considered to be a study of how information about the past is lost. Lacking a time machine, one can only obtain an incomplete or distorted record of the past. Nevertheless, some researchers who deal with invertebrate paleontology have argued that more information about the past is available than usually recognized (Martin, 1999). In general, taphonomy addresses theoretical arguments about the nature of the fossil record, how faithfully fossils testify to the life of the past, and even about whether major unknowable gaps exist in the record.

Consider the Eocene Messel site discussed above. During Eocene times, the site was a deep lake in a volcanic crater. One expects to find fish, turtles, crocodiles, and amphibians in such a site, and they do occur there. Yet, land mammals, birds, and bats are also found. What taphonomic processes account for this? Modern volcanic lakes in West and Central Africa sometimes emit dense plumes of carbon dioxide from the bottom of the lake, as carbon dioxide is degassed from magma below the earth's surface, and deep water in the lake rapidly turns over. These massive plumes reach the lake surface, and spread upward and laterally. Oxygen-breathing animals around and above the lake, including humans and their domesticated animals, are quickly asphyxiated as they inhale pure carbon dioxide. Humans living around Lake Kivu, in the eastern Congo, are aware of this.[6] If they experience sudden breathlessness or fatigue, they know they must run away from the shores of the lake, rather than sit down until the episode passes. If they sit or lie down, they will quickly die. In 1986, 1,700 people and many animals died from an explosive release of carbon dioxide around the shores of Lake Nyos, in Cameroon (Jones, 2010). Gas trapped in deep water near the bottom of the lake is suddenly ejected by events that disrupt the water, such as landslides or heavy rain. A similar scenario is thought to have occurred in Eocene times at the Messel lake. Birds and bats flying above the surface were instantly suffocated by carbon dioxide, and plummeted into the lake. Wide varieties of land mammals living around the lake margins also quickly died, and were incorporated into the soft lake sediments. Scavenging by other land mammals and insects was minimized, because everything died during this catastrophic event. Primates are found, but

---

[6] I was told that swimmers in Lake Kivu must also take care. Pockets of carbon dioxide may lie above the lake's surface. Human swimmers surfacing for a breath of air may get a lungful of carbon dioxide, instead. Low areas of land surrounding Lake Kivu can also collect carbon dioxide. Visitors to the Tongo Reserve, near Lake Kivu, are warned of the dangers of a sudden influx of carbon dioxide. Chimpanzees in Tongo are not immune to this danger. Tongo has low depressions that contain dense concentrations of chimpanzee bone.

Figure 3.7. Hadrosaur (duck-bill dinosaur) limb bones from the Ruby site, South Dakota, Hell Creek Formation, 66 mya. One of these bones has been jacketed in plaster, prior to removal. The bones are disarticulated, and show a preferred orientation. They were deposited as lag in an ancient river channel.

they were rare at Messel, even though they are frequent at other European Eocene sites. Primates were obviously living in the tropical rainforest around the lake, but primates may have been concentrated inland, and were relatively rare visitors to the forest immediately surrounding the lakeside margins. This discussion of catastrophic die-off, exceptional preservation, and differential fossilization is a good example of how taphonomic processes influence interpretation of the fossil record.

Knowledge of taphonomic processes can be used to explain peculiar completeness or incompleteness of fossil specimens. For example, the astonishing completeness of a 3.3-my-old juvenile hominid individual from Dikika, Ethiopia (Alemseged *et al.*, 2006), can be explained by burial occurring while the skin was still intact, forming a "casing" for the bones like the casing for a sausage. Whether the corpse was fresh or mummified is immaterial—what mattered was the containment of the bones by the skin. Similarly, four fossil human specimens from the 1.98-my-old South African site of Malapa have intact, virtually complete skeletons with articulated bones. These specimens, which are attributed to the species *Australopithecus sediba*, fell into a vertical, tunnel-like sinkhole carved out by water in soft limestone bedrock. These individuals died at the base of the sinkhole without being disturbed by carnivores, and their mummified skin kept all parts of the skeleton intact and in order.[7] Another example of startlingly complete primate fossils is found in the Miocene of Kenya. Remarkably complete limbs of fossil apes of the genus *Proconsul* are explained as remnants of kills made by a carnivore and cached in its living quarters—a den in an ancient hollow tree (Walker & Shipman, 2005). The hollow (with contents intact) was later filled by sediments.

Incompleteness is the norm in the fossil world. Either physical or biological processes can account for the incompleteness of specimens. Taphonomic processes are responsible for whatever remnants of a once intact corpse are left. For example, single elements such as jaws, limb bones, or teeth are removed from a cadaver and moved along a flooding river. Even if these elements remain intact, they are seriously abraded, just as a stone or cobble would be abraded by sediment load in moving water. Even though point bars in ancient rivers are often areas where bones accumulate, fossils recovered from point bars preserve a record of their battering in water that has long since disappeared. Many of the primate fossils recovered from the Eocene/Oligocene Egyptian Fayum localities show such abrasion. Because they are the earliest undoubted anthropoid primates (Chapter 12), reconstruction of their anatomy and lifeways is affected by their taphonomic history.

The fragmentation, transport, and abrasion of carcasses have been experimentally studied by paleontologists. Conditions that promote concentration and preservation have also been studied. Sedimentation is important, because scavenging, transport, and breakage are significantly slowed once remains are buried. The physical and chemical effects that occur after burial are called diagenesis. These effects include flattening and distortion, as well as chemicals in solution penetrating the rocky matrix containing fossils.

Stratigraphy and taphonomic history are also essential components of archaeological studies. They form a separate sub-discipline called geoarchaeology. In recent years, microscopic analysis of thin sections of sediment has been used to reconstruct minute details of ancient environments. In many cases, this microscopic analysis reveals the impact of taphonomic processes that would otherwise be invisible. For example, at the Middle Pleistocene English site of Boxgrove, which contains both archaeological materials and hominid fossils, thin sediment sections reveal earthworm excrement indicating bioturbation (Goldberg & Macphail, 2006:Fig. 3.16). Ancient humans were knapping flint. A thin section with a flint flake embedded in silt and mud demonstrates that sedimentation by tidal wash sealed the flakes *in situ* (Goldberg & Macphail, 2006:Fig. 7.1d). Thus, an episode of ancient human activity is preserved in place exactly where it occurred.

---

[7] It is possible that samples of actual soft tissue are preserved from the Malapa hominids. Lest any crucial comparisons be neglected, suggestions for studying this irreplaceable material are being solicited on an international scale via the internet. The website of the Malapa Soft Tissue Project is http://johnhawks.net/malapa.

Can one trust the fossil record?

Given the foregoing discussion about taphonomy, it is clear that the fossil record has major gaps and distortions (Jablonski, 1999). Indeed, it is known that the record is only formed during bursts of sedimentation, because the abundance of fossils is directly proportional to the volume of sedimentary rock examined. Marine mollusks yield the most complete fossil record, because of the constant sedimentation in the ocean basins, and because of the preservation of hard tissues. Even so, an examination of the record over the last 60 my shows that the quantity and diversity of marine mollusks is quite simply affected by the amount of sediments examined, or the amount of rock outcrops that contain fossils (Crampton et al., 2003; Smith, 2003). Thus, in a very fundamental way, perceptions about fossil abundance and diversity are affected by the availability of sediments containing fossils. Of course, one surveys and excavates in areas where fossils are abundant. Why would one search for fossils in areas where they are not known to exist? Yet, the fossil record retrieved from known localities does not reflect the full picture of contemporary life.[8] Areas not represented by sediment basins, such as mountainous regions, are woefully unrepresented by fossils. Researchers interested in fundamental problems about speciation, extinction, or ancient abundance and diversity have made statistical attempts to correct for known biases in the fossil record, such as the rock volume bias. Species arise and become extinct all the time. When statements are made about peaks of speciation or extinction at certain points in geological time, scholars take the general background level of species origin or extinction into account. Paleontologists conclude that a significant episode of speciation or extinction is occurring only when these processes exceed the statistically expected level.

In theory, one might expect that fossils would be far more abundant from recent time ranges. The less the amount of time over which fossil loss can occur, the richer the fossil record should be. Normal processes causing the loss of fossils include chemical alteration, erosion by wind and water, weathering by temperature changes, compaction and distortion of strata, and pressure and heat deep within the earth that cause the metamorphosis (mineralogical, chemical, and structural changes) of sedimentary rocks that contain fossils. Paleontologists therefore automatically suspected that "the pull of the recent" would be very strong: recent time would be overrepresented by fossils. However, a statistical test of this idea using a very dense marine invertebrate fossil record finds that only 5 percent of Cenozoic diversity can be accounted for by "the pull of the recent" (Jablonski et al., 2003). Thus, the abundance of fossils is not skewed in favor of later time ranges.

---

[8] A similar problem afflicts archaeologists. Material is retrieved from sites, but this may not reflect the full complement of human activities in a region. The problem is particularly acute in the earliest archaeological record, where the data are sparse. Scholars working in this time range have heated debates about the boundaries of sites, and the behavioral differences that may exist between the archaeologically rich "patches" of sites and the impoverished "scatters" between the patches.

Questions about evolutionary rates and modes of speciation are also affected by the way in which the fossil record is formed. For example, the episodic nature of sedimentary bursts once led paleontologists to hypothesize that evolution proceeded according to bursts as well. This is the idea of punctuated equilibria, promulgated by two invertebrate paleontologists, Niles Eldredge and Steven J. Gould (1972). They argued that evolution—morphological change illustrated by fossils—only occurred in abrupt episodes, at species origins and extinctions. This hypothesis was actually an alternative to natural selection (Cachel, 1992, 2006, n.d.). Eldredge and Gould argued that natural selection was always weak, and was responsible only for microevolution (evolution within populations), not the grand procession of life (the evolution of higher taxa). Yet, natural selection has been demonstrated in the living world, and it is now known sometimes to be very strong and rapid. It can be studied under field biology conditions, where it is called "Contemporary Evolution" (Chapter 2). Furthermore, the application of new analytical methods has now allowed directional selection on the phenotype to be detected in the fossil record (Hendry, 2007; Hunt, 2007a, 2007b). It was necessary to develop new analytical methods, because the fossil record is rarely complete and dense enough to detect short-term trends. The idea of punctuated equilibria was widely cited in textbooks and the popular press. Yet, ironically enough, it never found favor with professional paleontologists publishing in refereed journals, and the level of interest in the concept, as documented by citations, appears to be declining (Ruse, 1999:150–151).

One of the earliest and best-known examples of punctuated equilibria has now been invalidated, and the facts refuting it come from a detailed analysis of ancient geography and stratigraphy. The dense and well-dated fossil record of Plio-Pleistocene mollusks from the Lake Turkana Basin in northern Kenya was originally interpreted as a "prima facie" case for punctuated equilibrium (Williamson, 1981). The evidence was considered sufficient by itself to establish the existence of this evolutionary process. The Turkana mollusks apparently showed three episodes of abrupt morphological change (punctuations) followed by long periods of stasis (equilibria). However, a re-analysis of this record demonstrates that the appearance of novel fossil morphology actually represents foreign animals invading from outside the Turkana Basin during periods of wet climate, and not rapidly evolving local lineages (Van Bocxlaer et al., 2008). This episode illustrates two important points. First, that both taxonomy and stratigraphic context are important for interpreting fossils. Second, that paleontology is not merely a matter of discovering, describing, and dating new fossils—important insights can come from the re-analysis of well-studied data.

One of the most important pieces of information to be gleaned about fossil taxa is their first and last appearances in the fossil record. That is, scholars record the First Appearance Datum (FAD) and the Last Appearance Datum (LAD) for a taxon. The time between these two points is the span of the taxon's existence (Figure 3.8). FADs and LADs appear straightforward, but there are difficulties associated with them. Mammalian land mammal stages are units of geological time given by

Figure 3.8. Deciduous teeth of the African fossil pig *Metridiochoerus andrewsi*. Its First Appearance Datum (FAD) is 2.0 mya, and its Last Appearance Datum (LAD) is 1.6 mya. The presence of this species at a site can therefore be used to date the site to within this time range.

FADs. However, mammal FADs are extremely disparate across the North American continent. This is probably caused by poor sample sizes in many localities, but it has negative implications for paleontological dating (Alroy, 1998). Immigrant taxa, which are important for assessing continental connections and faunal invasions, do not appear to be particularly widespread or abundant.

One of the current major problems in paleontology is the disjunction between the time of a lineage's origin documented from fossils (its FAD), and the estimated times of origin documented from DNA and molecular differences in living organisms. The DNA and molecular data always seem to indicate a more ancient divergence time for lineages than the fossils do. This is also true for primates (Chapter 7). The obvious explanation is that, given the inherent unlikelihood of fossil preservation, the first appearance of a lineage's fossils will always postdate its true origin, especially if taxa are rare, or are rare at the beginning of their fossil history. For example, the discovery of *Juramaia sinensis*, a 160-my-old placental mammal, in China has pushed back the fossil record of the divergence of placental and marsupial mammals by about 35 my (Luo *et al.*, 2011). This ends the incongruity between fossil and molecular data in documenting the separation of marsupial and placental mammals. However, there are sometimes gaps of hundreds of millions of years between the molecular time of origin and the first appearance of fossils. These huge gaps are unlikely to be caused by the absence of fossils. There appears to be a fundamental problem in using the molecular data to generate times of origin. The reason for this is not clear. Vertebrate animals are especially affected by these gaps, so artifacts of generation time and genome size might be involved. One solution is to invent computer software that estimates molecular divergence with new probability models (Peterson *et al.*, 2008).

Alternatively, a recent shift towards studying genes and genomes, rather than simple DNA nucleotide changes, seems to be generating data that resemble the traditional morphological data used by paleontologists (Telford & Littlewood, 2008). Finally, major gaps between a time of origin posited by molecular data and that indicated by the fossil record are sometimes shortened by the discovery of new and earlier fossils. This is the case for the earliest known fossil animals. Their origin has been extended back by several factors: the discovery of trace fossils, basal fossil animals such as sponges, and possible fossil animal embryos in the 600-my-old Doushantuo Formation, China (Budd, 2008).

Occasionally, the first members of a group may appear in unexpected habitats. There is an 18 million year gap between fossils of the first vertebrates with four limbs (tetrapods) and trackways of prints documenting the presence of animals with feet and digits (Niedźwiedzki et al., 2010). The transition from fish to tetrapod was long thought to have taken place in flooded forested environments surrounding rivers. However, the first tetrapod trackways occur in marine intertidal flats or lagoons. This contradicts the idea that tetrapods had a non-marine origin, and raises the possibility that paleontologists searching for the first tetrapod fossils were examining rock exposures from the wrong habitat. This explains the 18 million year gap or "ghost range"—a period during which tetrapods existed, but from which no fossils have been recovered.

A fundamental problem is that the First and Last Appearances (FADs and LADs) of fossils may ultimately be controlled by the sediments or rocks that contain them. Periods when fossils are rare may be times when life is actually recovering from a mass extinction event, such as occurred at the Cretaceous/Tertiary boundary (Chapter 7). Alternatively, one may expect fewer fossils during periods of high aridity and low sedimentation. This seems to have occurred during the Oligocene epoch. Nevertheless, the Oligocene appears to have represented a true bottleneck for mammal evolution, because mammal diversity was severely reduced during this time, and mammals have never recovered (Chapter 11).

The simple analysis of First and Last Appearances sometimes reveals organisms that are seemingly absent or out of place. Colorful and memorable new terms have been invented for these taxa. Lazarus taxa are species or genera that amazingly re-appear in the fossil record long after they were thought extinct. They emerge like Lazarus from his tomb, dead for 3 days but then resurrected by Jesus Christ. Sometimes taxa are thought to persist through time, but are then later discovered actually to have been extinct. They were simply misidentified. These are Elvis taxa —actually gone, but seemingly still around. Lastly, there are taxa that are the sole surviving representatives of once abundant lineages or clades. They, too, are doomed to extinction, although they linger for a while. This is the Dead Clade Walking phenomenon.

Besides the known filtering effect of taphonomic processes, there are other fundamental biases affecting the fossil record. Paleontologists collect fossils at known fossil sites, and these areas therefore produce abundant evidence. Countries with a long tradition of paleontological survey and study are also overrepresented by fossils. These fossils are also used in biostratigraphy, and in identifying rock units, or formations. Ultimately, both paleontologists and geologists tend to believe that there is a common cause behind both the volume of rock and the abundance of fossils at any given time. Plate tectonic movements and global sea levels are generally identified as the common factors behind the correlated rock and fossil records (Peters, 2005).

4 The world of the past

The origin of continents and oceans

The title of this section is a translation of a book first written by the meteorologist Alfred Wegener in 1915, and revised and translated many times afterwards. Wegener proposed that a giant supercontinent (Pangaea) had once existed. Portions of this supercontinent split apart and moved away, creating ocean basins and the modern geography of the earth. This hypothesis was called continental drift. However, although Wegener amassed data on the similarity of continental coastlines (especially South America and South Africa), geological strata, and animal and plant distributions to support his hypothesis, he had no workable mechanism to explain the movements of the fragmented supercontinent. Most geologists considered Wegener to be an eccentric crank, although biologists appreciated the fact that continental drift eliminated the necessity for invoking land bridges to explain animal and plant distributions; and biologists further applauded the fact that continental drift rendered massive parallel evolution unnecessary. Wegener's hypothesis was resurrected in the 1960s, which witnessed a revolution in geological thought. Using new evidence about earthquake and volcanic belts, paleomagnetism, and sea-floor spreading, geologists outlined a new mechanism of continental movement called plate tectonics.

Movements (tectonics) affect plates of lithosphere. Convection currents deep within the mantle of the earth are the ultimate cause of these movements. These convection currents are driven by heat. Radioactive materials exist within the mantle. Their natural decay processes emit heat, and thus a molten iron core lies underneath the mantle. Movements within the mantle drag the overlying crust, and upwelling mantle plumes create local hotspots of volcanic activity, as observed in the Hawaiian Islands. These deep mantle plumes and hotspots can be tracked by volcanic islands and submarine sea mounts as plates pass over the plumes. These plumes can also affect the speed of plate movements and the intensity of volcanic eruptions. For example, the Réunion plume in the Indian Ocean is responsible for the simultaneous rapid motion of the Indian plate and slowing of the African plate beginning 67 mya (Cande & Stegman, 2011). The enormous Indian Deccan flood basalts are adjacent to the Réunion plume, and begin erupting at the same time. These massive volcanic eruptions contribute to the major extinctions that occur at the K/T boundary.

Tectonic plates can separate from each other, slide past each other, or converge. Plates separate from each other when new lava is extruded at their boundaries. This can be observed on land today in the Afar Triangle of Ethiopia. When plates separate at the central oceanic ridges, sea-floor spreading occurs, and ocean basins are formed. The denser lava sinks, and the lighter continental crust sits high on the surface of a plate. Most of the ocean floor is young. About 50 percent of it is no older than 65 my, the length of primate evolution. We primates are equal in age with much of the ocean floor! The rate of sea-floor spread can be quick (6 cm per year at the San Andreas Fault, which is sliding part of California north past the North American Plate), moderate (3 cm per year at the South Atlantic Ridge), or catastrophic. The entire island of Sumatra was shifted by 10 meters during the colossal magnitude 9.3 earthquake of December 26, 2004. The fault line created by this earthquake extended 1,500 km, the surface of the entire planet was raised and lowered by at least 1 cm, and the earth rang for weeks with free oscillations caused by the earthquake's seismic forces. Plate tectonic movements can therefore sometimes be quick and violent. When plates converge, a plate that carries oceanic crust subducts or descends into the earth's mantle, and creates a series of volcanic islands—an island arc—as the descending plate deforms the mantle. When plates carrying continental crust collide, the lithosphere is driven upward, and mountains are formed. The Andes Mountains and the Himalayas result from this type of plate movement.

The movement and dispersal of land animals is obviously affected by the opening of water gaps and ocean basins. G. G. Simpson (1940a) first examined the nature of land mammal dispersal by showing the decreasing likelihood of land mammals moving through land bridges or corridors, passing climatic or geographic filters, and crossing substantial water gaps in a sweepstakes fashion (Chapters 10 and 13). To this list of decreasingly less probable movement (corridors, filters, and sweepstakes dispersal), McKenna (1973) added two new elements that were made possible by plate tectonic theory: Noah's Arks and Beached Viking Funeral Ships. In the Noah's Ark scenario, a lithospheric plate moves and bears a cargo of living land animals. In the Beached Viking Funeral Ships scenario, a moving plate bears fossils. If the plate eventually contacts another plate, the fossils are now fixed to another continent. However, these fossils are "immigrants," and have no biogeographic relationship to fossils or living animals found in the new land. Because India was a Noah's Ark for much of the Tertiary, and did not contact Asia until the Miocene, primate evolution and dispersal in India and Asia is affected by plate tectonic theory.

Climates of the past

Detailed climatic records for over 150 years illustrate how mammals react to climatic change. Different species respond in different ways (Blois & Hadly, 2009). At the local level, seasonal changes might influence diet, mating, home range size, migration, or hibernation. Increasing seasonality might lead to shifts in diet, habitat, and geographic range. At the regional level, differences between populations could lead to divergence and speciation. Prolonged climatic change is likely to be associated with immigration, speciation, or extinction of species.

Up until about 34 mya, the climate of the earth was distinctly warmer than at present. Mean global temperatures were much higher, and there was no ice, or virtually no ice, at the poles. The atmosphere of the earth had higher concentrations of greenhouse gasses. Volcanic eruptions were frequent from about 60 to 40 mya, which partly accounts for this higher concentration. Methane emissions from shallow ocean sediments bubble up to the atmosphere and also affect greenhouse gas concentrations (Chapter 8). On the other hand, beginning at 34 mya, Antarctic ice becomes permanent. Ephemeral ice sheets appear in the northern hemisphere at 10 mya, during the Late Miocene. Northern ice sheets become permanent, and reach a continental scale at about 3 mya. The growth and recession of the great continental ice sheets has been a focus of intense study. Geologists have been debating the cause of the mysteriously increasing and diminishing northern ice sheets for nearly 170 years (Ramo & Huybers, 2008). Major plate tectonic events clearly set the stage. Land surfaces sitting over the poles, or polar ocean surrounded by land, allow ice to build up in high latitudes. Land surfaces that restrict oceanic circulation are also important. With the rise of the Isthmus of Panama at 3.5 mya, warm equatorial water could no longer flow around the entire globe. This appears to have been the final trigger for the advent of permanent ice sheets in the northern hemisphere.

It has been difficult to fine-tune our understanding of the activity of these northern ice sheets. Besides the difficulties of dating their presence and extent, models that explain their waxing and waning are proving difficult to understand. From about 3 to 1 mya, glacial cycles were regular, and lasted about 41,000 yrs (Ramo & Huybers, 2008). This matches the hypothesis of Milutin Milanković, who argued in the 1930s that variations in the earth's orbit are responsible for glaciation. When the northern hemisphere receives less solar radiation during the summer, snow lasts throughout the year, and gradually builds up into continental ice sheets. Known astronomical variations led Milanković to predict a 41,000 yr glacial cycle. Beginning at about 1 mya, and continuing to the present, Ice Age climate appears to show 100,000 yr cycles (Ramo & Huybers, 2008:Fig. 1). Questions about solar radiation, the coupling or decoupling of northern and southern hemisphere glaciation, the geographic extent of early ice sheets, and the thickness and ablation of ice sheets are currently being debated. In particular, climatologists are still uncertain about how changes in solar radiation that affect the earth's upper atmosphere are translated into drastic variations in ice volume at ground level. Hypotheses about the relative length of ice growth and melt seasons are now being tested.

For the Cenozoic, at least, the paleoclimatic record is detailed enough to show that the carbon cycle is intimately linked to climate. At the end of the nineteenth century, the American geologist Thomas C. Chamberlain hypothesized that fluctuations in the carbon cycle were the trigger for global ice ages. He argued that the weathering of high mountain ranges and the reduction of oceanic calcium carbonate would reduce atmospheric CO_2 levels, causing lower global temperatures. Low rock weathering rates and high oceanic carbonate production would increase atmospheric CO_2, causing higher global temperatures. But carbon dioxide levels in the atmosphere fluctuate in a more complex way. These levels are affected both by influxes from volcanic out-gassing and by the sequestering or release of organic carbon. Changes in continental ice volume are strongly correlated to the amount of CO_2 in the atmosphere. As the partial pressure of atmospheric CO_2 changes, paleobotanical data also change to reflect either tropical or temperate shifts. Atmospheric carbon dioxide thus affects climate, but other greenhouse gasses also force climate change. Atmospheric methane has a greater effect than CO_2 does. Eruptions of methane into the atmosphere from sea-floor sediments were apparently responsible for the terrific rapid upward spike of global temperature at the Paleocene/Eocene boundary.

Marine sediments are currently emitting oil and gasses, including methane. The rate of this emission is not constant. A number of factors affect the rate of hydrocarbon ooze. During periods of Pleistocene deglaciation, tar deposits occur in sediments off the California coast, indicating greater seepage of hydrocarbons as the climate ameliorated (Hill *et al.*, 2006). Incorporation of methane into the atmosphere would further intensify the natural climatic warming.

Much of the fine-grained analysis of climate change in the Late Cenozoic is conducted through geochemistry. Measurements are made of the ratio of heavy to light oxygen extracted from sea-floor sediments and terrestrial ice cores. Reconstructing fluctuations between cold and warm global climate is based on these oxygen isotope ratios.

Habitat reconstruction

It is not enough to outline the broad details of geography or climate. When one considers the life of the past, one desires knowledge of vegetation, precipitation, and seasonality before limning in the fossil animal species of an ancient community. Pollen has an outer surface that is not only extremely durable, but also has a characteristic morphology for genera. Pollen can sometimes be retrieved from sites. Impressions of leaves, fruits, flowers, stems, and sometimes whole tree trunks of fossil wood—macrobotanical remains—can be recovered from sites. Thus, one can gain some understanding of plant species that were present at fossil sites. By measuring the topography of leaf edges that occur in different climates today, paleobotanists have established a link between leaf edge structure and mean annual temperature. The simpler the leaf margin (e.g. palm leaves), the higher the temperature. Complicated leaf margins (e.g. oak leaves) occur in more temperate conditions. Empirical regression lines between the degree of leaf evagination (i.e. the total length of leaf margin) and mean annual temperature are generated from living plant species (Wolfe, 1978, 1993). The margins of fossil leaf assemblages retrieved from ancient sediments are then measured, and ancient temperatures are inferred from the results.

Many Cenozoic sites have plant genera and species that are alive today, and this allows botanists to reconstruct an ancient habitat in some detail. Often, the plant taxa are out of place. For example, tropical trees occur in what are now desolate badlands in the American West during the Paleocene and Eocene. These ancient trees occur in what is obviously a very different habitat from the modern landscape. Sometimes taxa occur in suites that are not observed in modern times. This situation is more difficult to interpret, because it implies that community structure and species interactions were different in the past. These comments also apply to animal genera and species. A thorny complication in both cases takes place when an ancient taxon has no living representatives, which occurs with increasing frequency as sites become more ancient. The solution for enigmatic plant and animal taxa is to investigate their anatomy in detail in an attempt to infer their lifeways. For example, does the plant have woody tissue? What is the bone density and cross-sectional area of an animal limb? Chapter 5 will highlight ways in which the lifeways of extinct animals can be inferred.

Ancient soils (paleosols) can sometimes be recovered from sites. Detailed knowledge of the temperature and rainfall regimes that create similar soils in the modern world allows specialists to reconstruct ancient habitats based on the depth and nature of ancient soil horizons. Miocene ape habitats in Africa and southern Asia have been reconstructed using paleosol data (Retallack, 1991), and habitats dating back to the Mesozoic are routinely studied using paleosols. Nevertheless, one would like even finer detail—an impression of the degree of open country or wooded area at a site. Is this possible? Yes. Correlating the carbon isotopes from thousands of samples of modern soil associated with satellite photographs of vegetation in the same area allows one to discover the relationship between modern soil temperature and ground cover. Carbon isotopes retrieved from ancient soils (paleosols) can then be used to reconstruct ancient soil temperatures and infer the degree of shade or tree cover that existed at an ancient site (Cerling et al., 2011; Feibel, 2011). Quantification of shade allows modern biologists to categorize habitats as grasslands, open woodlands, or dense forests. These paleosol researchers use a worldwide database of modern sites to define grasslands as having less than 40 percent tree cover; woodlands as having greater than 40 percent tree cover; and forests as having greater than 80 percent tree cover. Examining sites in East Africa back to 7 my reveals that most of the sites with fossil humans have less than 40 percent canopy cover (Cerling et al., 2011). They are thus defined as grasslands, which has important implications both for human evolution and for the evolution of Old World monkeys that are sympatric with humans at these sites (Chapter 15).

When coexisting or sympatric species are considered together, paleontologists explore the way that ecosystems function. In the modern world, a number of factors are known to affect the structure and function of ecosystems (Figure 4.1). Some of these operate at a local level; some of these operate at a continental level.

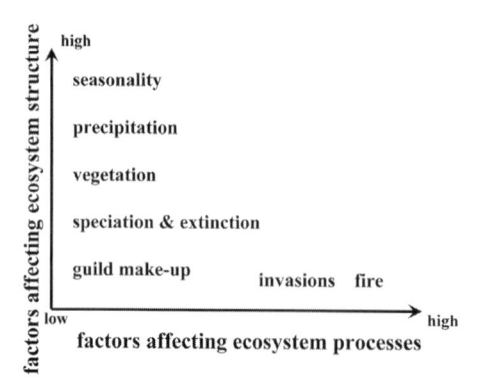

Figure 4.1. The relative ranking of factors that affect ecosystem structure and processes.

Community Assembly

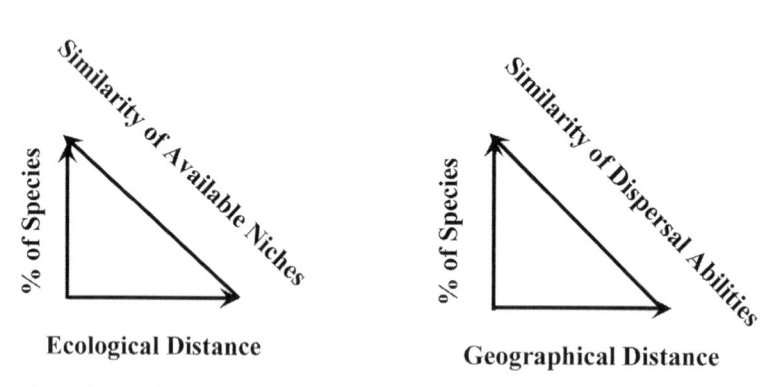

Figure 4.2. Ecology versus geography as a determinant of community assembly. Evidence indicates that a true equilibrium state (stasis) does not exist. That is, if perturbations occur, the community does not return to its initial state. At any given time, a community is an assemblage of species that are accreting or being removed in a partly random fashion.

Many researchers study living communities, and lament species loss within these communities. Ecological and geographical factors are known to affect species composition within communities (Figure 4.2). Comparison between communities can yield insights into the way that natural selection or random events affect evolution. Similarity between communities in the percentage of species present can be caused by a resemblance in available niches or by the dispersal abilities of taxa. However, community composition in the modern world appears to be very plastic or labile. Native species go extinct in a local area, and are easily dislodged or replaced by invasive species. If perturbations occur, the community does not return to its initial state. Evidence indicates that a true equilibrium state (stasis) does not exist. At any given time, a community is an assemblage of species that are accreting or being removed in a partly random fashion. An ecological niche does not need to be occupied by any given, single species, but by similar species with similar niches—members of a guild. The same processes occur in the fossil record. Paleontologists have falsified the idea of "coordinated stasis." Coordinated stasis is the idea that a network of species interactions is so strong that it resists change. New or invasive species therefore cannot penetrate an existing ecosystem. No evidence for this has been found in the fossil record.

What affects primate presence in a community? This is partly determined by how primates disperse. They cannot fly and they cannot swim. Another factor limiting primates is their nearly universal reliance on tropical or subtropical forest habitats. Tropical or low-latitude animals are much more sensitive to changes in the physical environment than temperate animals are. This is known as Rapaport's rule (Stevens, 1989; Cachel, 2006). Primate sensitivity to climatic or habitat change quickly removes them from an ecosystem. When living primates are studied, there is no evidence that they experience competition from sympatric primates. This may be caused by recent extinction events since the Late Miocene that removed many primate species. Primate survivors have a wealth of food and space that fossil primates—shoehorned into species-rich ancient communities—did not have. If living primates experience competition, it may come from frugivorous birds and bats or other arboreal mammals, such as squirrels or the small arboreal marsupials of South America.

5 The lifeways of extinct animals

Introduction

Georges (Baron) Cuvier (1769–1832), who worked in the late eighteenth century, is often identified as the Father of Vertebrate Paleontology. He achieved this eminence by virtue of two things. First, he recognized that some organisms had become extinct. The idea that species were not immortal was a revolutionary one. Species were traditionally held to have remained unchanged and immutable since the point of Creation. Cuvier believed that episodic global catastrophes accounted for these extinctions. A new creation would account for subsequent species. The history of life on earth thus became a series of revolutionary extinctions and creations.

Second, Cuvier recognized that fossil bones and teeth resembled those of living animals, and that these bones and teeth could be used to reconstruct the lifeways of extinct creatures. In fact, it was the fond expectation of Baron Cuvier that future anatomists would discover morphological rules or relationships that would allow an absolute knowledge of the lifeways of an extinct species from even a single bone or tooth. Animals were machines whose bodies worked through mechanical principles. It was the duty of functional morphologists to discover these principles. Cuvier seems to have expected that there would be an inevitable relationship between an animal's niche and its anatomy. In the same way that mathematicians have discovered the equations for an ellipse or a circle, Cuvier believed that future anatomists would discover the absolute relationships, as rigorous as equations, between form and function. To this end, Cuvier articulated a "law" of correlation of parts.

In a word, the shape of the tooth implies the shape of the condyle, that of the scapula, that of the nails, just as the equation of a curve implies all its properties; and just as by taking a property separately as the basis of a particular equation, one would find again and again both the ordinary equation and all the other properties, similarly the nails, the scapula, the condyle, the femur and all the other bones taken separately, give the tooth, or give each other; by beginning with any of them, someone with a rational knowledge of the laws of organic economy could reconstruct the whole animal. (Fortelius 1990:210).

This is an anatomist expecting to discover rules of animal design that would resemble the formal equations of analytical geometry. Cuvier also strongly implies that standard categories of animal design exist: the equivalent of the circles, ellipses, parabolas, and other shapes recognized in analytical geometry.

Biologists who study living animals are aware that the lifeways or ecological niches of these organisms can be determined, in theory, by a multitude of factors that shape the niche—what G. Evelyn Hutchinson called "an abstractly inhabited hypervolume." Nevertheless, field biologists routinely identify four factors that determine the major elements of a niche (Cachel, 2006). These elements are body size, diet, locomotion, and the temporal patterning of behavior—for example, is the animal diurnal or nocturnal? Does it breed seasonally? Does it migrate? Does it hibernate?

Body size

Nothing is more important in animal design than body size. Once one considers multicellular organisms, no biological traits are immune from the effect of body size. The regression line between body size and metabolic rate in endothermic animals has a slope of 0.75. The exact relationship and slope may vary according to taxon (McNab, 1978, 1980, 2012). Locomotion, ranging behavior, diet, digestion, metabolic rate, and reproduction are affected by body size. These factors interact in a complex way. Metabolic rate is affected by diet and the ranging and foraging needs of a particular diet. Yet, dietary quality affects physiological problems of digestion and energy budget. Metabolic rate subsequently affects reproduction and reproductive success. Thus, the rate of increase of a species in a particular habitat is influenced by body size, diet, and local ecological parameters. Adults of the same species may differ in body size because of their sex. This is called sexual dimorphism—the sexes occur in two morphs. In mammals, adult males are often larger than adult females, but the degree of sexual dimorphism can vary widely between species. Sometimes it is dramatically large; sometimes it is negligible. Within humans, the degree of sexual dimorphism varies between populations (Figures 5.1–5.3).

Metabolic rate and primate reproduction and reproductive success have been studied in some detail, because they have implications for the human condition. For example, orangutans have an astonishingly low metabolic rate for mammals. It averages 23 percent lower than expected for their body weight, and explains their delayed maturation and strikingly low reproductive rates (Pontzer *et al.*, 2010). This reflects the balancing of orangutans on a knife's edge of adequate overall caloric and protein intake. Female Bornean orangutans must be in a positive energetic balance before they can reproduce (Knott & Thompson, 2012). In human females, the energy content of breast milk is affected by economic status and the sex of the nursing infant. Because male infants grow larger, the milk that mothers provide them has an energy content that is 25 percent greater than that provided to female infants (Powe *et al.*, 2010). Human brain size at birth is affected by the metabolic demands placed on the mother by the fetus, and not by the size of the birth canal. Birth occurs when the metabolic demands of the fetus are about to surpass the mother's ability to meet both the energy requirements of the fetus and her own energy requirements (Dunsworth *et al.*, 2012).

Figure 5.1. Sexual dimorphism in adult humans, illustrated by a limestone statue of Nen-Khefet-Ka and his wife Nefer-Shemes, Old Kingdom, Egypt, 2414–2347 B.C. Courtesy of the Oriental Institute, University of Chicago.

This analysis ignores the fact that birth does not release mammalian mothers from energy constraints: lactation is far more costly to mammalian mothers than pregnancy is (Hayssen, 1984:Fig. 2). Lactation is associated with the greatest expenditure of energy by mothers because the quantity of milk produced must not only support survivorship of the young for an extended period of time, but must also support their continued growth. Costs to the mother increase as the young grow larger. One must also note that the brain of a newborn mammal can continue to grow after birth, given that the bones of the neurocranium are not fused at birth.

Any discussion of body size must take into account the sizes of different species and the relative impact of general body size on any trait being examined. This impact of body size is called allometry–the effects of differences in scale (Figure 5.4). Allometric simply means "different in shape." Isometric means "the

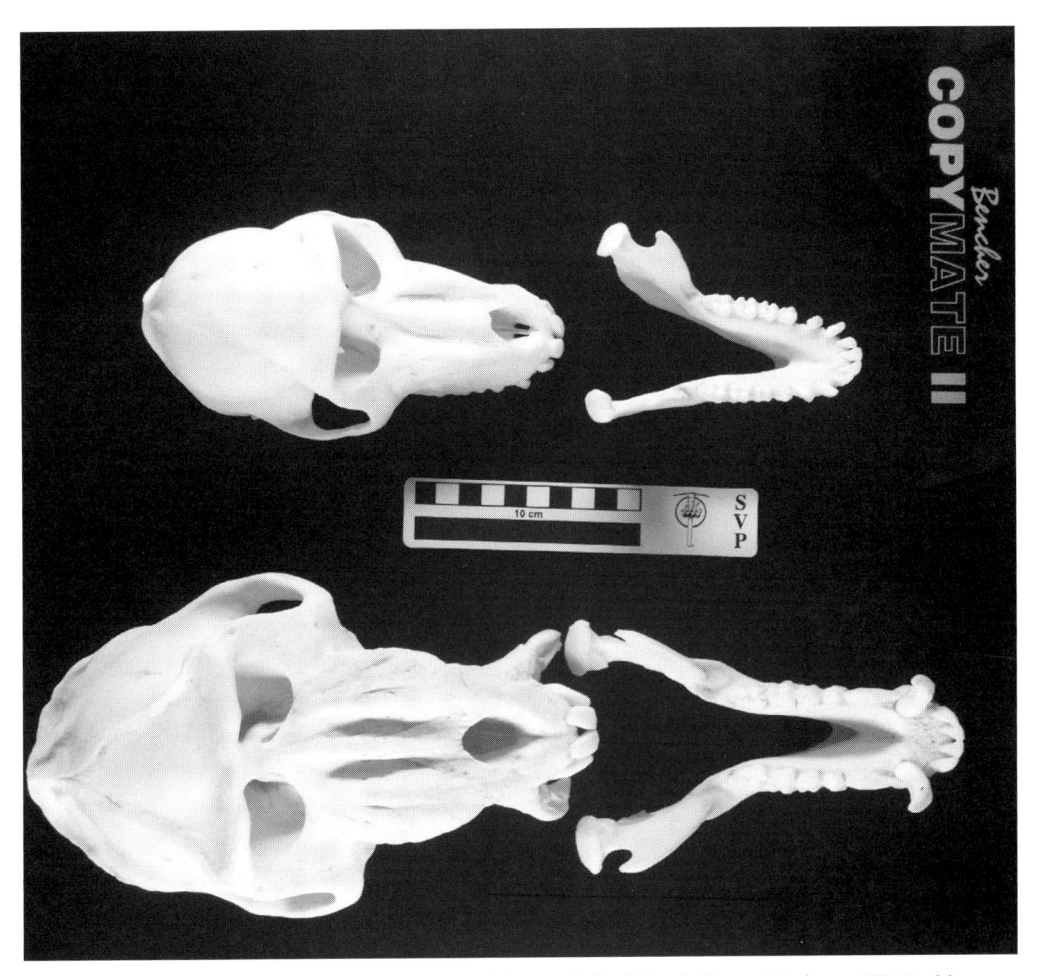

Figure 5.2. Sexual dimorphism in adult mandrills (*Mandrillus sphinx*), an Old World monkey species. The male (lower) is much larger than the female. Many Old World higher primates (catarrhines) exhibit profound sexual dimorphism, with males often being twice the size of females.

same shape." Isometry results when the same shape is maintained regardless of differences in scale—doubling the size of a femur, for example, while maintaining the same shape. Julian Huxley (1932) produced the first equation illustrating the impact of scale. Size can be compared over many orders of magnitude, from mice to blue whales. When the axes are geometrically (logarithmically) scaled, the relationship becomes a power function:

$$y = ax^b \qquad (5.1)$$

or

$$\log y = \log a + b \log x \qquad (5.2)$$

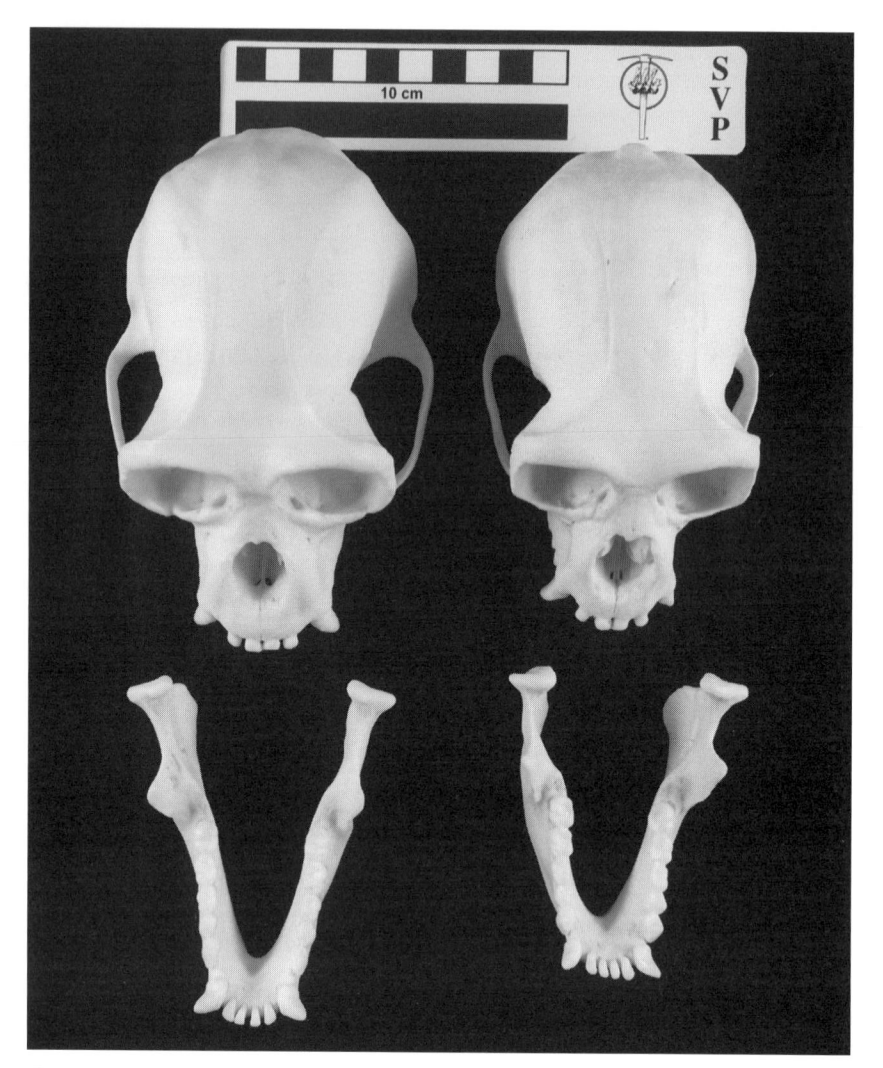

Figure 5.3. Sexual dimorphism in adult siamangs (*Symphalangus syndactylus*), a small-bodied ape species. Note that the male (left) is not much larger than the female. This reduced sexual dimorphism is unusual for catarrhine primates. Gibbons (genus *Hylobates*), small-bodied apes that are smaller than siamangs, show an even greater reduction in sexual dimorphism.

When the axes are linear:

$$y = bx + a \qquad (5.3)$$

y represents some trait (e.g. femur length) measured in species of different sizes; x represents body mass or some other measurement of body size; a (where the line intercepts the y axis) and b (the slope of the regression line) are

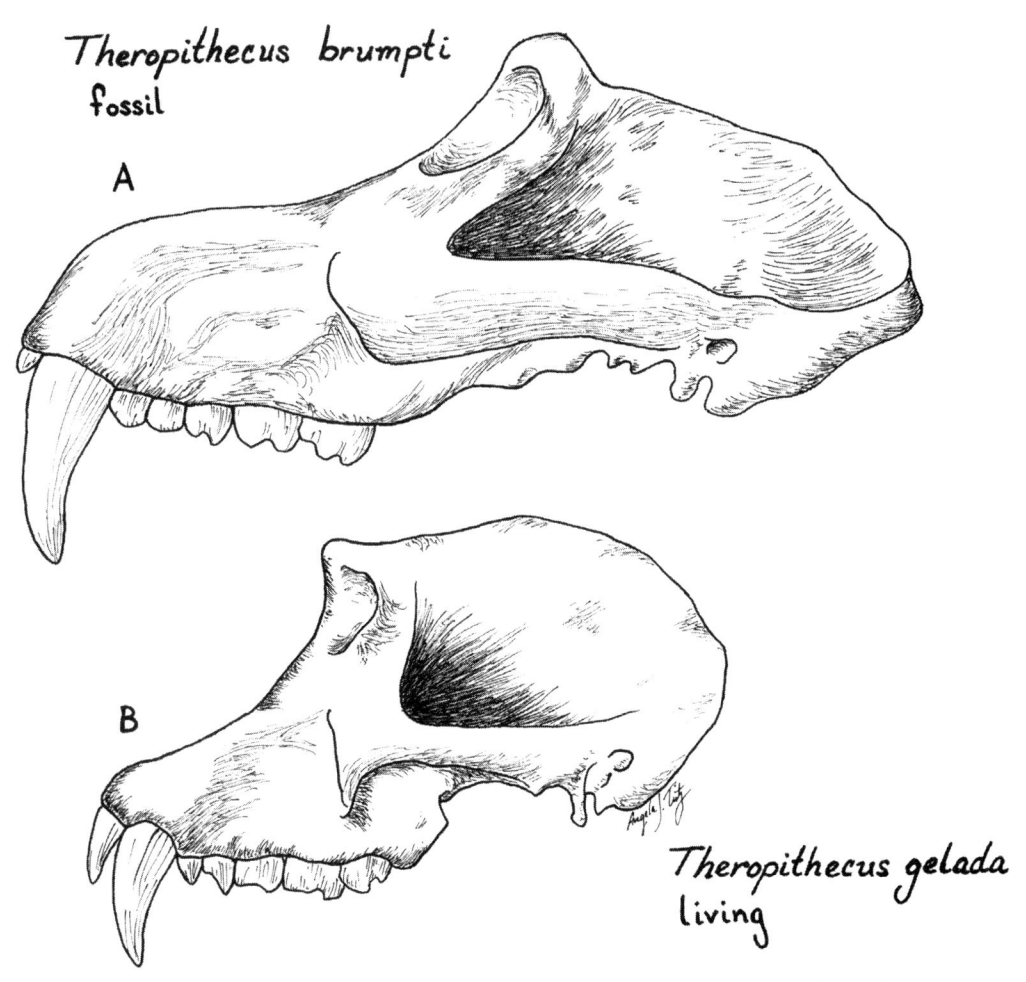

Figure 5.4. How allometry affects the shape and relative proportions of cranial structures. A fossil species of gelada baboon (*Theropithecus brumpti*, A) is contrasted with the modern gelada baboon (*Theropithecus gelada*, B). Note that the relative size and shape of the neurocranium, snout, and orbits are affected, and that the orbits of *T. brumpti* appear to be angled upward very strongly. These differences merely reflect allometry. They therefore do not indicate higher-order (e.g. genus-level) differences between these two taxa. Illustration by Angela J. Tritz.

constants. Logarithmic scaling is useful when size contrasts cover several orders of magnitude.

Estimates of body size are crucial for an understanding of the life of a fossil animal. But how can one generate an estimate of body size for a fossil species? Because teeth may be the only information available for a fossil species, the relationship of tooth size and body size has been studied in living mammals. How big was a fossil taxon if only its teeth are known? The worst-case scenario is having only a single fossil tooth. Can one say anything about body size from only a single tooth? Here is the methodology. Dental measurements (usually crown areas generated by multiplying length and width measurements) are taken in living species. The relationship between dental measurement and body weight is graphed, and an equation is generated that relates a particular measurement (e.g. lower M1 area) to body weight in a group of living species (carnivores, primates, etc.). If a fossil lower M1 is discovered, the crown area of this tooth is measured, and then plugged into the primate equation. Individual equations have been generated for each permanent tooth in the upper and lower jaws. The tooth crown areas of the lower postcanine teeth—and especially M_2—appear to have the most robust correlation with body weight in living primates, 0.964 and 0.968, respectively (Gingerich & Smith, 1985:Table 1). The area of the M2 always accounts for about one-third of the total molar area in mammals (Kavanagh et al., 2007). This linchpin function of the M2 may account for its high correlation with body weight. Conroy (1987) has developed regression equations to estimate body weight in fossil primates using the area of M_1.

Sometimes problems develop if teeth in living animals are specialized. For example, species in the living marsupial genus *Caluromys* have specialized molars that are smaller for a given body weight than in other marsupials. Including these species in the analysis between marsupial molar size and body size reduces the strength of the correlation (Gordon, 2003). This is not a trivial fact. *Caluromys* species have been used in studies about primate locomotion and primate origins (Chapter 7). These animals (called woolly opossums) also appear in studies reconstructing the lifeways of Paleocene primates (Chapter 8). It is therefore interesting that using their molar tooth size would lead to underestimates of body size. The opposite problem occurs with studies of fossil humans. Early fossil humans called australopithecines are megadont—their teeth are larger than expected for their body size. Dental measurements alone therefore overestimate body size in this fossil human group.

When fossil limb bones are available, their length can be used to estimate body mass. However, the limbs of living primates are longer than expected when contrasted to other mammals of the same body mass. This is especially true for the femur, but primates also have additional length in the humerus, tibia, and ulna bones (Alexander et al., 1979; Alexander, 1985). When researchers are interested in both body size and locomotion, the relative size of joint surfaces that are weight-bearing can be used to infer body size. Weight-bearing joints, of course, support body mass during both static posture and during active, complex locomotor activities that may create compressive, tensile, and shear forces. If cranial bones are available, cranial length can be used in body size estimates of fossil species. Researchers interested in the relationship between brain and body size in fossil species advocate that the diameter of the foramen magnum in the cranium—the aperture through which the spinal cord passes to the brain—may be the best indicator of body size when examining brain evolution. In summary, there is no best option for inferring body size in fossil mammals. If estimates of body size are available from two or more body areas, the optimal strategy would be to combine these estimates in order to infer body size.

The sex of fossil mammals, including primates, is usually inferred from bimodal distributions of traits. For example, if teeth from the same species occur in two sizes, the larger teeth are presumed to come from males, and the smaller teeth are presumed to come from females. There are several confounding factors: age, geographic variation, and species identification. Deciduous and permanent teeth cannot be compared, and animals from different populations may be different in size because of local habitat differences. Species identification is the ultimate confounding factor—different species can be different in size, and therefore teeth that are different in size may represent different species, and not males and females of the same species. Miocene primate genera such as *Proconsul*, *Dryopithecus*, and *Sivapithecus* are affected by considerations of sexual dimorphism and species identification. Even if teeth come from the same locality, they may not come from the same species. Teeth may have collected at a site from different areas through taphonomic processes such as river transport. Site formation processes must always be studied.

Bergmann's rule, known since the nineteenth century, and confirmed for the vast majority of living endotherms, states that, if a genus is widely distributed, larger species are found in cold environments, and smaller species are found in warm environments. This is also true within a species, if it is widely distributed. It is difficult to establish whether Bergmann's rule operates in fossil mammals, because their remains are generally rare, and not well dated. However, during an abrupt, dramatic rise in global temperature in the Paleocene/Eocene Thermal Maximum at 55 mya, a number of mammal lineages become smaller. This is exceptionally well confirmed for the earliest horses, which decrease their size by about 30 percent over the first 130,000 yrs of the Paleocene/Eocene Thermal Maximum, and then afterwards increase their size by about 76 percent (Secord *et al.*, 2012; Smith, 2012). Pleistocene land mammals dramatically increase their size during global temperature drops, when many living mammals had megafaunal equivalents.

In the 1880s, the American paleontologist Edward Drinker Cope empirically demonstrated that body size tends to increase with time within a lineage. This became known as Cope's rule, and it is a well-established feature of mammal evolution. However, paleontologists have argued about why Cope's rule exists. Because larger body size increases competitive ability within a species, competition has always been thought to drive body size increase. But larger body size has costs—for example, a larger body needs more food. If food is scarce or unpredictable, a larger animal may have more difficulty surviving and reproducing than a smaller counterpart of the same species. A general study of Cope's rule using 554 species of fossil mammal species distributed throughout the Cenozoic finds that the explanation is more complicated than intra-specific competition at any given point in time. Size increase occurs with more specialized diets, and when global temperature declines (Raia *et al.*, 2012). Size increase also comes with a heightened risk of extinction, because larger animal species reproduce more slowly and need larger home ranges. Thus, body size can be startlingly labile over the duration of a given lineage. In living mammals, it can shift from generation to generation, depending on changes in the local environment.

What can comparative genomics reveal about mammalian body size? Domesticated dogs have the greatest range of body size in living mammals, because the largest breeds overtop the smallest breeds by 57 times. This variability in size doubtless exists because humans have enthusiastically selected for body size among the different breeds. It is now known that a single gene variant (allele) of the *IGF1* gene is principally responsible for body size in all small dog breeds (Sutter et al., 2007). Body size in dogs thus has a major genetic underpinning in this one gene. Because this gene occurs in other mammals (including humans and non-human primates), variability in body size within a species or between closely related species may have a relatively straightforward genetic explanation. In humans, who are the only primates where body size has been intensively investigated, many genes and allelic variants of genes are known to influence stature, with each gene having a very small, additive effect. Height and body weight are also influenced by sexual dimorphism. The degree of sexual dimorphism in humans, however, is different between human populations. In general, the larger the average adult stature of a population, the greater the difference in statural size dimorphism (Cachel, 2006).

Diet

The primatologist Richard Kay empirically demonstrated that a dietary threshold associated with body size exists in living primates (Kay, 1975). Living primates that are under 500 g consume insects and gums and resins. Primates above 500 g begin to supplement their diet with leaves. The 500 g separation has become known as Kay's Threshold. The threshold exists because the cellulose and hemicellulose in leaves requires a longer gut and more complex digestion than insects or gums do. Digestion of cellulose is often dependent on the activity of gut bacteria. A longer gut equates to a larger body size. Animals above the threshold cannot acquire solitary insects at the rate and density that they need to support their larger size. Social insects (e.g. termites, ants) are an exception. Fruit is quicker to digest, and can be consumed by animals both above and below the threshold. Kay's Threshold appears to operate throughout the mammals.

Major categories of diet in mammals can quickly be discerned from the morphology of the mandible, the relative size of the anterior and posterior teeth, and the height of the molar crowns (Maynard Smith & Savage, 1959). In fact, mastication in animals with very different diets can be examined using biomechanics (Figure 5.5).

Vertebrate paleontologists have used teeth and tooth crown morphology to infer the diet of extinct organisms since the time of Baron Cuvier. Living organisms with known diet provide the database from which inferences are drawn—an

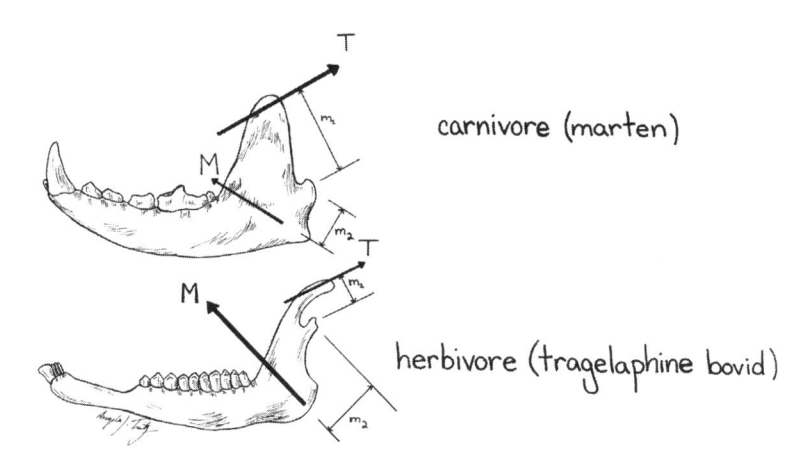

Figure 5.5. Basic mammalian jaw mechanics. This figure illustrates how relative size of the anterior and posterior dentition, height of the molar crowns, height of the coronoid process, and height of the mandibular condyle can differentiate between a specialized carnivore and a specialized herbivore. Because the lower jaws are scaled to the same size, the contrasts are more visible. The size and vector of the temporalis muscle (T) and the masseter muscle (M) are also indicated. These are major muscles of mastication whose attachment areas can be clearly seen on intact bone. The moment arms, which measure the biomechanical efficiency of these muscles (T = m_1, M = m_2), are also shown: m_1 spans the length between the area of attachment of the temporalis muscle on the coronoid process and the mandibular condyle; m_2 spans the length between the area of attachment of the masseter muscle on the angle of the mandible and the mandibular condyle. The relative size of these moment arms also differentiates carnivore and herbivore. After Maynard Smith and Savage (1959). Illustration by Angela J. Tritz.

anatomical use of uniformitarianism. There is thus a long paleontological tradition of using teeth to infer diet. And, because teeth are the most well-preserved elements in the fossil record, they are the major focus of paleontological research, and are also used to diagnose taxa.

The general morphology of the molar crowns has long been thought to be related to the texture of an animal's food, and has figured in some of the earliest attempts to infer diet in fossil primates (Gregory, 1916). Quantification of molar shear crests (which cut through vegetation or insect bodies) relative to a primate's size has been used objectively to infer diet in fossil primates. This methodology was initiated by Richard Kay (1975). The following example illustrates its utility. Molar shear crests and body size from 13 Late Eocene primate taxa from the Fayum Depression were closely examined to infer diet (Kirk & Simons, 2001). Generalized dietary categories were created. Anthropoid species appeared largely frugivorous, regardless of differences in molar crown morphology. Prosimian species had more dietary diversity, ranging from insectivory to folivory, with intermediate categories of generalized frugivory to frugivory/insectivory.

The association between diet and dental morphology has recently been examined in a novel way, through the use of geographic information systems (GIS) analysis, in two orders of mammals (Evans *et al.*, 2007). These are carnivores and rodents, which are both diverse and long separated by evolutionary history. Three-dimensional laser scans are made of both upper and lower postcanine tooth rows, and data are processed into digital elevation maps (the GIS format). Tooth rows are scaled to the same size, so that differences reflect surface complexity, and not differences in scale. Examined in this way, the dentition of the two mammal groups is surprisingly similar. Carnivores and rodents differ in body size, tooth classes, number of teeth, tooth replacement, cusps and crests on tooth surfaces, and range of chewing motion. Nevertheless, carnivores and rodents in the same dietary class have the same degree of tooth shape complexity; only hyper-carnivores (one of five dietary classes) are different (Evans *et al.*, 2007). Thus, tooth complexity reflects the texture of the food that is eaten, irrespective of ancestry. Natural selection trumps phylogeny, because lineages that are very different independently evolve similar dental occlusal surfaces to process similar dietary items.

Microwear on the crowns of fossil teeth reflects the texture of food items in the diet. However, analysis of living primates demonstrates that a microwear signature can be removed within a week, so that microwear reveals nothing more than the last meal. This is the equivalent of the traditional method of gauging diet in mammals—through analysis of the stomach contents of trapped or shot individuals. Taphonomic processes affecting fossils can also alter microwear, e.g. when grains of sediment slide against teeth incorporated within a deposit. Another line of investigation examines plant phytoliths. These are durable silica fragments incorporated within the body of a plant. Phytoliths can have a distinctive morphology that sometimes allows for plant family-level or genus-level identification. If found in ancient soils or sediments, phytoliths allow vegetation to be reconstructed, but they can also be found embedded in tooth crowns or the calculus (tartar) on the sides of a tooth.

The chemistry of enamel from fossil tooth crowns is undoubtedly the best window into ancient diets. Enamel is resistant to diagenesis, because it is crystalline and almost free of organic material. Stable isotopes of carbon, nitrogen, and oxygen reveal the trophic level of a fossil species within its community, because different isotopes from dietary items and water are incorporated into the developing tooth crown. Predators that eat herbivores are higher in the food chain; i.e. they are at a higher trophic level. Their isotopic signatures will reflect those of the animals that they eat, but these signatures will be sequestered into smaller and smaller packages, depending on the height of the trophic level that they occupy. Different photosynthetic pathways utilized by plants allow diet to be examined more finely (Cerling *et al.*, 2010; Lee-Thorp & Sponheimer, 2013). Carbon isotopes are incorporated into the tissues of the animals that eat them. Animals that rely on C_3 plants (leaves, bark, other plant parts, resins, nuts, and fruit from nearly all trees, bushes, shrubs, and forbs) can be separated from

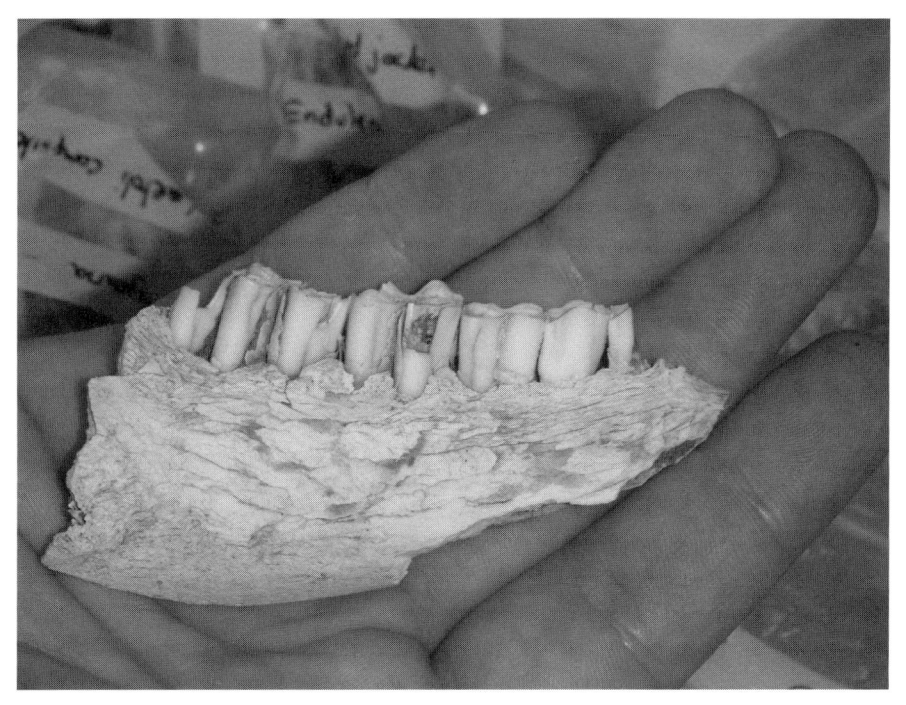

Figure 5.6. Fragmentary mandible of a living zebra, collected from the Serengeti. Collagen will be extracted from this bone, and the relative abundance of carbon and nitrogen isotopes will be determined. This will position the zebra within the trophic grid of the modern Serengeti ecosystem.

animals that rely on C_4 plants (tropical grasses and some sedges), or the CAM[1] photosynthetic pathway (relatively rare arid succulents). Animals that eat other animals mirror the diet of their prey. Omnivores eat both plants and animals. Nitrogen isotopes yield evidence of reliance on marine resources, and oxygen isotopes yield evidence of the degree of reliance on water from actively drinking or water obtained indirectly from leaves or other ingested plant parts (Figures 5.6 and 5.7).

Fossil hominids can demonstrate the utility of these methods. Stable carbon isotopes have been used intensively to study the diet of fossil hominids (Ungar & Sponheimer, 2011). The South African australopithecine species *Australopithecus africanus* and *Australopithecus robustus* were higher in the food chain than living non-human primates, because they had diets that included meat from vertebrates that were eating C_4 grasses. They did not resemble living or fossil baboons from

---

[1] CAM is an acronym for crassulacean acid metabolism. This refers to a third, alternative pathway for carbon metabolism in plants. These are succulents found primarily in arid environments, such as euphorbias.

Figure 5.7. Rutgers graduate student René Studer-Halbach examines the readouts of carbon and nitrogen isotopes after collagen samples have been combusted within the adjacent mass spectrometer. This work is taking place at the Research Laboratory for Archaeology and the History of Art, Oxford University.

South Africa, or living chimpanzees, including those from modern savannah or open environments. However, australopithecine species were diverse in diet. *Australopithecus* [*Paranthropus*] *boisei* in East Africa had an extraordinarily high C_4 signal—it resembled the modern gelada baboon, whose diet consists almost entirely of grass. Yet, two specimens of *Australopithecus sediba* in South Africa had no C_4 signal at all. This is unexpected, because they therefore most resemble the far more ancient hominid *Ardipithecus ramidus*. *Australopithecus sediba* relied on C_3 food items, such as fruit and herbaceous plants. Dental microwear reveals that it ate hard objects. Plant phytoliths embedded in dental tartar show that it ate tree bark and woody tissue or cambium (Henry *et al.*, 2012). This exegesis into australopithecine diet demonstrates not only that closely related fossil species may vary in diet, but also that a fossil species may be like nothing in the modern world. A fossil species may have no exact matches among living organisms.

Stable isotopes can help to differentiate the niches of closely related sympatric species. For example, working at the Research Laboratory for Archaeology and the History of Art (RLAHA) at the School of Archaeology, Oxford University, my advisee Mr. René Studer-Halbach is compiling reference data on carbon and nitrogen isotopes from bone collagen in living mammals from Laetoli and the Serengeti ecosystem in Tanzania. The relative abundance of these isotopes can be determined by mass spectrometry (Figures 5.6 and 5.7). Mr. Studer-Halbach anticipates studying isotopes from the tooth enamel of six fossil cercopithecoid monkey species from the Pliocene site of Laetoli, Tanzania. He will examine resource competition and niche separation in these species. The Plio-Pleistocene site of Makapansgat Limeworks in South Africa contains 12 sympatric species of cercopithecoid monkeys, separated into three genera (Chapter 15). Investigation of stable isotopes allows one not only partly to separate these fossil species by diet, but also to question incongruity between dietary data and morphological data (Fourie et al., 2008).

How uniform or how variable was an animal's diet through time? This can be investigated using carbon isotopes. Delicate laser ablation of tooth enamel within single teeth of *Australopithecus robustus* was performed at successive intervals along the tooth crown that approximated the intervals between ancient seasons or years. This work demonstrates that individuals could ably shift their diet from season to season, or from year to year (Sponheimer et al., 2006). Members of this species were therefore not specialized in diet.

The most profound shift in diet that occurs in the life of mammals takes place during weaning, when a juvenile is no longer dependent on its mother's milk. Weaning can occur very abruptly or in a more gradual fashion. Data from living humans and rhesus macaques reveal that weaning can be discerned from the enrichment of barium relative to calcium in the enamel of deciduous teeth (Austin et al., 2013). Applying these results to a single fossil Neanderthal permanent M1 from Scladina, Belgium, indicates that caretakers began to introduce adult food to this child at the age of 7 months. At the early age of 1.2 years, the Neanderthal child's diet was composed exclusively of adult food. This technique for inferring the age of weaning can be applied to non-human fossil primates. Because it can reveal the abruptness of weaning, it has the potential for discerning fine-grained behavioral interactions between ancient mothers and their offspring. Were youngsters suddenly pushed and slapped away from their mothers as they attempted to nurse? Or was weaning more gradual, and therefore less traumatic to the youngsters?

Bone collagen can sometimes be retrieved from fossil species that are not very old, or that lived in temperate or cold environments, where the degradation of collagen was halted or hindered. Stable isotopes from collagen can yield information like that from enamel. Collagen from fossil mammals and humans from Late Pleistocene sites in Europe has allowed the trophic interactions in ancient communities to be examined (Lee-Thorp & Sponheimer, 2006). Neanderthals appear to have been largely carnivorous. They fed on terrestrial mammals, and their diets were similar to those of contemporary wolves, hyaenas, and cave lions. Anatomically modern humans, who succeed the Neanderthals, were also carnivorous, although their diets were more diverse, and included fish.

In spite of new work into the chemistry of prehistoric enamel or bone collagen to reconstruct diet, there has been a long-standing tradition of studying specifics of molar crown morphology to infer fine details of diet in fossil species or to elucidate ancestry. Nevertheless, tooth crowns may not reveal details of diet or the secrets of phylogeny. This statement represents a revolution in mammal paleontology. Simple genetic changes underlie tooth number and defects in enamel formation. *Pax9* gene dosage influences the number of teeth in mice and other mammals, including humans (Kist *et al.*, 2005). In humans, a single mutant allele results in the congenital absence of most molar teeth, and sometimes second premolars and incisors. A further reduction in normal *Pax9* alleles in mutant mice results in the loss of third molars, small or missing lower incisors, and defective enamel in the ever-growing incisor teeth. Thus, a minimum number of normal *Pax9* alleles are needed for the standard development of mammalian dentition.

The number and size of the molar teeth are often critical to mammalian species diagnosis. What controls the growth and size of individual molar teeth? In light of the importance of mammalian molar teeth in taxonomy, it is important to note that evolutionary development has revealed the factors that trigger the formation of molar teeth, and the factors that determine the relative size of molar teeth in the distal tooth row. Through mouse experimentation, researchers discovered a mechanism of activation and inhibition in the sequential development of molar teeth (Kavanagh *et al.*, 2007). One feature of this mechanism is that the second molar always accounts for one-third of the total molar area. Using this algorithm, predictions were made about what the relative size of the first and third molars should be in 29 mouse species that differed in diet. These predictions were confirmed, which demonstrates that evolutionary development has the capacity to generate and test hypotheses. This is important, because research in evolution and paleontology is often thought to lack the ability to falsify hypotheses by experiment.

A simple genetic basis underlies major differences in molar crown shape and even molar number in mammals. Dental traits are not necessarily determined by a complex genetic basis. Experiments with mutant mice demonstrate that differing expression of the protein ectodysplasin can simultaneously affect many dental traits (Kangas *et al.*, 2004). These traits include tooth number and size, cusp number and shape, and both longitudinal and transverse cresting. The dental phenotypes can be so dissimilar that taxonomists would probably sort them into separate genera, in spite of the fact that different levels of only a single protein (ectodysplasin) shape the phenotype. Furthermore, the dental traits that respond most strongly to ectodysplasin signals are the least variable traits. These are the traits that taxonomists preferentially select for unequivocally sorting out taxa. This research suggests that the diversity of mammalian tooth shape has been overemphasized. It may not reflect fundamental differences in diet, or be the ultimate window into ancestry (Cachel, 2006).

Currently, 29 separate gene mutations are known to affect molar development in the mouse (Harjunmaa *et al.*, 2012). Most of these cause molars to be lost or simplified. This agrees with what is abundantly known about mutations that affect human molar development. However, there are numerous examples in the fossil record of molar cusps increasing in many mammal lineages. Experiments with mouse embryonic teeth demonstrate that cusp numbers can be greatly increased without altering tooth size, to the point where the crown topography resembles the complex third molar of the giant panda—a phenotype that is completely alien to the normal mouse molar. This result was achieved by simultaneously manipulating three different developmental pathways. There may be a genetic bias against increasing molar cusps, and this may occur at slower rates through evolutionary time than decreasing complexity. Nevertheless, selection pressure for masticating fibrous plant foods (as in the giant panda processing bamboo) has been strong enough to create multi-cusped molars independently many times over during the course of mammal evolution (Harjunmaa *et al.*, 2012).

The postcanine teeth of mammals scale isometrically with body size—they fall on the regression line when compared to body weight, and are exactly as large as expected for body size (Fortelius, 1990). This is also true for primate postcanine teeth (Gingerich & Smith, 1985). Primate masticatory muscles are also isometric with respect to body size (Cachel, 1984). Isometry of postcanine teeth and masticatory muscles presents a conundrum for classical biomechanics. It implies that occlusal stress is independent of body size. But large mammals need to eat more food, and can eat lower-quality plant food to fuel metabolic processes. Processing of this food should require a greater postcanine surface area and larger masticatory muscles. That is, larger animals should have positively allometric postcanine teeth and masticatory muscles. Lucas (2004) explains the strangely small molar teeth of large mammals by arguing that they ingest larger food particles that fracture at lower stresses, and which therefore do not demand large occlusal surfaces. I have argued that the fractal geometry of gut absorption area or total capillary area might explain the isometry of postcanine teeth and masticatory muscles (Cachel, 2006). The area of the gut that digests food and the area of the capillaries that transports nutrients to cells scale like a three-dimensional structure or volume.

Inferring behavior from morphology

Locomotion

The shape of organisms is plastic. The concept of phenotypic plasticity—the ability of a given genotype to produce different phenotypes, depending on the environment—illustrates the possibilities of organic form. Function is a manifestation of the possibilities of form. Shape is plastic, but function is supreme. If natural selection occurs on traits that perform the same biological function, they will be highly correlated and will covary. This pattern of morphological integration can be used to demonstrate the existence of natural selection on functional complexes (Olsen & Miller, 1958). Yet, how is function itself studied, especially in extinct organisms?

The skeleton and its muscles are the first line of evidence for inferring behavior in fossil animals. The skeleton is alive in living animals. It needs a constant blood supply, and its mineral components, particularly calcium, are in constant flux with the blood stream, because critical serum levels of calcium are necessary for nerve function. The skeleton responds to weight-bearing and activity by constant remodeling—i.e. bone is deposited or resorbed according to rules known since the nineteenth century. The thickness and internal trabecular organization of cancellous bone reflects its biomechanical function as the living bone responds to stress generated by posture (weight-bearing itself) and movement. In eight genera of higher primates (including humans and three other hominoids), a constellation of features in trabecular bone under the heads of the femur and humerus can be used as signatures of locomotor groups (Ryan & Shaw, 2012). These features include trabecular number and density, trabecular connectivity and orientation, and the occurrence of rods or plates of trabecular bone. Wolff's Law, dating to 1884, states that bone is locally built up (deposited) where it is needed, and locally removed (resorbed) where it is not needed (Ruff et al., 2006). Tension leads to bone deposition; compression leads to bone resorption. Muscle contraction generates tension, and bone surfaces are marked by areas of muscle origin and attachment, called entheses or muscle scars (Benjamin et al., 2006). The surface area of muscle scars in fossil animals can be digitized to yield a quantitative approximation of area. Muscle strength, which depends on the muscle weight or the physiological cross-section of a muscle, remains unknown. Nevertheless, there are ways of studying muscle strength in fossil animals—to study bite force, for example. These are dependent on examining bone deformation or strain when bone is stressed. Bone strain in fossil specimens can be studied either directly through minor surface striations called Sharpey's fibers, deformation of a load-bearing cast of the specimen modeled in a photoelastic substance, or through finite-element analysis of a 2D or 3D computer digital image of a CT-scanned specimen (Bell et al., 2009). Regions in a fossil bone that experienced compression, shear, or tension caused by posture and mode of locomotion during life can be identified by these methods.

Muscles produce electrical signals when they are active. This fact has been known since the late nineteenth century, when the French physician Guillaume Duchenne discovered that the atrophied leg muscles of patients with muscular dystrophy did not generate the electrical signals that the muscles of normal people did. During the 1960s, John V. Basmajian (1963, 1972, 1974) began to produce a body of work on the pattern of electrical signals produced by the muscles of human beings engaged in normal behavior. This is called electromyography or EMG (Figure 5.8). One of Basmajian's innovations was to use fine-wire electrodes implanted directly into individual muscles. Thus, the electrical record could definitively be associated with a single muscle—it was not a diffuse record that might be produced by several muscles. Basmajian then collaborated with the primatologist R. H. Tuttle to compare the EMG patterns in the forelimb and hindlimb of humans and great apes during posture and locomotion that is normal for each species—e.g. suspensory arm posture, knuckle-walking, bipedal walking (Tuttle, 1994; Tuttle

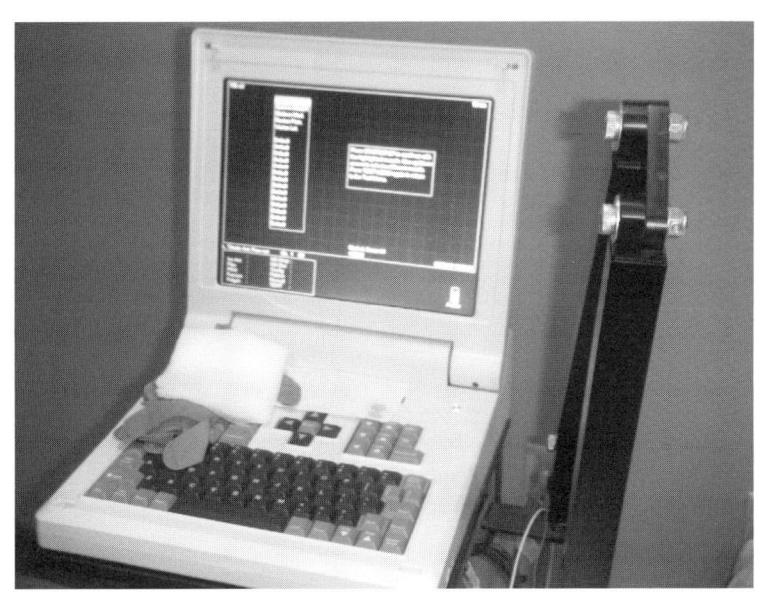

Figure 5.8. Electromyography is routinely used in medicine to diagnose muscular and neurological problems. In this standard medical EMG apparatus, surface electrodes with short cables pick up the electrical activity of muscles, and the record of this activity is displayed on the computer screen.

et al., 1979). EMG analysis is now routinely and extensively used in studies of primate locomotion, as well as in studies of primate mastication. Remote telemetry currently allows EMG signals to be recorded wirelessly. There is no long tail of cables leading from electrodes to the recording device that might interfere with normal activity, or to hinder movement. Surface electrodes can now pick up signals from underlying muscles without invasive implantation. Electromyography is therefore now used to detect the muscles that are active when animals move, including the patterns of activity of different muscles, and the recruitment of muscles at different phases of movement.

Gait, speed, and joint angles can be examined in living organisms to extract detailed information about locomotion and kinematics. Data gleaned from these analyses can be used to infer locomotion in extinct species, the impact of different substrates on locomotion, or to model transitions in locomotor mode. For example, my advisee Ms. Melanie Crisfield is using high-speed three-dimensional motion-capture video on living human subjects whose center of gravity, trunk mobility, and other variables have been altered as they move across dry or wet sand (Figure 5.9). This explores the effect of anatomical features seen in early humans on bipedal gait and joint angles. Such analysis allows one to model the transition from arboreality to human terrestrial bipedality, even though the fossil evidence for this transition is absent or ambiguous.

Locomotion can be inferred even if postcranial bones are completely missing in a fossil taxon. Cranial bones that preserve the semicircular canals of the inner ear

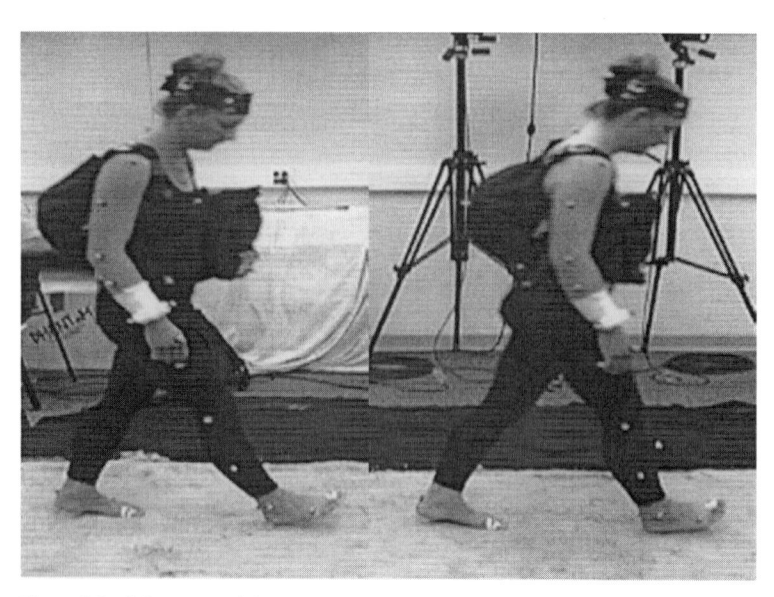

Figure 5.9. A human subject traverses an experimental wet sand trackway at the Computational Biomedicine, Imaging and Modeling Center (CBIM), Rutgers University. High-speed video cameras capture motion in three dimensions. The backpack and wrist weights alter the center of mass to mimic body shape in australopithecines, an early hominid group ancestral to genus *Homo*. Australopithecines had a heavier thorax and longer arms than modern humans do. The subject carries a weight equivalent to the weight of an ancient infant. Reflective discs attached at joint surfaces allow complex joint movements to be analyzed from the video record. Courtesy of Ms. Melanie Crisfield, M.A.

relating to balance can be used to reconstruct locomotion in fossil primates (Ryan *et al.*, 2012). The inner ear or otic capsule is extremely hard—it is formed from the rock-like petrosal bone—and it is likely to be preserved even when more fragile elements of the cranium are lost. The semicircular canals contain otoliths which react to movements of the head. Because the semicircular canal system tracks angular rotations of the head as an animal moves, and coordinates posture and movement using visual, proprioceptive, and otolithic cues, the degree to which an animal engaged in fast, acrobatic movements can be gauged. A study of 16 fossil anthropoid primates, ranging in age from the Late Eocene to the Late Miocene, reveals significant changes in locomotion over 35 my.

Ranging behavior

Every animal that has ever been studied in the wild has a home range—an area of land that the animal traverses during its normal activities. Social animals live together within a home range. Field biologists can monitor an animal moving within its home range, either through direct observation or by remote telemetry. Researchers can map the boundaries of a home range, because animals do not venture beyond its limits. This is important, because animals do not wander at random through the environment. They know the resources contained within their home ranges, and they know its dangers. Areas in the home range that are very highly frequented because they contain crucial resources are called core areas. A home range whose boundaries are covertly or overtly defended is called a territory. Biologists track the movements of an animal through time, and then use software to construct the irregular dimensions of a home range on a map. There are a number of ways to estimate home ranges, based on the way that animals disperse, data points are analyzed, and the intensity of land use (Samuel & Garton, 1985; Meretsky, 1987). Paleontologists are beginning to explore something similar to a home range. Using large fossil databases and well-dated sites whose paleolatitudes and paleolongitudes are known, paleontologists can glean some idea about the range size of a fossil species (Carotenuto et al., 2010; Raia et al., 2012). There are obvious problems. For example, the discontinuous nature of the fossil record, uneven distribution of fossils through succeeding time intervals, and ancient lake and ocean margins can cause range sizes to be overestimated. Another problem is that paleontologists use minimum convex polygons to lay out range sizes. The outermost points of a distribution are connected together to form these polygons. Field biologists consider this to be a simple way to estimate home range, because it is very sensitive to outlying data points, and incorporates areas that are not actually used. However, it is difficult to see how paleontologists have any alternatives to this method. The important point is that they are beginning to investigate ranging behavior in fossil animals.

Elements of ranging behavior in individual fossil animals can sometimes be retrieved. Isotopes of the element strontium that are retrieved from ancient tooth enamel preserve a history of ranging behavior. Strontium enters the enamel of a growing tooth crown from the bedrock of a particular area. Strontium signals that are foreign to a local landscape indicate that an animal has moved into the area from a different landscape than the one in which its tooth crowns were formed (Figures 5.10 and 5.11). This method has been used to examine migration patterns in Pleistocene mammals in Europe. It also has the potential to differentiate movement patterns between sexes in the same species. Thus, strontium isotope signals from two ancient South African human species (*Australopithecus africanus* and *Australopithecus robustus*) are different in males and females (Copeland et al., 2011). Females of these species have foreign strontium signals, indicating that they dispersed into an area from outside; males have a local strontium signature. Males therefore did not range as far as females, and did not disperse from their natal landscape. Note that these strontium results imply that males remain within the group to which they were born, but females range more widely, and join foreign groups. Thus, strontium isotopes can be used to reconstruct the mating behavior of fossil organisms. Unlike the majority of living social primates, where females are philopatric or remain within their natal group, while males disperse, these fossil australopithecine species had females dispersing to find mates in foreign groups.

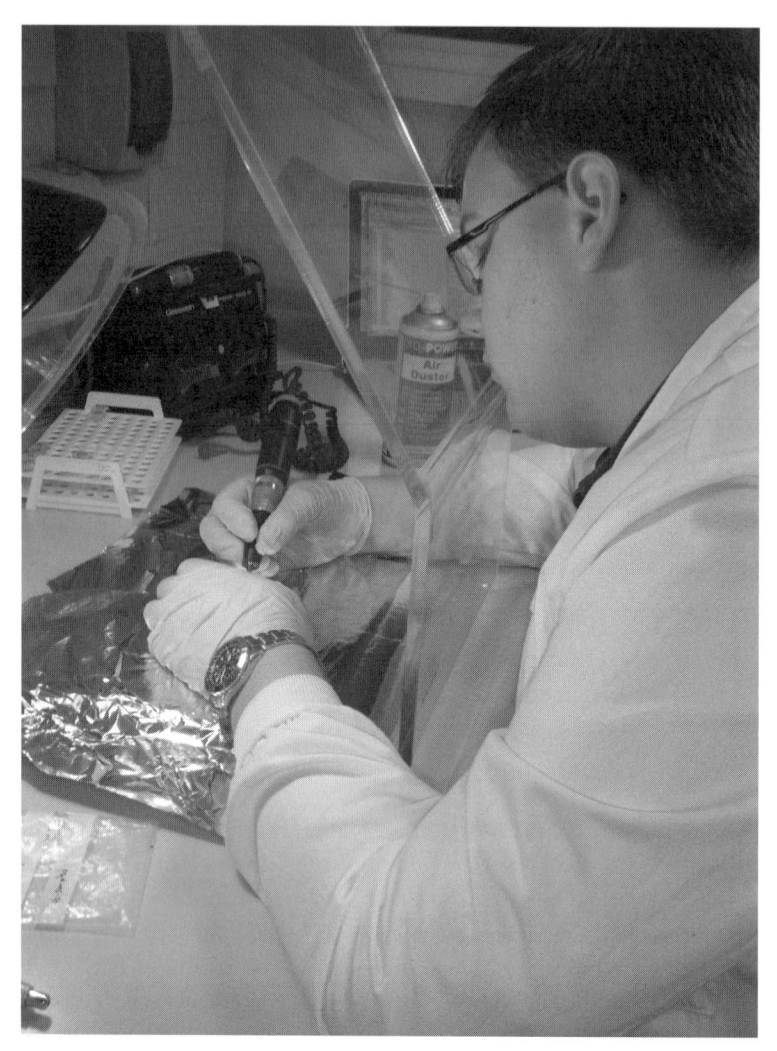

Figure 5.10. Rutgers graduate student René Studer-Halbach serially samples enamel from mammal teeth for variation in strontium isotopes, which allows migration across a landscape with varying bedrock to be discerned. This work is taking place at the Research Laboratory for Archaeology and the History of Art, Oxford University.

Temporal patterning of behavior

Was a fossil animal diurnal? Was it nocturnal? Sometimes a reasonable case can be made for a diurnal or nocturnal lifestyle in a fossil organism. Living animals, including some primate species, are also known to have a mixed pattern of behavior, depending on local circumstances—that is, they are cathemeral. It will probably never be possible definitely to establish cathemeral behavior in fossil organisms, although it may be implied by the absolute size of the eyes (see below).

The size of the eyes relative to body size is routinely used to infer diurnal behavior versus nocturnal behavior in fossil animals. Eye size is given by the size

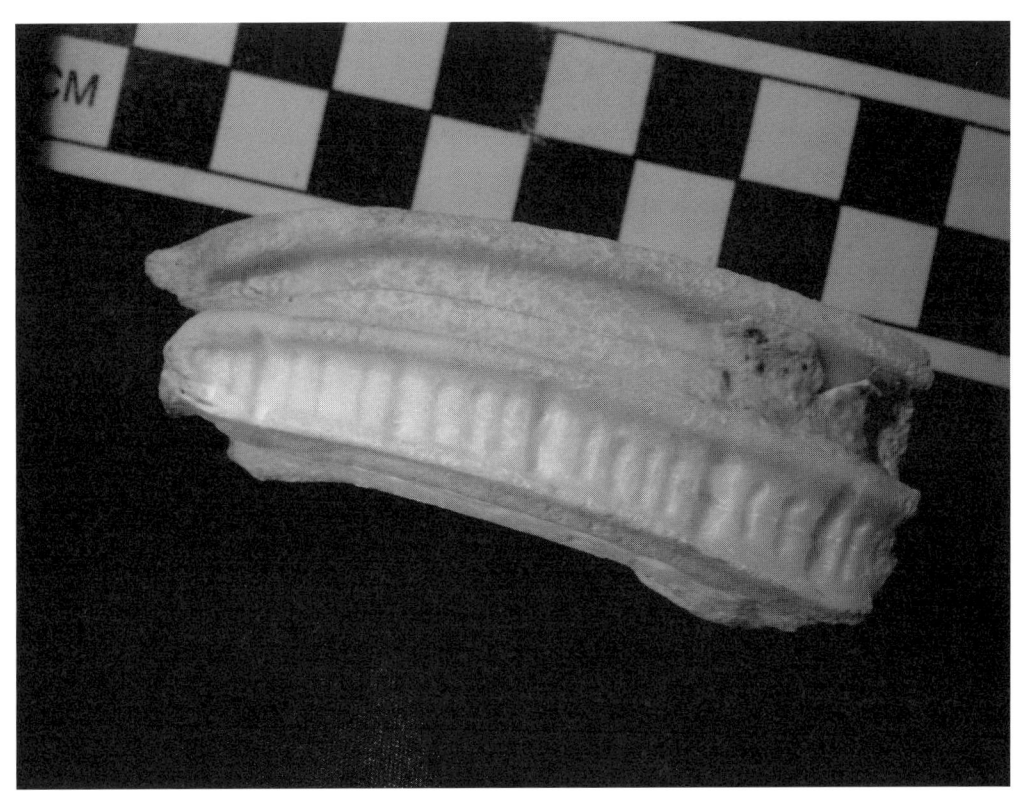

Figure 5.11. Tooth from a modern zebra that has been serially sampled. The longer samples of enamel will be tested for carbon and nitrogen isotopes. The smaller, intervening samples will be tested for strontium isotopes. The scale is in centimeters.

of the orbit (usually the diameter of the orbit), and body size is given by a number of variables. Cranial length or skull length is usually taken for the estimate of body size, particularly since a good portion of the cranium must be present if orbital size can be examined. A large sample of modern species in the same group is needed for the comparative database, and the modern species must include both nocturnal and diurnal forms. If the orbital size is large relative to body size, the fossil is declared to be nocturnal, and if the orbital size is small relative to body size, the fossil is declared to be diurnal. A famous vertebrate example is the theropod dinosaur *Troodon*, found in the Late Cretaceous of Alaska. Its very large eyes relative to body size have been used to infer that this genus was a permanent resident, and did not migrate south as sunlight diminished in the high latitude winter. A primate example is *Teilhardina asiatica*, whose small eyes relative to body size have been used to infer that this species was diurnal (Chapter 9).

Paleontologists strive to extract as much information as they can about lifeways from the feeble fossil evidence. It is important to note that living primates have similar eye sizes relative to head size, regardless of whether they are nocturnal, diurnal, or cathemeral (Kirk, 2006). Furthermore, living non-primate mammals show only a moderately weak correlation between relative eye size and activity pattern (Ross & Kirk, 2007).

Eye size in mammals may reflect more than body size or activity patterns. Maximum running speed may influence eye size, with faster mammals having absolutely larger eyes than their slower relatives. This association is called Leuckart's law, and is based on the idea that species that can reach very fast speeds need large eyes to support visual acuity, and to avoid colliding into objects when they run. An eye that is absolutely larger will yield greater image resolution than a smaller eye, irrespective of the body size of the animal. This is because a larger image can be projected onto the retina. This larger image would encounter a greater number of photoreceptors, which would enhance visual acuity. Leuckart's law was first used to explain the relatively large eyes of birds, but has now been investigated in 50 mammal species spread across 10 orders (Heard-Booth & Kirk, 2012). Three primate species were examined: a New World monkey (*Cebus capucinus*), an Old World monkey (*Erythrocebus patas*), and the gorilla (*Gorilla gorilla*). The Old World patas monkey has the fastest running speed of any living primate, having been clocked at 55 km/hour, and has the largest absolute eye size of examined primate species. Its eyes are, in fact, 25 percent larger than the eyes of other Old World monkeys in the same size range. Leuckart's law does seem to operate in primates.

Absolute eye size is also significantly larger in cathemeral mammals (Heard-Booth & Kirk, 2012). These species have the largest absolute eye size, when compared to nocturnal or diurnal species. Cathemeral species also tend to have a larger body size. These findings might be used to infer cathemeral behavior in fossil organisms.

Reliance on sensory modalities can also be used to infer lifeways. Relative expansion of visual centers of the brain versus centers emphasizing processing of information from olfaction or sound would lead one to infer diurnality, or vice versa. For example, CT scans of the cranium of *Parapithecus grangeri*, an early anthropoid primate from Egypt, allow one to examine the size of the olfactory bulbs in this specimen, which are larger with respect to the rest of the brain than one sees in living anthropoids (Bush *et al.*, 2004).

Additional behavioral inferences

Behavior can be inferred not only from the details of musculoskeletal structure. A subtle study of morphology can yield evidence of the brain, the senses, and activity of fossil species. For example, in tyrannosaur dinosaurs, study of the brain, neurocranium, and ear region shows that these dinosaurs had an enhanced sense of smell, were sensitive to low-frequency sounds, and had superior reflexes coordinating quick movements of the eye and hand (Witmer & Ridgely, 2009). Age at death can yield information about life history variables, if the sample size from a contemporary fossil assemblage is large enough. Most dinosaur species may have experienced high mortality before they reached adulthood. Only a few survivors reached full adult size and maximal lifespan (Erickson *et al.*, 2009). This implies that attaining reproductive maturity was associated with physiological costs or increased predation pressure.

The question of dinosaur physiology—Were they ectothermic? Were they endothermic?—has generated extensive study about the relationship between physiology and bone growth in living mammals. The definitive answer to this relationship is now known. Cross-sections of the femur in wild ruminant mammals from Eurasia and Africa were studied. Ruminants are especially constrained by the stable temperature demanded by the digestive fermentation processes taking place within their four-chambered stomachs. Histological analysis demonstrates that ruminant bone growth is cyclical. Bone growth stops during an unfavorable season when food resources are limited and metabolic rate declines (Köhler *et al.*, 2012). This study disproves the idea that only ectothermic animals show seasonal arrests in bone growth. Endotherms also show seasonal arrests, providing that they are growing slowly enough to record the impact of several seasons. The ruminants studied were collected from habitats stretching from polar to tropical regions. Hence, seasonal variation in both temperature (high latitudes) and precipitation (low latitudes) affected bone growth. The ecological correlates of the histological lines signifying halted bone growth can therefore be recognized. This has the potential for unraveling the impact of seasonality on bone growth in fossil mammals.

Migration can be inferred in fossil animals. As tooth crowns are formed in living mammals, the growing enamel incorporates strontium from the local bedrock. Because bedrock has varying strontium isotopes, serially sampling enamel from the teeth of living species reveals patterns of migration that cross through different bedrock areas (Figures 5.10 and 5.11). This methodology has been used to infer seasonal migration in Ice Age mammals, as well as different patterns of dispersal away from their natal area in male and female fossil humans.

6 Evolutionary processes and the pattern of primate evolution

What drives evolution? Physical environment versus biological factors

In the modern world, large geographic regions are distinguished by characteristic assemblages of animals and plants. This creates distinctive biogeographic areas. A. R. Wallace was the first scholar to appreciate this fact (Wallace, 1876), thus creating the discipline of biogeography. Later scholars applied a biogeographic approach to paleontology, trying to discern ancient realms of faunal similarity. G. G. Simpson created a straightforward statistic to measure faunal similarity, and applied it to mammal faunas through Cenozoic time (Simpson, 1943). This Simpson Coefficient is:

$$\frac{C}{N_1} \times 100$$

C is the total number of taxa at a certain level (species, genera, families, orders) that are held in common between two faunas; N_1 is the total number of taxa at the same level that are present in the smaller faunal sample. Applying Simpson's Coefficient to Early Eocene mammals reveals that North America and Europe were part of a single faunal province, but Asia was much different, presumably because a broad ocean gap (the Tugai Sea) separated Asia from Europe, and a similarly broad ocean gap (the Bering Strait) separated Asia from North America. Ellesmere Island, high in Hudson Bay, was part of the integrated North American/Europe faunal province (Flynn, 1986).

Why is biogeography important? Suites of coexisting animal and plant species can affect the evolution of a taxon or taxonomic group by affecting resource availability and intra-specific and inter-specific competition. Early Eocene primates, for example, are part of the integrated North American/Europe faunal province, which includes areas now high in the Canadian Arctic, like Ellesmere Island (Chapter 9).

In addition to geography, other factors in the physical environment can affect living organisms. Many current researchers favor climatic factors as the ultimate driving force in biological evolution. This idea has even received a special name—the "Climate Pulse" model of Elisabeth Vrba, generated by analyses of mammal evolution (Vrba, 1992, 1995). Yet, is it true that evolution would stop if there were no perturbations within the physical environment? The evolutionary ecologist

George Evelyn Hutchinson once fiendishly posed this impossible exam question to his hapless students at Yale University: "[I]magine an utterly isolated island in a constant environment. How big would it have to be to permit indefinitely ongoing evolution?" (Jolly, 2006:148). Hutchinson pioneered the study of community ecology, especially the study of competition and niche differentiation between sympatric species in a community (Slack, 2011). Evolutionary ecology can direct study to an even grander scale—not only to the coexistence and interaction of species in a community, but to the coexistence and interaction of species across an entire continent (Brown & Maurer, 1989). In either case, questions are asked about how species divide up necessary resources like food and space.

The idea that biological interactions themselves can drive evolution, irrespective of the physical environment, lies behind the Red Queen's hypothesis (Van Valen, 1973a). It seems obvious that the probability of a species' extinction should be related to the length of its existence. After all, the probability of an individual organism's death in any given year is related to its chronological age—the older an organism is, the more likely it is to die. However, using mammal fossils from the beginning of the Cenozoic, Van Valen examined the relationship between extinction and duration using FADs and LADs. He discovered that the duration of a species' existence had no relationship to its probability of extinction. Competition between species drives evolution, and those species that are competitively inferior go extinct. Thus—just as the Red Queen told Alice that a person in Wonderland has to run as fast as he can just to stay in the same place—Van Valen argued that a species has to evolve as fast as it can just to persist. The Red Queen's hypothesis has recently been tested and is supported. Nineteen terrestrial mammal lineages with rich Cenozoic records that are either extinct or in decline (e.g. elephants and horses) demonstrate that species survivorship is just as dependent on high rates of origination or speciation as on the ability to sidestep extinction caused by the changing physical environment (Quental & Marshall, 2013). Note that the category of organisms examined (terrestrial mammals) also encompasses primates, which indicates their similar vulnerability. Survival by itself is not enough. A lineage needs to generate new species, in order to resist the impact of speciation from sympatric competitors. A corollary follows from these observations: there is an everlasting arms race between species. Life is a vast arena of competition. The Red Queen was right.

Another corollary of the Red Queen's hypothesis is that there is no equilibrium condition for communities. A community has no natural equilibrium condition—i.e. a steady state in which the number and abundance of species remains static. A number of factors, including ecology and dispersal abilities, affect the species composition of a community. In addition, higher-level factors, such as seasonality and the presence of fire regimes, affect the structure of ecosystems (Figures 4.1 and 4.2). The instability of communities and ecosystems is something that conservation biologists do not recognize. They continually strive to maintain a suite of species with their concomitant relative abundances in a certain area. Species that were present and their abundance when biologists first surveyed an area are considered the "natural" state. Conservation biologists act to preserve these species and their relative abundance in a modern reserve. These biologists want to preserve the "natural" state or equilibrium condition by protecting species that are becoming rare and removing invasive species. Yet, if the Red Queen rules, there is no equilibrium condition.

A third corollary of the Red Queen's hypothesis is that the existence of genuine or authentic natural communities is called into question. A whole realm of biology—community ecology and conservation biology—is threatened. Many scholars study primate communities and strive to maintain a steady-state condition: the condition in which explorers first encountered primates in a certain geographic area. The Red Queen undermines the philosophical underpinnings of this endeavor.

Examination of rates of morphological change in placental and marsupial mammals over the last 165 million years demonstrates that rates vary significantly between orders (Venditti *et al.*, 2011:Fig. 2). A great burst of change occurred in mammals 90 mya, when flowering plants diversified and spread. Another great burst occurred at 65.5 mya, after the Cretaceous/Tertiary mass extinction opened up many new niches previously occupied by terrestrial dinosaurs, flying reptiles, and water reptiles. Yet, no overall slowing in evolutionary rates took place, even after lineages diversified. Rates can increase, decrease, or remain stable, and the most speciose orders (bats and rodents) do not demonstrate continually high rates of morphological change. In mammals, ecological niches at any given time are not stocked full. They are continually replenished as species compete for resources or coevolve. This confirms the Red Queen's hypothesis. Species-rich orders experience the same difficulty occupying niches as orders represented by a handful of species do, because niches are constantly shifting. Thus, niches or adaptive zones continually evolve through time, depending on a multitude of factors, including interactions within and between species.

Nevertheless, the physical environment does change. Are some species more susceptible to climatic change? Yes. Tropical species in lower latitudes are more frail in terms of changes to the physical environment (e.g. temperature, precipitation, seasonality). Even altitude has a greater effect on tropical species (Janzen, 1967). In short, species in lower latitudes are more susceptible to changes in the physical environment, and react more to their occurrence. This is known as Rapaport's rule (Stevens, 1989; Cachel, 2006). Rapaport's rule is confirmed by worldwide analysis of the latitudinal ranges of species in equatorial regions versus temperate or Arctic regions. Rapaport's rule is also confirmed by physiological responses to temperature change. Even though only a slight change of ambient temperature occurs in the tropics, metabolic rates in tropical species show far greater shifts than in those from the Arctic (Dillon *et al.*, 2010).

It is known that drastic environmental changes have occurred on a global scale over the course of the last 550 million years, a time interval called the Phanerozoic. These include major geographic changes relating to the position of continents and oceans, caused by plate tectonic movements. The average global temperature, degree of seasonality, and global sea levels have altered through time. The earth of the past may have not resembled the present earth at all. These changes were entwined with biological evolution. The latest physical variable to be investigated is the concentration of atmospheric oxygen. Major changes in atmospheric oxygen have occurred through the Phanerozoic. These are caused by the geochemical cycling of carbon and sulfur. The burial of organic matter creates a spike in photosynthesis, releasing oxygen. The greatest percentage of atmospheric oxygen occurred 300 mya, during the height of the Carboniferous. Large trees and forests evolved to cover the land surface. Experiments with animal embryos demonstrate that rate of development, body size, bone composition, and other variables are affected when oxygen levels shift from 16 to 35 percent, which is the range seen during the Phanerozoic (Berner et al., 2007). Rising atmospheric oxygen during the Tertiary is therefore thought to be associated with an increase in mammalian body size.

Natural selection and adaptation demonstrated

At the level of the genome, natural selection acting on single genes is often assumed to be neutral—i.e. no selective advantage is conferred by the gene. This is especially the case when a molecular clock is used to describe the evolutionary divergence between lineages, given a scarcity of fossil evidence. Genetic differences are marked by the establishment of new mutations, and fossil evidence is used to anchor these differences to some point in time. The equation for the molecular clock is:

$$K = 2NuP \quad (6.1)$$

where K is the rate at which mutations become fixed or permanent, u is the rate at which mutations occur, N is the population size, and P is the probability that a mutation will become fixed. If mutations are neutral, then every mutation has an equal likelihood of becoming fixed. That is, P is $1/2N$, or:

$$K = 2Nu1/2N \quad (6.2)$$

This reduces to:

$$K = u \quad (6.3)$$

That is, the rate at which mutations become fixed is the rate at which they occur. Genetic differences are therefore directly proportional to evolutionary time. Note the fundamental assumption of selective neutrality. Investigations of living organisms demonstrate that mutations are not always neutral. However, many researchers believe that the utility of the molecular clock overrides its false assumption of selective neutrality, especially when fossil evidence is rare.

Yet, paleontologists largely examine morphology, and not genetic changes, except in the rare cases where ancient DNA can be extracted from fossils. It is now known that rates of natural selection and subsequent adaptive response can vary. In general, the rates of change are inversely proportional to the length of evolutionary time over which they are observed. Phenotypic change measured in fossils over millions of years appears very slow—a fact that mistakenly led some researchers to argue that fossil species remained static in morphology (Eldredge & Gould, 1972). Rates of phenotypic change observed in the wild today under conditions of "contemporary evolution" are 10,000 to 10 million times greater than in the fossil record (Cachel, 2006). This is caused by the way in which the geological record is formed—through bursts of sedimentation. Whether a fossil record exists at all depends on sedimentary processes. Sedimentation itself is episodic or erratic, and erosion by water and wind, gravity, soil formation, bioturbation, and tectonic processes disrupt the stratigraphic record. Paleontologists interpreting the fossil record are thus confounded by time-averaging: the cumulative effect of these disruptive processes through geological time.

What about variability in paleontological datasets? Paleontologists will often use the coefficient of variation (CV) to examine variability in their data. The CV measures the variability in a measurement with respect to the mean of that measurement in a population. It is given by the following equation:

$$CV = 100 \times (S.D.)/x \qquad (6.4)$$

This is the ratio of the standard deviation (S.D.) to the mean (x), multiplied by 100. One advantage of this ratio is that it is dimensionless—it is independent of the unit of measurement. It can thus be used to compare datasets with different units of measurement or widely different means. Thus, one might use the CV to examine endocranial volume in cubic centimeters versus the CV of femur length in centimeters or the CV of postcanine tooth area in centimeters squared in a fossil species. The value of this exercise is that it allows one to compare the variability in different regions of a fossil organism's body. Because natural selection can only operate when a phenotype is variable, one can therefore assess the potential for evolutionary change between these different regions. Is postcanine tooth area or femur length more variable? That is, which CV has the higher value? The CV with the higher value has the potential for greater evolutionary change. Similarly, one might compare the CVs of endocranial volume between different fossil species, and therefore assess the relative probability for change in brain size across these different species.

Modern rapid phenotypic changes are caused by a variety of factors, such as climatic change, habitat shifts, invasive species, or predation. Very rapid changes in morphology can mimic the extinction of a species if intervening morphological stages are not preserved. The transformation in morphology is invisible to paleontology, unless the fossil record is very dense and well dated. The transformation of morphology within an evolving lineage from ancestral to descendant species is termed anagenesis. It is rarely observed in the fossil record of land animals, but it has been documented in primate evolution. Fossil sites in the American West preserve examples of anagenesis for some lineages of Paleocene and Eocene primates (Chapters 8 and 9). Not only can natural selection be fast, but adaptation can also be fast. Field experiments demonstrate swift and meaningful morphological response to changes in natural selection. For example, David Reznick and his colleagues have been artificially transplanting native guppies from lowland areas of four rivers in Trinidad upstream into the headwaters of these same rivers. They examine the local ecology of each stream, predation pressure, the growth of each fish, its ranging and foraging patterns, and its reproductive success, using genetic data. The results demonstrate that significant evolutionary change can occur in these fish in as little as 4–11 years (Reznick *et al.*, 1997; Pennisi, 2012). In another example, anole lizards artificially seeded onto seven tiny, sparsely vegetated Bahamian islands quickly evolved significantly shorter limbs as they adapted to narrower arboreal substrates (Kolbe *et al.*, 2012). Hence, local ecology can drive significant evolutionary change, and the adaptive response to selection can be very rapid. This is Contemporary Evolution.

However, the morphology of living organisms reflects trade-offs between performance on a number of tasks, not just locomotion, as in the example above. Several well-studied examples of morphological differences between closely related species (including the Galápagos finches studied by Darwin and bats) demonstrate that a range of morphologies exist, but some theoretically feasible morphologies are absent. This reflects a balancing act between performances on a number of different tasks important to the organism. Each task has an optimal morphology, which is not the optimal morphology for other tasks. Because trade-offs exist, the values of the different traits are correlated, and a distinctive morphospace is generated for the group of related species. Variation in morphological traits within a population is of the same nature as variation in traits between species. The problem of maximizing fitness when a number of tasks are involved has been studied both theoretically and experimentally (Shoval *et al.*, 2012). Tasks can be inferred from measured phenotypes within the morphospace, and the relative importance of a trait in terms of fitness can be demonstrated. This has implications for paleontology, where only morphology can be studied. Morphospace has been examined in detail by invertebrate paleontologists, who use it to examine adaptation in fossil organisms (McGhee, 2007). The absence of morphologies that are theoretically possible is particularly interesting—most of the morphospace is empty, illustrating that groups of related organisms are experiencing similar evolutionary trade-offs between traits as fitness in a particular habitat is maximized. A shift to another area of the morphospace signals that a novel shift in adaptation has occurred, and that a novel adaptive zone has been opened up.

The fossil record demonstrates many instances where many closely related species evolve in a relatively brief period of evolutionary time. These events are termed adaptive radiations. The primate fossil record consists of six adaptive radiations: the Paleocene radiation, the Eocene radiation, a radiation of hominoids during the Miocene, a radiation of Old World monkeys during the Plio-Pleistocene, the platyrrhine primate radiation, and the Malagasy prosimian radiation. Ecological triggers to adaptive radiations have been examined in a range of living animals. Famous examples of adaptive radiations include the anole lizards discussed above, where the genus *Anolis* contains nearly 200 species, and stickleback fish, which exhibit explosive speciation over the last 11,500 years. In both these cases, novel habitats are being colonized: the anoles are colonizing Caribbean islands, and the stickleback fish are colonizing freshwater lakes and streams left behind after the disappearance of continental glaciers. The platyrrhine and Malagasy primate radiations can similarly be explained by the invasion of new lands. South America was an island continent for most of the last 65 million years, and the ancestors of the Malagasy prosimians were seeded onto an isolated island. Given the importance of adaptive radiations in primate evolution, characteristics that define radiations in other groups may also be found in primate radiations. One of these characteristics is an ecological shift that opens new adaptive zones or lifeways to organisms. These organisms respond to new opportunities through phenotypic change. Some researchers argue that hybridization is important during radiations (Seehausen, 2004). New, closely related species rapidly radiating within the same genus may hybridize. Adaptation is promoted because beneficial traits are transferred between the hybridizing species. This has been proven by detailed genetic analysis of the rapidly evolving butterfly genus *Heliconius* (The *Heliconius* Genome Consortium, 2012). Hybridization here has fixed genes that are involved in mimicry, which is an important anti-predator defense.

Another well-known example of adaptive radiation is that of cichlid fishes in the great lakes of the East African Rift system. However, these fish are widespread throughout Africa, and are found in many lakes where they do not undergo radiation. This invites the question as to why this group sometimes becomes speciose, and why it sometimes does not diversify. There are features of the physical environment and features of cichlid biology that contribute to the likelihood of adaptive radiation. Lakes where cichlids are radiating are deep and old, and receive more intense sunlight; cichlid species that are radiating are sexually dimorphic in color (Wagner *et al.*, 2012). Applying these factors to primates yields the following possibilities: one would expect that primates—being generally limited to tropical and subtropical forests—would radiate in areas where forest habitats are old and relatively undisturbed. Sexual selection involving dimorphism in color or courtship behavior involving colorful visual signals would also be important. Both of these factors operate on the Old World monkey group known as guenons, members of the genus *Cercopithecus*. The guenons evolve in tropical rainforest nuclei that expanded and contracted during the Pleistocene. Courtship behavior involving coat colors and patterns on the head and hindquarters exists. Gibbons are speciose when compared to other hominoids, and they occur in dipterocarp rainforests of Southeast Asia. Although body size is similar in male and female gibbons, species are often dimorphic in coat color.

Systematics, evolutionary trees, and homoplasy

Groves (2001) discusses a very peculiar taxonomic idea that had been proposed by Hennig (1966). A taxonomic rank should be equivalent across kingdoms of organisms, irrespective of whether one is examining bacteria, flowering plants, molluscs, insects, or mammals. This is not true. An average family of flowering plants is not the same as an average family of insects or mammals (Van Valen, 1973b). Hennig opined that there should be a universal link between geological age and taxonomic rank. He thus argued that taxonomic ranks are equivalent because they were ultimately established by the origin of the rank in geological time. Groves suggests that this idea might be resurrected for well-known groups, although age-adjusted for each group, given that "molecular clocks" can now supplement fossil evidence. Of course, these molecular clocks are ultimately also based on the fossil record. Comparing primate ranks with those of other mammal groups, Groves suggests a relationship between primate ranks and geological time (2001:Table 1). Infraorders date to the Middle Eocene (45–40 mya), families date to the Middle to Late Oligocene (28–25 mya), and genera date to the Late Miocene (11–7 mya). However, a number of primate families and genera violate this concept. It is no accident that even the most ardent supporters of Hennig's ideas have sheepishly abandoned the notion that a universal link exists between taxonomic rank and geological time.

If cladistic methods are applied to living organisms at the level of populations, a major problem appears. Species boundaries vary depending on what traits or characters are used in the analysis. Increasing the number of traits only leads to a finer level of lineage resolution (Avise, 2000). Ultimately, one is left with a single individual. The problematic application of cladistics to living organisms should make one hesitate before applying this methodology wholesale to the entire primate record.

Morphological, behavioral, physiological, or genetic traits can be used to construct phylogenetic trees. In each case, shared traits that arise in the last common ancestor are used to identify their descendants. Many software packages exist allowing researchers to construct phylogenetic trees. The multitude of possible trees is a major problem. Despite almost two decades of hope, anticipating the arrival of a magic software program that could pick the correct phylogenetic tree, even complex algorithms do not allow identification of the most likely tree (Whitfield, 2007). Homoplasy caused by convergent evolution is a major culprit, here. Homoplasy is the evolution of traits that evolved independently in remotely related organisms through convergent evolution. An alternate problem is that traits may evolve, but then be lost, as is the case when parasitic species lose complex structures. Furthermore, several phylogenetic trees may be equivalent in their explanatory power. Most frustrating of all is when different morphological or genetic traits from the same group of organisms yield different phylogenetic trees.

Much to the chagrin of researchers who are dedicated to the cladistic methodology, homoplasy is rampant in the animal world. Organisms with simple anatomy are particularly problematic, because their minimal morphology may have independently evolved several times, and thus offers no clues as to evolutionary relationships. Sometimes comparative genomics can be used to resolve difficulties, as is the case with problems about the relationships between the animal phyla (Whitfield, 2007). Yet, convergent evolution is ubiquitous. Some researchers have been so impressed by convergent evolution, and by the apparently small number of morphological solutions to adaptive problems, that they argue for the existence of unknown factors that constrain or limit morphological possibilities (Conway Morris, 1998).

There are numerous documented examples of parallel and convergent evolution, which create homoplasy. In one case, the genetic origin of the homoplasic trait is known. Stickleback fish have undergone an extensive Holocene radiation since the end of the Pleistocene. Regulatory changes affect the *Pitx1* genetic locus, so that expression of the gene is absent in animals with reduced pelves (Shapiro *et al.*, 2006). This occurs in different stickleback genera, as well as in mice and manatees. Thus, pelvic reduction mediated by the *Pitx1* gene is widespread among vertebrates. Profound morphological changes that, at first guess, might appear to be underlain by profound genetic differences unique to each group are actually all caused by regulation of the product of the *Pitx1* gene. Because marine stickleback fish have repeatedly invaded freshwater lakes and streams over the last 11,500 years, their ability successfully to colonize and adapt to novel freshwater environments has been intensively examined by ecologists and geneticists (Schluter, 2000a, 2000b). Field experiments with parallel speciation in sticklebacks have also been conducted in small freshwater ponds (Schluter, 2000a). The genomes of a large sample of marine and freshwater sticklebacks from North America and Eurasia have been sequenced, in order to discover the genetic bases for the phenotypic variation that is undergoing selection pressure. The results show that existing genetic and chromosomal variation is repeatedly tapped as freshwater is invaded. Changes in genes that code for proteins occur, but changes in regulatory genes are more important in explaining the recurrent adaptive radiations of these fish (Jones *et al.*, 2012). The genetic basis for repeated parallel evolution is therefore known for this vertebrate group.

Ancient DNA extracted from bone collagen might be thought to yield a royal road to solving questions about the species status or evolutionary relationships of extinct organisms. DNA degrades quickly, however, particularly under warm and wet conditions, so information is rapidly lost. And it has recently been shown that ancient DNA degrades even under museum conditions, because standard museum conservation practices that mandate cleaning fossil material result in a great loss of retrievable DNA (Pruvost *et al.*, 2007).

Evolution and development

The late Leigh Van Valen once famously argued that "Evolution is the control of development by ecology. Oddly, neither area has figured importantly in evolutionary theory since Darwin, who contributed much to each. This is being slowly repaired for ecology … but development is still severely neglected …" (Van Valen, 1973c:488). Over the last 20 years, the study of evolutionary development has exploded, and is still growing in importance.

In fact, it is now known that ecology, development, adaptation, and natural selection interact in a circular fashion. Ecological factors affect embryonic growth, and thus affect the newborn phenotype. The phenotype of newborns, juveniles, and adults is subject to adaptation by natural selection, which is affected by the local ecology of a species at any given time. This process is epitomized by the effect that nutritional deficiencies during embryonic life have on adult body size—i.e. how ecological factors control development and therefore adult body size and form. Starvation during embryonic life leads to smaller adults, although not all organs shrink accordingly. Variation in tissue response to dietary restriction illustrates phenotypic plasticity, or the degree to which a genotype can yield different phenotypes. In *Drosophila*, male genitalia are resistant to size reduction. Differential organ response to nutritional insult is regulated by the degree to which a certain transcription factor (FOXO) is expressed (Tang *et al.*, 2011). It therefore now appears that development has a central position in morphological transformations through evolutionary history. Evolutionary development promises to illuminate paleontological transitions (Thewissen *et al.*, 2012).

Gene regulation is emerging as a major evolutionary process, affecting diversity at a molecular level. This is true even within the same species. In humans, for example, about 25 percent of the examined genes differ significantly in terms of protein expression when Japanese and European databases are compared. Yet, developmental differences also affect the grand procession of life. Research in evolutionary development ("evo-devo") is demonstrating that major morphological changes can occur through modifications taking place during embryonic life. It is now known that a suite of genes (*Hox* genes) is responsible for laying down the fundamental blueprint of the embryo. These genes have remained virtually intact through 500 million years of animal evolution. These crucial developmental genes are conserved throughout the realm of bilaterally symmetrical animals. The last common ancestor of all bilateral animals lived more than 500 mya, when an explosive radiation of animal phyla occurred during the Cambrian. In fact, one explanation for the abrupt appearance of these phyla in the fossil record (the "Cambrian Explosion") invokes the final organization of these developmental genes into their present configuration after a period of genetic experimentation (Marshall, 2006). The interactions of *Hox* genes explain morphological similarity within major groups of organisms.

There are 26 presacral vertebrae in virtually all mammals. The numbers of thoracic and lumbar vertebrae are variable, but inversely so. The differentiation of mammalian vertebrae into different types (cervical, thoracic, lumbar, sacral, and coccygeal) along the long axis of the spinal column is ultimately caused by the interaction of *Hox* genes. Virtually all mammals, with a handful of exceptions, possess seven cervical vertebrae, regardless of the length of their necks. Sloths are an exception, but the extra cervical vertebrae of sloths are accounted for by *Hox* interactions. The rigidity of cervical vertebrae number in mammals indicates that powerful developmental constraints are occurring. About 1 percent of living humans are born with only six cervical vertebrae—the vertebra that would normally be the last one in the neck bears either a single rib or a pair of ribs. Miscarried fetuses with six cervical vertebrae are far more frequent. Newborns with this phenotype have an 80 percent probability of death before reaching the end of their first year of life, and children with cervical ribs have a likelihood of developing certain types of cancer that is 120 times greater than normal (Myers, 2007). Thus, it is clear that *Hox* genes patterning the cervical vertebrae in mammals are fundamentally linked to genes protecting against developmental abnormalities and cancer. When the *HoxC6* gene is expressed in experimental mouse models, ribs appear all along the spine, from neck to pelvis. This snake-like transformation illustrates the antiquity and unchanging nature of *Hox* genes, which clearly link together all vertebrates, and are also found in invertebrates, as well.

When enhancers of *Hox* genes are experimentally altered by multiple minor mutations, there can be a dramatic change in morphology (Frankel *et al.*, 2011). Each mutation has a subtle effect, but the total result is remarkably different. Thus, regulators of *Hox* genes can cause morphological change. A corollary to this statement is that, under normal conditions, when single enhancer mutations occur, morphological change is gradual. This is another reason to suspect the mechanism of punctuated equilibria as an explanation for evolutionary change (Chapter 3).

Study of the relative length of the forelimb and hindlimb or the relative length of limb segments (e.g. proximal forelimb to distal forelimb) has been a constant refrain in primate evolution (Le Gros Clark, 1971). Prior to the mid 1960s, major researchers like Adolph Schultz, William Straus, Jr., and Sir Wilfred Le Gros Clark argued that the long forelimbs of apes were highly specialized. They argued that this made it unlikely that humans had evolved from an unknown ape—Old World monkeys, whose limbs were approximately equal in length, were more likely ancestors of humans. Yet, what is the probability that relative limb length or limb segment length can change? Do genes and development highly constrain limb length? Experiments with fetal mammals reveal the origin of anatomical differences in relative limb length. A fetal mouse model with a mutation in the *Prx1* gene has a distal forelimb (forearm) that is 6 percent longer than normal—the elongated long bone in its forearm approaches that of a fetal bat (Cretekos *et al.*, 2008). Thus, the shift from generalized mouse forearm to specialized bat forearm is based partly on a single gene mutation that enhances the expression of the *Hox* gene *Prx1*. Selection pressure on phenotypic variation within a population can therefore result in macroevolutionary change: the origin of bat wings and powered flight in mammals (Cooper & Tabin, 2008). This study implies that shifts in limb length or limb segment length can occur relatively quickly, and do not depend on a radical reorganization at the genetic level. Differences between primate species in relative limb length or limb segment length need not reflect profound genetic differences or great evolutionary divergence. This holds true within lineages, as well. For example, the fossil species *Homo erectus* differs from australopithecines, an earlier fossil human group, because the legs (lower limbs) are lengthened, and the arms (upper limbs) are shortened. This shift in relative limb length need not imply a profound genetic reorganization.

Hand and foot structures are of particular importance to primate evolution. The first digit of the hand (pollex or thumb) and foot (hallux or big toe) are large and capable of grasping in most primates. The *HoxD13* gene is expressed in the first digit, but three other *Hox* genes are active in the remaining four digits. Various primate species reduce the pollex (colobine and ateline monkeys) or both the pollex and hallux (orangutans) independently of the other digits. This is explained by the different *Hox* patterning of the first digit. The human pollex is the largest among primates. Nevertheless, what is presumably a *HoxD13* mutation can transform a human pollex into a digit like any other on the hand. This "bear paw" phenotype is associated with three phalanges on digit I, rather than the normal two characteristic of a pollex. Furthermore, muscles and tendons specific to the pollex are absent, and the pollex is not opposable (Held, 2010:Fig. 2B). Thus, a single *Hox* mutation can transform a human hand into one resembling that of a generalized mammal.

Comparative genomics

Comparative genomics can reveal gene function, including not only coding for protein production, but also coding for the design of embryos. One might be skeptical about the use of model organisms such as *Drosophila* flies or mice to illuminate gene function, or the genetic bases for phenotypic traits in humans or other primates. However, homologous genes with the same function (called orthologs) are widespread in the animal world, and occur even in yeast and *E. coli* bacteria. A major study of 13 model organisms that tested the degree to which orthologs are more similar in function across species discovered that they were significantly more similar than paralogs—homologous genes with a different function (Altenhoff *et al.*, 2012). Retention of the ancestral function occurs significantly more often than change in function. The null hypothesis therefore becomes similarity of function for these shared genes, even if the function of a protein is not known for a particular species. Discovery of how a protein functions in yeast, *E. coli*, and mice can illuminate how it functions in humans or other primates.

Comparing the genomes of living species also reveals the degree of genetic divergence between lineages, rates of evolutionary change, and genetic changes unique to species. The genomes of several catarrhine species have been examined: the rhesus macaque (*Macaca mulatta*), the two orangutan species, the two gorilla species, the common chimpanzee, and humans. Many changes involve gene duplications. The vast majority of these duplications (80 percent) occur after the separation of hominoids from Old World monkeys at about 25 mya; some duplications occur at the approximate separation of the orangutan lineage, at about 16–12 mya; and human and the African great ape lineages show a great surge in gene duplications at about 10 mya (Marques-Bonet *et al.*, 2009). The rate of change is not constant, and the duplications are not random. General catarrhine changes involve traits like amino acid metabolism and the release of neurotransmitters, which affect nervous system function. Genomes of the two orangutan species demonstrate a significant slowing of evolutionary rates when compared to other hominoids (Locke *et al.*, 2011).

The comparative study of mammalian genomes is also revealing changes specific to human origins. Humans possess a unique version of the *FOXP2* gene, which underlies the motor control of language, and thus promotes the easy and fluent production of language. Neanderthal humans also possess this unique variant, suggesting that they, too, were capable of using language like modern humans. The *FOXP2* gene in other mammals is crucial in normal mouth and tongue movements occurring when the young are nursing. Gene duplications very late in time herald differences in human diet that occur after the domestication of grains that are high in starch. Because salivary amylase immediately breaks down starch as soon as starchy food items are put in the mouth, there is great variability in copies of the *AMY1* gene, or salivary amylase gene, depending on whether human diets emphasize domesticated grains or plant foods gathered from the environment (Perry *et al.*, 2007). Common chimpanzees show no copy variation in *AMY1*.

A number of gene duplications occur about 2–3 mya, at the point where genus *Homo* diverges from australopithecine ancestors. These include two human-specific duplications in *SRGAP2*, a gene that is active in the neocortex of the developing and adult human brain (Charrier *et al.*, 2012; Dennis *et al.*, 2012). One of these duplications (*SRGAP2C*) has a high level of protein expression, and is essentially fixed in all human populations (Chapter 14). Mouse experimental models show that the effect of these human-specific duplications is to hinder the function of the ancestral mammalian *SRGAP2* gene. This slowing of function creates more immature dendrites on the existing neurons. This subsequently increases neuronal migration, branching, and density in the neocortex.

7 Primate origins

The Cretaceous world

The Cretaceous Period is notable for documenting the evolutionary success of dinosaurs as dominant land animals. The end of the Cretaceous provides abundant evidence of this, particularly in North America. Novacek (2007) refers to the terminal Cretaceous as a "Dinosaur Camelot," which poignantly invokes lost glories in forgotten landscapes. Flying and aquatic reptiles were also dominant at this time. The rise of angiosperms or flowering plants also occurred during the Cretaceous. As related later in this chapter, at least one researcher, Robert Sussman, believes that the origin of primates was linked to the rise of angiosperm plants. The paleogeography of the earth during the Late Cretaceous and at the Cretaceous/Tertiary boundary is well known (Figures 7.1 and 7.2)

Morphological, molecular, and genetic data have been used to study the relationships of living and fossil placental mammal orders. The plate tectonic separation of Africa and South America that began about 100 mya has been used to explain the divergence of placental mammal orders. However, a recent analysis using both morphology and molecular evidence refutes the notion that the creation of the Atlantic Ocean had anything to do with the divergence of placental mammal orders (Asher et al., 2003).

Nevertheless, ever since the existence of plate tectonic movements became known, the separation of land masses has been used to explain the differentiation of primate groups, e.g. the origins of platyrrhine and catarrhine primates through the opening of the South Atlantic (Hershkovitz, 1977). A more extreme version is presented by Heads (2010), in which ancestral Euarchontans (Chapter 1) are widespread throughout the Supercontinent of Pangaea. Allopatric speciation is caused by plate tectonic movements, as animals are passively conveyed on lithospheric plates like the animals in Noah's Ark. Continental rifting and opening of new ocean separates plesiadapiform primates in the north, euprimates in the south, and colugos and tree-shrews on terrain that eventually becomes Southeast Asia. A counterclockwise rotation of Africa/Arabia towards Asia leads to an overlap of plesiadapiform primates and euprimates in the north, and colugos, tree-shrews, and euprimates in Southeast Asia. Volcanism and faulting in Africa during the Early Jurassic at 180 mya separates the ancestors of prosimians and anthropoids. The formation of the Mozambique Channel during the Middle Jurassic at 160 mya separates lemurs from galagoes and lorises. Finally, platyrrhine and

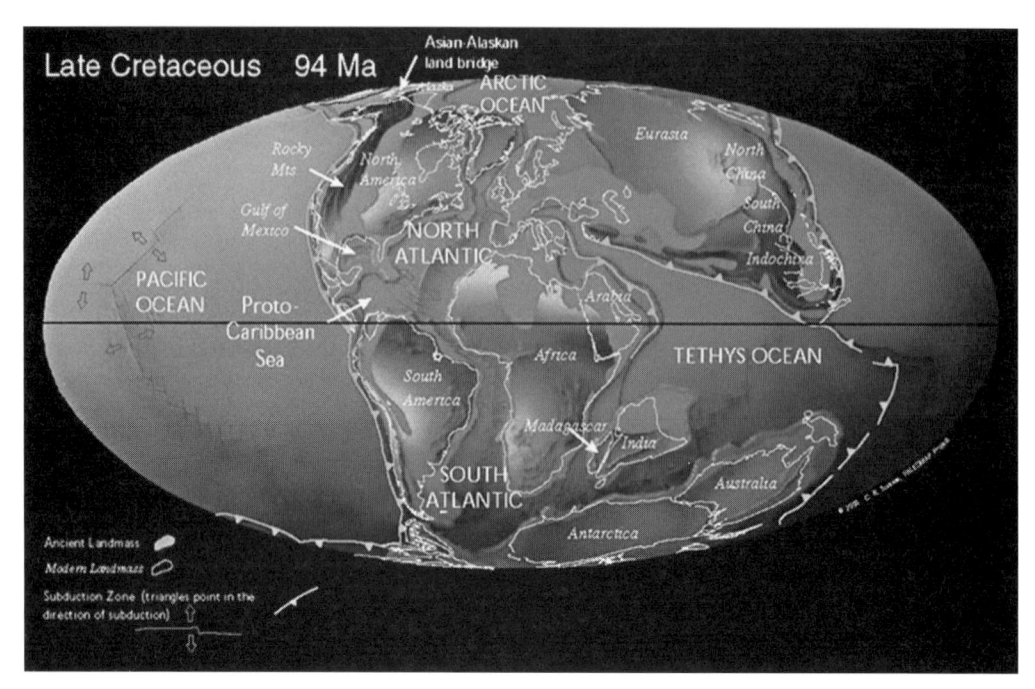

Figure 7.1. Paleogeography of the earth during the Late Cretaceous, 94 mya. Courtesy of Dr. Christopher Scotese.

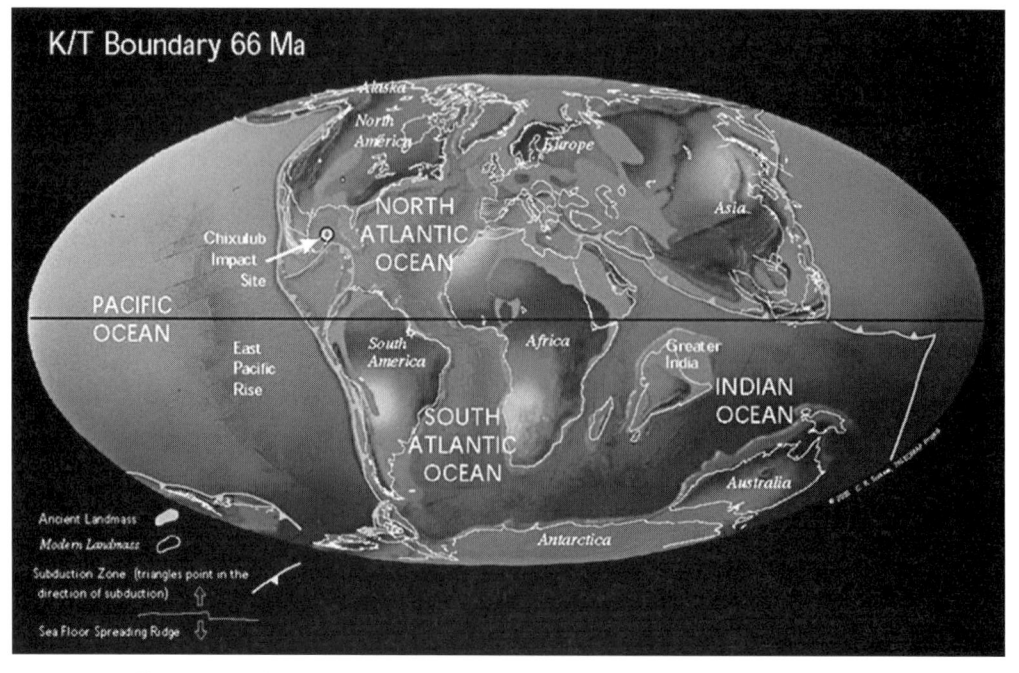

Figure 7.2. Paleogeography of the earth at the Cretaceous/Tertiary (K/T) boundary, 65.5 mya. Courtesy of Dr. Christopher Scotese.

catarrhine primates are separated by the formation of the South Atlantic during the Early Cretaceous at 130 mya. Heads (2010) thus explains the origin of major primate groups by plate tectonic movements and vicariance, and does not need to invoke land animals dispersing over ocean barriers. A major flaw in this vicariance model, however, is that it pushes the origin of the Euarchontans back to the Triassic. The origin of their descendants (plesiadapiforms, euprimates, prosimians, anthropoids, galagoes and lorises, platyrrhines and catarrhines) is also pushed far back in time. There is no support for these dates in the fossil record.

The Cretaceous/Tertiary mass extinction

Primates emerge in numbers directly after the Cretaceous/Tertiary mass extinction. It was one of the five great mass extinction events in earth history. After the catastrophic die-off occurred, ecological vacuums were created that could be exploited by opportunistic survivors. The cause of the mass extinction event at the Cretaceous/Tertiary boundary is commonly considered to be an asteroid that impacted at the Chicxulub Crater in the shallow ocean waters off the Yucatán coast. Certainly, there is a global signature of this impact, in the form of a narrow sedimentary boundary layer found at multiple locations. This layer is enriched in the element iridium, which is rare on earth, but more abundant in extraterrestrial sources.

Despite the consensus opinion that an asteroid impact was responsible for the Cretaceous/Tertiary mass extinction, other factors may have contributed to the event. In particular, gigantic volcanic flood basalt eruptions known as the Deccan traps occurred in the latest Cretaceous. Enormous, slow-moving lava eruptions took place, creating piles of lava that were millions of cubic kilometers in extent. These basalts covered significant geographical areas. The gasses released by these flood basalts may have affected global climate. Stable carbon and oxygen isotopes have been extracted from Cretaceous paleosols from Texas in order to estimate atmospheric carbon dioxide. The data show two major upward spikes in global temperature during the latest Cretaceous (Nordt *et al.*, 2003). Hence, both terrestrial and extraterrestrial factors seem to have contributed to global events at the Cretaceous/Tertiary boundary.

Land plants also suffered a major extinction at the Cretaceous/Tertiary boundary. Angiosperms were the dominant land plants during the Late Cretaceous of North America. Although ferns, cycads, ginkos, and conifers were present, they made up less than 10 percent of the flora. Study of over 22,000 macrobotanical specimens from 161 localities in North Dakota has shown that 57 percent of the species go extinct at the Cretaceous/Tertiary boundary, but the ferns rise immensely directly after the mass extinction (Wilf & Johnson, 2004). North America was heavily devastated, as one might expect, given the proximity of the Chicxulub impact crater off the Yucatán coast and the fact that the angle of the impact veered to the northwest. However, New Zealand also experienced both deforestation at the Cretaceous/Tertiary boundary and an upward spiking of ferns immediately afterward (Vajda *et al.*, 2001). Empirical evidence of how North American ecosystems reeled after the asteroid impact is given by fossil leaves and the insects that fed on them. During the latest Cretaceous, fossil leaves show high species diversity and damage from different types of insect feeding. This is normal in tropical forest conditions, and illustrates the intricate nature of animal and plant coevolution in tropical food webs. After the Cretaceous/Tertiary boundary, one 64.4 mya Montana site has low plant diversity, and high leaf damage by a particular insect feeding mode (leaf-mining); a site in the Denver Basin of Colorado dated to 63.8 mya has a highly diverse flora, but little leaf damage (Wilf *et al.*, 2006). This illustrates aberrant ecosystem function. Food webs in the Western United States did not recover until the Late Paleocene, after the passage of 5–10 million years.

Defining primates

Some orders of mammals (e.g. even-toed hoofed mammals [artiodactyls] or bats) possess keystone features that ineluctably allow one to identify either living or fossil members. For artiodactyls, a double-pulley astragulus bone in the ankle is the keystone feature—it even links artiodactyls to the earliest fossil whales, allowing one to identify from which group the four-legged, terrestrial ancestors of whales emerged. For bats, specializations for powered flight are keystone features. Alas, living primates possess no keystone features. This accounts for the changing composition of the primate order. Over the last 40 years, both tree-shrews and fruit bats have been first included and then excluded from the living primates. The lack of a keystone feature makes the diagnosis and definition of fossil primates even more problematic. Fossil species may be based on only fragmentary teeth, and soft-tissue evidence is entirely gone. Furthermore, the adaptive zone of early fossil primates may not resemble that of modern primates. Questions about primate origins and the identification of the first primates are affected by both the absence of evidence and changing adaptive zones through time. In the following sections, I identify 12 features that generally allow one to separate living primates from other mammals (Cachel, 2006:4–11). However, there are exceptions to these features among living primates, which exemplifies the lack of a primate keystone feature. Problems of primate identification are accentuated with increasing antiquity. Fragmentary fossils located in Deep Time are always likely to remain problematic.

(1) Primates have a generalized limb structure and five digits on each of the extremities, and they retain the ancient mammalian clavicle. (2) Primates have grasping extremities. This is caused by their freely mobile digits, particularly the first digit of the hand (pollex) and the first digit of the foot (hallux). Primate digits are long relative to the length of the bones in the palm (metacarpals) or the sole (metatarsals). This is especially evident in a ratio comparing the length of the proximal phalanx to the length of the metacarpal (metatarsal). (3) Primates have flattened nails, instead of sharp, compressed claws. At least one digit in every living primate is nailed—the hallux. The gripping surface of every digit ends in a

fleshy pad that is extremely responsive to touch. The naked skin on the grasping surface of the hands and feet (and prehensile tail, if one occurs) has intricate three-dimensional patterns. These patterns (dermatoglyphics) provide friction, and lessen the probability of slipping. The dermatoglyphics are associated with Meissner's corpuscles, which are complex nerve endings sitting beneath the high ridges of the dermatoglyphics. In contrast to the nerve endings of other mammals, Meissner's corpuscles lie naked and unprotected by a surrounding membrane. This renders them exquisitely sensitive to pressure or touch.

(4) The primate snout or muzzle is reduced, although it may be secondarily elongated in some species whose ancestors had short faces, such as the baboons. This secondary elongation is caused by selection pressure for longer postcanine tooth rows. (5) Primates are strange among mammals, because they deemphasize the sense of smell. Thus, primates have olfactory bulbs that are small in volume when compared to the endocranial volume or body mass. Living anthropoids have smaller olfactory bulbs than living prosimians. The accessory olfactory bulbs that process signals from the vomeronasal organ are also small in primates that possess this organ (Smith *et al.*, 2007). The vomeronasal organ collects chemical cues from the environment about sex and sexual cues. It is a very important organ among mammals generally, but is absent in higher Old World primates.

(6) Vision is the most important sense in primates. The eyeballs or orbits become increasingly convergent, rotating around to the front of the head. Nearly all primates have the plane of the orbits oriented to the front of the head, rather than angling upwards or downwards. Orbital convergence and frontality allow for an overlap of visual fields from both eyes. Binocular vision thus occurs. A remarkable transformation of the retina, retinal projections, and visual centers of the brain allow stereoscopic or true three-dimensional vision to take place. The brain integrates data from the overlapping visual fields. Besides the systematic representation of the visual field of each eye projecting to the opposite optic tectum (which is found in all mammals), one-half of each visual field also projects to the opposite optic tectum. This necessitates a complete re-wiring of the retina, retinal projections back to the brain, and the visual centers of the brain itself (Allman, 1982). At first, this retino-tectal system was considered to be a unique, keystone primate feature, but fruit bats were later found to have this specialization. Are they flying primates? No. They evolve a primate-like retino-tectal system through convergent evolution.

Anthropoid primates have a relatively small cornea, which supports greater visual acuity or image resolution in these animals (Kirk, 2004). That is, they can detect the fine spatial details of a visual stimulus. Further indication of specialization for vision in primates is seen in the density of cells within the primary visual cortex of the brain. In carnivores, the density of cells within the hippocampus and primary visual cortex scale isometrically and correlate with each other, but the densities of cells within these two areas are not correlated in primates (Lewitus *et al.*, 2012). This indicates that mosaic evolution is occurring in the primate brain, with the primary visual cortex becoming specialized.

(7) In primates, there is some bone separating the globe of the eye from the anterior temporalis muscle. This muscle lies directly behind the eyeball, and is separated from it by an intervening fat pad. The bony separation in primates may be either a postorbital bar of bone or a complete bony plate or postorbital septum. This condition is not unique to primates, because other mammals may develop a postorbital bar (e.g. tree-shrews, horses). Heesy (2005) argues that the independent evolution of the postorbital bar in different mammalian lineages depends on complex relationships between relative orbital size, orbital orientation, relative size of the neurocranium, and the degree of angular deviation between the temporal fossa and the orbit. In general, the postorbital bar may help to strengthen or brace the lateral wall of the orbit. Without the bracing effect of a postorbital bar, large-scale eye movements would deform the eye, and, to a lesser extent, movements of the temporalis muscle and its fascia would deform the eye when an animal chews (Heesy, 2005:Fig.12).

(8) The tympanic bulla (the tympanic floor) encloses the middle ear in mammals, and is a uniquely mammalian feature. It contains the malleus, incus, and stapes bones that function in the sensitive mammalian auditory system (Chapter 1). The tympanic bulla can be formed from a number of bones, and the pattern can be used as a taxonomic key—e.g. in carnivore classification. The tympanic bulla on the base of the primate cranium is formed from the petrosal bone. The absence of a petrosal bulla was responsible for removing tree-shrews from the primate order (Van Valen, 1965). Tree-shrews (Order Scandentia) have an entotympanic bulla. The skeletal elements that form the auditory bulla in mammals are diverse. They can be used to define various mammalian groups, but there is no obvious pattern of evolutionary transformation. In fact, the bulla can remain partly cartilaginous. Entotympanic bullae are widely distributed in mammals, although an entotympanic bulla was probably not the primitive eutherian condition (Novacek, 1993). Primitive eutherians likely had an ectotympanic bulla. Thus, bulla composition is diverse in mammals. Bulla composition does not seem to affect auditory acuity.

(9) The molar teeth of primates have a very simple occlusal surface, with no intricate crests or enamel folds (Figure 7.3). These molar teeth are quadrangular. Even the earliest primate genus (*Purgatorius* from 66–61.7 mya) has quadrangular molars with low, rounded cusps that indicate generalized herbivory. Primates reduce the number of incisor teeth, and lose one premolar from the front of the tooth row at an early date. All living prosimian primates (with the exception of *Daubentonia* [the aye-aye] and *Tarsius* [the tarsiers]) possess a tooth-scraper or tooth-comb at the front of the mandible. It is also present in tree-shrews, developing through convergent evolution. The principal function of the tooth-scraper is to collect exudates (gums, saps, or resins) from damaged vegetation. Thus, the tooth-scraper allows new dietary resources that are rich in carbohydrates and calories to be consumed.

(10) Placental tissues become increasingly more elaborate in primates. An intricate contact is established between maternal and fetal blood supply.

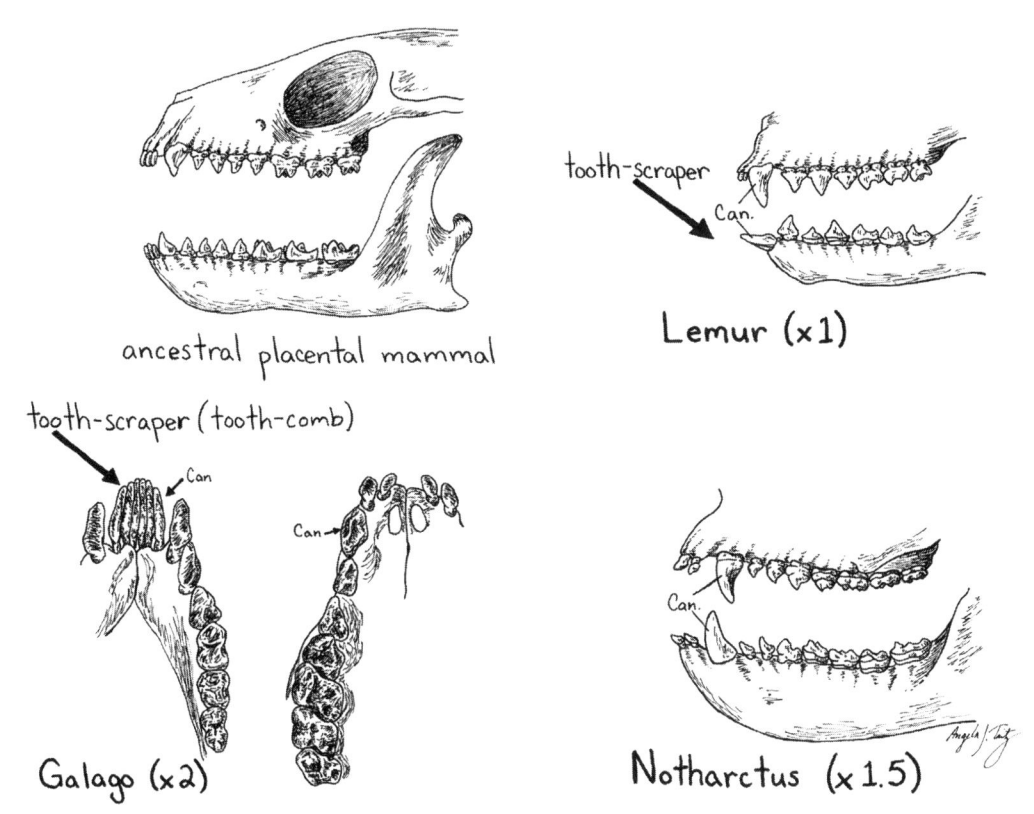

Figure 7.3. The dentition of an ancestral placental mammal is contrasted with the dentition of *Notharctus*, an Eocene prosimian primate, and two living prosimian genera (*Lemur* and *Galago*). The ancestral mammal has a dental formula of 3:1:4:3. Note that the primate incisors have been reduced to two, and that one premolar has been lost from the front of the tooth row. The tooth-scraper (tooth-comb) can be seen in the mandibles of nearly all living prosimians. It is composed of the lower incisors and canine, which become elongated, narrow, and procumbent. The tooth-scraper is not present in the Eocene primate *Notharctus*. *Lemur* is shown at normal size (×1); *Galago* (×2) and *Notharctus* (×1.5) are enlarged to illustrate details. Illustration by Angela J. Tritz.

(11) In comparison with other mammals of the same body weight, primate growth or ontogenetic periods are expanded. (12) The primate brain—especially the neocortex—is large relative to body size.

In contrast to other mammals, primates have a major distinction between hindlimb and forelimb function. The hindlimb serves to grasp and the forelimb serves to manipulate (Gregory, 1920). Although all of the limb bones are relatively elongated in primates with respect to body size, primates have an especially long femur, and the muscles of the thigh and foot are also relatively large (Alexander *et al.*, 1979, 1981; Alexander, 1985). The hindlimbs are also the principal brakes and accelerators of primate movement—forces generating movement are not shared equally between the four limbs. This situation contrasts with that of other mammals, where biomechanical analysis indicates that all four limbs carry equal amounts of weight and account for equal amounts of force responsible for movement. This can be demonstrated in living animals in a locomotion laboratory utilizing force plates. It can also be demonstrated by observing the articulated vertebral columns of living mammals. The neural spines on the dorsal surface of the vertebrae slope either forward or backward. There is one vertebra (the anticlinal vertebra) that stands nearly upright, and that marks the shift in slope between the neural spines. This anticlinal vertebra is easy to identify in most mammals—each limb carries an equal amount of weight, but large neck muscles move the head. It is very difficult to identify the anticlinal vertebra in primates, because of hindlimb dominance in locomotion.

Because of the dominance of the primate hindlimb in locomotion, the hindfoot stays in contact with the substrate for a disproportionately long time. Primates exhibit a peculiar footfall pattern when they walk: left hindfoot, right forefoot, right hindfoot, left forefoot. This preferred walking gait is called the diagonal sequence diagonal-couplets gait. This gait may be the result of primates originating in arboreal settings with small, discontinuous branches (see below).

Primates also demonstrate a wider range of locomotion than other mammals do. For example, although most primates are arboreal quadrupeds, they evolve highly specialized arboreal leapers and slow climbers, as well as terrestrial quadrupeds and bipeds. This diversity of locomotion may be caused by a lack of morphological integration in the primate pelvis, which renders it more susceptible to evolutionary change (Lewton, 2012).

Primate origins

One vexing problem in modern paleontology is the disjunction between the dates of origin yielded by actual fossils and the dates of origin yielded by the molecular clock (Chapter 3). The molecular clock always estimates the origin of a group at a far earlier date than the fossils indicate. The gap between these dates is sometimes jarringly large, indicating that it is not simply caused by a failure to discover the first fossils for a group. Primates are no exception to this problem. The first euprimates appear at about 56 mya. The first plesiadapoid primates appear in the latest Cretaceous, about 66 mya. Yet, the average date estimated for primate origins by the molecular clock is 82 mya. One explanation for the age gap given by primate fossils and estimated by molecules is that there is a very strong inverse correlation between rates of molecular change in primates and three variables: body size, absolute endocranial volume, and relative endocranial volume (Steiper & Seiffert, 2012). Rates of molecular change are fast when these three variables are small. The last common ancestor of all primates was small-bodied, and had both an absolutely and relatively small endocranial volume. Primate lineages later independently develop an increase in body size and an increase in absolute and relative endocranial volume through convergent evolution. These changes signal a decrease in the rate of primate molecular evolution. When this decrease is accounted for, the corrected molecular timescale indicates an origin of primates near the Cretaceous/Tertiary boundary, or even more recently in the Paleocene. Thus, the gap between primate origins as indicated by the molecular clock and indicated by fossil evidence disappears. It is not known whether the disparity between molecular and fossil timescales can be resolved in the same fashion for other mammalian groups (Steiper & Seiffert, 2012). A slowdown of the molecular clock in humans has been recognized for some time (Li & Tanimura, 1987), and a general hominoid slowdown is widely accepted. However, the corrected molecular clock timescale for primates indicates that convergent slowdowns occurred many times in primate evolution, because of selection pressure on different lineages to develop the extended life histories generated by larger body size and brain size.

In addition to the disparity between primate origins as indicated by the molecular clock and primate origins as indicated by the fossil record, there has been an effort to reinterpret the primate fossil record as showing continually increasing diversity through time: that is, to show that the number of primate species expand with time in an inverted cone, and never undergo any significant extinction events. This is not supported by the fossil record. However, the primatologist Robert D. Martin (1986, 1990) argues that this type of primate diversification must have occurred, and that the incompleteness of the fossil record in Deep Time alters the true pattern. Martin and his colleagues (Tavaré et al., 2002) later produced a similar primate diversification curve, and estimated that primates had originated in the Late Cretaceous at 81.5 mya. Because this date was close to the date estimated using molecular clock evidence, Martin and his colleagues believed that this coincidence of dates confirmed their reinterpretation of the primate fossil record. These elaborate attempts to replace the actual pattern of primate evolution with a recalculated artificial pattern are actually attempts to rescue primates from the specter of extinction events. Yet, if primates were spared from such extinction events, and only continued to expand their numbers, primates would be unique among vertebrates.

An arboreal lifestyle is ancient in primates. It is now known that *Purgatorius* was arboreal. *Purgatorius* is the earliest primate genus. It dates to 66–61.7 mya, and is found in northeastern Montana, USA, and Saskatchewan, Canada. *Purgatorius* is known to be arboreal because of the discovery of characteristically primate-like tarsal bones for this genus (Kaplan, 2012; Milius, 2012). The ankle joint was extremely mobile, allowing movement to occur in many dimensions. However, the tarsal bone was not elongated, as it is in primates that are specialized for leaping. Hence, *Purgatorius* scampered and scrambled in the trees, engaging in flexible and nimble movements that selected for ankle flexibility.

Adaptations for arboreal life evolve independently in many mammal groups, and appear among the oldest therian mammals. The earliest known placental mammal comes from the Jurassic of China, and dates to 160 mya (Luo et al., 2011). It has a generalized dentition and a scansorial forelimb, modified for arboreal climbing, scrambling, and leaping. Thus, traits associated with the primate order, such as arboreal locomotion and limb modifications for arboreal life, are already documented in the oldest of placental mammals.

A number of different ideas have been advanced about primate origins. I will detail five of them here, in the chronological order in which they were presented: (1) the arboreal life per se (Wood Jones, 1916; Elliot Smith, 1924); (2) generalized herbivory in an arboreal environment (Van Valen & Sloan, 1965; Szalay, 1968); (3) vertical clinging and leaping (Napier & Walker, 1967); (4) visual predation (Cartmill, 1972, 1974b, 1975); and (5) the evolution of flowering plants (angiosperms) as a trigger for primate origins (Sussman & Raven, 1978; Sussman, 1991). A sixth idea is beginning to emerge—the idea that a number of major primate features evolve as specializations for arboreal locomotion among discontinuous and mobile small branches (Cachel, 2006).

The idea that primate features evolved because of arboreality per se is the earliest suggestion about primate origins (Wood Jones, 1916; Elliot Smith, 1924). Traits such as grasping hands and feet, nailed digits, convergent orbits, an emphasis on vision, and an enlarged brain evolve as specializations for an arboreal life. Primates develop these traits because they had been arboreal since the dawn of the Cenozoic, longer than any other order. The implication was that, if a group persists in an arboreal lifestyle long enough, primate traits would emerge. These traits are the ultimate adaptations to arboreal life, and primates are the consummate arboreal animals. This idea was falsified in the early 1970s by observing that many mammals are arboreal without bring primate-like. Consider the gray squirrel, a common denizen of urban backyards. It has no grasping hands and feet, nails, convergent orbits, visual specializations, or an enlarged brain. But the gray squirrel is certainly not deficient in its command of arboreal locomotion, because it easily climbs, scampers, and leaps in the trees, and it can amble across telephone wires with the ease of a tightrope walker. Paleontologists in the 1960s argued that the first primates were small, generalized herbivores living in trees (Van Valen & Sloan, 1965; Szalay, 1968). The dietary assessment was based on the quadrangular molars with low, rounded cusps that are found in the first primates. However, many other Paleocene mammals were both arboreal and generalized herbivores without being primates. Primates were not unique occupants of Paleocene forests. When field observations of wild prosimian primates commenced in the 1960s, a new category of primate locomotion—vertical clinging and leaping—was recognized. The researchers who first documented this category argued that it was the most ancient form of primate locomotion, because it was observed in prosimians, and was responsible for such primate traits as grasping hands and feet, nailed digits, convergent orbits, visual specializations, and hindlimb dominance (Napier & Walker, 1967). This idea was falsified by the revelation that anatomical traits associated with vertical clinging and leaping locomotion show no morphological integration. They therefore evolve independently in many lineages.

The visual predation hypothesis of primate origins argued that convergent eyes oriented in the frontal plane, strong, grasping pollex and hallux, and nailed digits result from hunting active insect prey in arboreal settings where branches are fine and thin (Cartmill, 1972, 1974b, 1975). It was later stipulated (Cartmill, 1992) that the first primates were nocturnal insect predators, because a narrowing of the

Figure 7.4. A living loris actively moving across arboreal supports that are small in diameter, compared to the body size of the loris. The orbits are highly convergent, and in the frontal plane. The hands are extremely specialized for grasping small-diameter supports. The thumbs are highly enlarged, and are rotated about 180° away from digits III–V. The second finger is reduced, so that the hand resembles the biological equivalent of a pair of pliers. The feet have the same morphology, although they are not illustrated. The loris eyes and hands and feet are adaptations for hunting of active insect prey in arboreal settings where the branches are thin and fine. The loris is thus a living exemplar (though highly specialized) of the visual predation hypothesis of primate origins. Illustration by Angela J. Tritz.

visual field to enhance visual acuity is more important for a nocturnal animal than a diurnal one. A panoramic visual field is not useful for a nocturnal animal. Figure 7.4 illustrates a highly specialized living primate that exemplifies the visual predation hypothesis. Yet, the earliest primate fossils demonstrate no indications of an insectivorous diet, convergent orbits, or nailed digits (Chapter 8). Cartmill (1974b, 1975) argued that they should be removed from the Order Primates.

While many researchers have complied with this suggestion, there is increasing indication that these Paleocene taxa were indeed primates. Furthermore, the fossil evidence indicates that primate-like traits begin to accrete in some of these lineages with time (Chapter 8).

A fifth idea about primate origins arose in the late 1970s (Sussman & Raven, 1978; Sussman, 1991). This idea argued that primate emergence is triggered by the origin and diversification of angiosperm plants. The first primates were coevolving with these plants, becoming specialized angiosperm pollinators. This idea is unlikely, for the simple reason that angiosperms emerge and undergo a major radiation during the Middle Cretaceous. This date far precedes the first fossil evidence of primates. Furthermore, animal pollinators must be small enough not to damage the reproductive parts of the plant as they feed on nectar and passively transfer pollen to other plants. Insects are therefore the principal pollinators of angiosperm plants. Bats can be pollinators of night-blooming flowers, because they can hover in the air while they delicately lap up nectar. But primates are likely to smash or tear flowers if they feed on nectar. Neither do specialized animal pollinators eat flowers or pollen—activities which living primates do. Sussman and his colleagues have recently attempted to resurrect the angiosperm hypothesis (Sussman *et al.*, 2013). However, rather than focusing on the origin of angiosperms as the launch-point for primate origins, the hypothesis now argues that the first primates emerge in the Late Paleocene when they begin to exploit the fruits and seeds of flowering plants. Sussman *et al.* (2013) specifically identify *Carpolestes simpsoni* as the first primate. Yet, this taxon is generally considered to be a plesiadapiform primate that is beginning to acquire euprimate traits (Chapter 8). And the plesiadapiforms, although forbidden primate status by some researchers, first appear at the Cretaceous/Tertiary boundary.

Arboreal locomotion in small branches

Early Paleocene forests may have been lush and impenetrable, because mammalian herbivores that were large enough to damage trees and create patches of open woodland did not yet exist. Dense forest structure and deep shade made sapling growth difficult. This may have led trees to develop new reproductive strategies that emphasized large, energy-rich seeds. Tree species needed to attract animals that could consume fruit and seeds without damaging the seeds through digestion (Flannery, 2001). After animals defecated the intact seeds, the seeds were dispersed far away from the parent plant, and were therefore not in competition with the parent plant for sunlight and soil nutrients.[1] Saplings could grow. This

---

[1] Scholars can infer the presence of missing seed predators in modern environments. In the Pleistocene Neotropics, many large mammals such as gomphothere proboscideans, ground sloths, and horses fed on large, predator-dispersed fruit, such as the giant seed pods of the guanacaste tree (*Enterolobium cyclocarpum*). These ancient megafaunal species became extinct at the end of the Pleistocene, creating a landscape with abundant giant seeds and no seed predators. Human introduction of domesticated horses partly re-establishes the ancient ecology (Janzen & Martin, 1983).

reproductive strategy on the part of plants meant that mammals specializing in fruit and seed predation could evolve for the first time. Primates would be among those mammals.

All living primates and even the earliest fossil primates have long digits. When relative digit length is measured by a ratio of digit length/metacarpal or metatarsal length on the same ray, living primates are characterized by long digits. Paleocene plesiadapoid primates like *Plesiadapis*, *Dryomomys*, and *Carpolestes* also have long digits, as do Eocene fossil euprimates like *Darwinius*. The long digits of primates are thought to have evolved for grasping small branches (Hamrick, 2012). This small-branch setting was the background for primate emergence.

The earliest primates used all four limbs and all four extremities when traversing the arboreal environment. However, the hindlimbs were more engaged in strongly grasping supports that were already tested, while the forelimbs were engaged in searching out new supports. Locomotion was often halting and disrupted, because arboreal supports were discontinuous. The smaller the primate, the more discontinuous the arboreal setting becomes. And early primates were small. Although light in weight, they would often have encountered dangerous gaps in the canopy. Locomotion in small, terminal branches involves movement across mobile, swaying supports, and animals often need to bridge gaps that loom from branch to branch (Figure 7.5). Cartmill *et al.* (2002) argue that the characteristic quadrupedal walking gait of primates—diagonal sequence diagonal-couplets (otherwise very rare for mammals)—evolved in a small-branch setting. This is a pattern of footfall that runs thus: left hindfoot, right forefoot, right hindfoot, left forefoot (technically described as Lh Rf Rh Lf). Note that the hindfoot always precedes the forefoot in touchdown, and that weight is transferred to the contralateral or opposite side of the body after the hindfoot has touched down. The hindfoot precedes the forefoot because the hindfoot grasps more powerfully, and is responsible for both more acceleration and more braking forces than the forefoot. The transfer of weight to the contralateral side of the body creates more stability among the small branches, because the center of gravity passes close to the line connecting the diagonally opposite feet (Cartmill *et al.*, 2002:Figs 6 & 16). The animal is better able to maintain its balance. Prolonged contact of the hindfoot with the substrate increases the diagonality of the walk. The hindfoot grasps a known, tested support just as the contralateral forefoot comes into contact with an unknown, untested support. Given the grasping specializations of the hindfoot, a primate may retreat or regain its balance if the new support contacted by the forefoot begins to break or sway. Common marmosets (*Callithrix jacchus*) are separate from other primates on the Cartmill *et al.* data plots. They have a different walking gait, because they use claws to climb and cling to tree trunks and large branches, and their hallux is reduced. This is a secondary condition, however. The ancestors of marmosets had a typical primate diagonal sequence diagonal-couplets walking gait. They have reverted to the normal mammalian pattern, because they have lost the grasping predominance of the hindfoot and re-developed claws from nails (Chapter 13).

Comparative studies that shed light on primate origins

One should consider colugos when addressing primate origins. Colugos are the mammal group most closely affiliated with the Order Primates (Figure 1.5). Modern colugos have peculiar, comb-like incisor teeth, with as many as 20 little

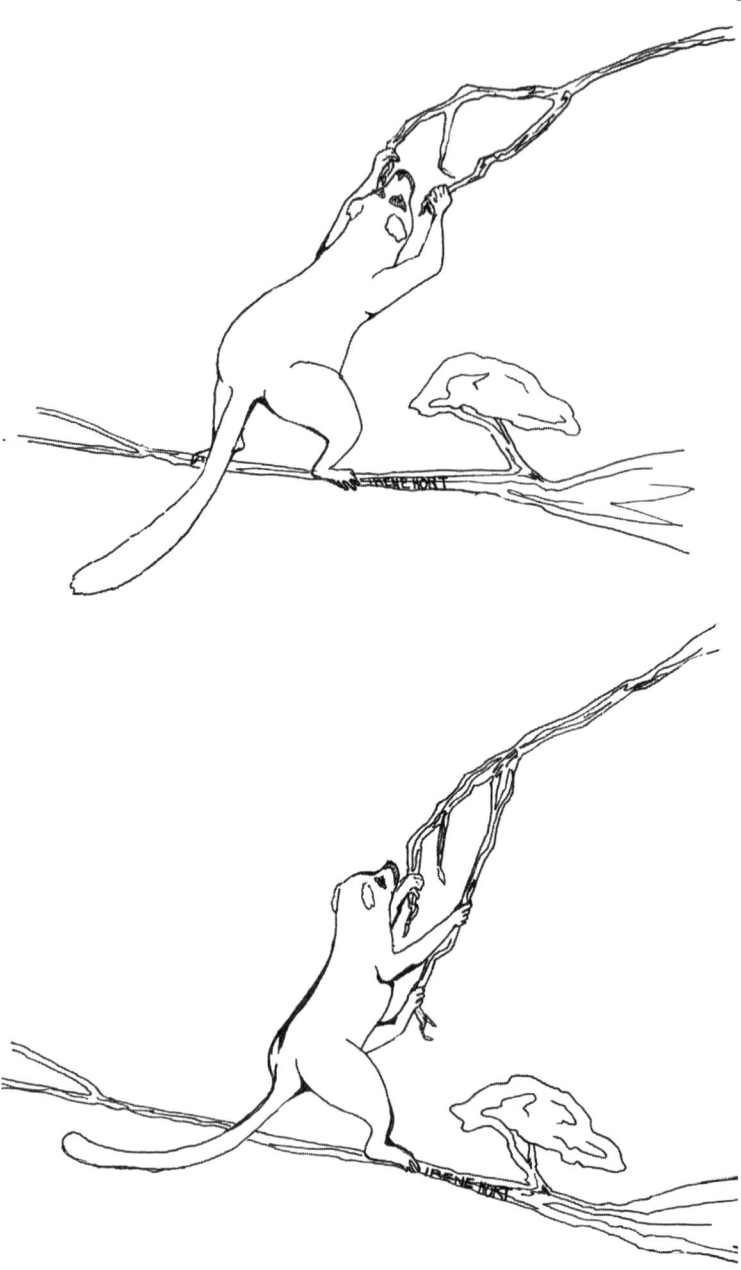

Figure 7.5. Arboreal locomotion using discontinuous, mobile small branches. "Bridging" behavior is shown, as the primate transfers its body weight between different, mobile branches. Note that the feet support the body from tested branches, while the hands search out new, untested branches. Illustration by Irene V. Hort.

tines on each incisor. Yet, although colugos are arboreal, they are poor climbers. They have no opposable first digits (pollex or hallux) to grasp branches, and support their rather heavy body weight with claws alone. Their weak limb musculature mandates that they climb upward slowly. However, they do suspend themselves well underneath arboreal supports. They are fantastically specialized gliders. Hence, colugo arboreal traits involve suspension and gliding between high canopy supports. Neither suspension nor gliding are traits discernible in the first primates. And the poor climbing abilities and lack of grasping first digits in colugos also vitiate their use in modeling primate origins.

Marsupial phalangers (Family Phalangeridae), which can have prehensile tails, a nailed hallux, and grooming claws on their feet, do seem to foreshadow primate traits. Chapter 8 includes a discussion of phalangers and other marsupials that illuminate the lifeways of extinct Paleocene primates.

Regardless of how one identifies the first primates, they coexisted with multituberculates, members of the Order Multituberculata, the longest-lived of any known mammalian order. Multituberculates appear in the Early Jurassic and persist until the Early Oligocene. If one measures evolutionary success by survival, this is the most successful order of mammals that ever lived. They were the first mammals to become accomplished herbivores, which accounts for their evolutionary achievement. They thrived until the appearance of the true rodents. Multituberculates developed striking similarities to plesiadapoid primates. The skull of *Taeniolabis* of the Lower Paleocene has the chisel-shaped incisors and diastemata seen in many plesiadapoids, as well as the shearing premolars and ever-growing incisors of the most specialized of the plesiadapoids (Kermack & Kermack, 1984).

Now, consider size. Body size is inextricably entwined with energetic requirements, and energetics can therefore harshly constrain the evolution of a species (Brown *et al.*, 1993). Some researchers (Gebo *et al.*, 2000; Gebo, 2004) have argued that the earliest primates were very small, equivalent to shrews in body size. Because these researchers consider the first primates to be euprimates, their arguments involve species that first appear in the Eocene. The genus *Eosimias*, weighing between 67 and 179 g, has advocates for its being a possible anthropoid primate. Two 45-my-old primate calcanei from China are less than half the size of the calcaneus of *Microcebus*, the smallest genus of living prosimian, which averages about 65 g in weight. These Chinese fossils may be the smallest known primates. Estimates indicate a body size of 17.2 g based on the larger calcaneus and 10.6 g based on the smaller calcaneus (Gebo *et al.*, 2000). This is about one-tenth the size of the living pygmy marmoset genus (*Cebuella*), the smallest living anthropoid, and within the size array of living shrews. An order of magnitude also separates the average living prosimians and anthropoids, and this argues against the anthropoid status of the Chinese fossil calcanei. Furthermore, primates, unlike many other mammals, increase their body size during a dramatic upward spiking increase in global temperature at the Paleocene/Eocene boundary (Chapter 8). I think that shrew-sized early primates are unlikely, because of a variety of problems that shrews experience, purely because of their tiny body size.

Shrews are among the smallest living mammals. They are currently tied with bumblebee bats for the distinction of tiniest living mammal. Body size affects a number of important variables in the life of an animal. For example, the average lifespan of a shrew is a year and a half. The average lifespan of a rat is 2 years. Lifespan is directly proportional to body size. Thus, any primate that is shrew-like in size has different life history variables than larger mammals. Several predictions based on the life histories of living primates would have to be revised. One could not expect that a shrew-sized primate would have the relatively prolonged ontogenetic periods found among living primates.

Relative food consumption is higher for smaller animals. They cannot tolerate long periods without food. They therefore pay a relatively higher price for maintaining a stable core body temperature above the temperature of the ambient air than large animals do. They must breathe in oxygen, oxidize food, and expend energy as calories at a higher rate to maintain the physiological processes necessary for life. Metabolic rate is increased in small animals. This is expensive. Diminutive animals like hummingbirds, little bats, and shrews sometimes allow their body temperature to drop for several hours when they are not active. This saves on the oxidation of food to fuel physiology. Very small animals therefore frequently enter periods of torpor (McNab, 2012). For example, when food is scarce, small marsupial carnivores drop their body temperature by 10 °C, and enter a period of torpor that can last for 10 hours. The smallest shrew (*Suncus etruscus*), which weighs an average of 2.2 g, frequently enters torpor. The shrew thus avoids spending the energy required for maintaining a continual high body temperature at such a low body mass. A shrew-sized primate would also need to experience episodes of torpor in order to conserve energy.

A fast metabolic rate mandates a high-quality diet. A shrew-sized primate either would be highly carnivorous or would feed on nectar. Any degree of seasonality would absolutely force a shrew-sized mammal into torpor or hibernation. Shrews in temperate areas hibernate during the winter, but drop their internal body temperature down to almost freezing. The size of their body and internal organs even regresses, a well-known occurrence known as the "Dehnel effect." Body length and neurocranial dimensions shorten, and major organs (e.g. brain, liver, kidneys) are reduced during hibernation, but expand to their normal size afterwards. These changes occur because it is impossible for shrews (unlike bears) to store enough calories as fat to last through a winter without eating. However, surviving seasonal episodes of cold or lack of food might not be a problem during the warm and equable global climate of the Early Cenozoic.

What sensory systems are utilized by small mammals? Nocturnal mammals depend upon sound and olfaction, and not vision. In order to locate objects accurately at night, sound wavelengths must be of the same order as the size of the animal's head. If not, sound would diffract around the animal's head, and accuracy would be lost. Small mammals utilize high sound frequencies that travel in straight lines. Insectivorous bats actively home in on their prey utilizing sonar, a sensory system also utilized by tenrecid insectivores. If shrew-sized primates were nocturnal, their sensory modalities would have emphasized olfaction and high-frequency sound. This runs contrary to the visual specializations seen in primates, which is another reason to doubt that the earliest primates were shrew-like in size.

8 The Paleocene primate radiation

The plesiadapoid primates

What events took place 65.5 mya at the beginning of the Cenozoic Era, which is colloquially known as "The Age of Mammals?" This is the Cretaceous/Tertiary (K/T) boundary, and it is now clear that most placental mammal orders originate after this boundary (O'Leary et al., 2013). This realization was achieved after integrating over 4,500 morphological traits from living and fossil placentals with molecular data. If this analysis is corroborated, the typical age gap between phylogenies derived from fossils and those from DNA has been eliminated. In practical terms, what this means is that the search for unknown, ghostly Cretaceous ancestors for primates and other placentals will cease. Ideas about primate origins that are based on a Late or even Middle Cretaceous genesis are vitiated (Chapter 7). The oldest known placental genus is only 91 mya. In addition, skeletal, dental, brain, uterine, and other soft-tissue traits of the hypothetical common ancestor of all placentals can be reconstructed (O'Leary et al., 2013: Fig. 2). Most astonishing is the documentation that about 10 speciation events separating placental orders occurred in as little as 200,000 years—a fantastically high rate of morphological evolution. As a group, placental mammals clearly experienced an incandescent evolutionary radiation as they began to occupy niches left vacant by the extinction of terrestrial dinosaurs and reptiles that were specialized for life in the water and air. What place did primates hold in the great placental radiation?

A general examination of placental and marsupial mammal evolution over 165 million years demonstrates that rates of evolutionary change were low and steady for the first 60 million years. Rates then doubled about 90 million years ago, when a major placental mammal divergence occurred (Venditti et al., 2011) between the Laurasiatheria (e.g. bats, carnivores, artiodactyls, and perissodactyls) and the Afrotheria (e.g. elephants, sirenians, and hyraxes). There is no general pattern in evolutionary rate for mammals as a whole. Species do not always evolve rapidly after the appearance of a lineage, and then experience slow evolutionary change. Rather, evolutionary rates can vary from a 3-fold decrease to a 52-fold increase. The majority of mammal species—including the most speciose orders (bats and rodents)—have no significant and lasting increases or decreases in rates of morphological evolution (Venditti et al., 2011). But primates have a fairly irregular history of morphological change. The primates only experience their first

Primate Morphological Evolution

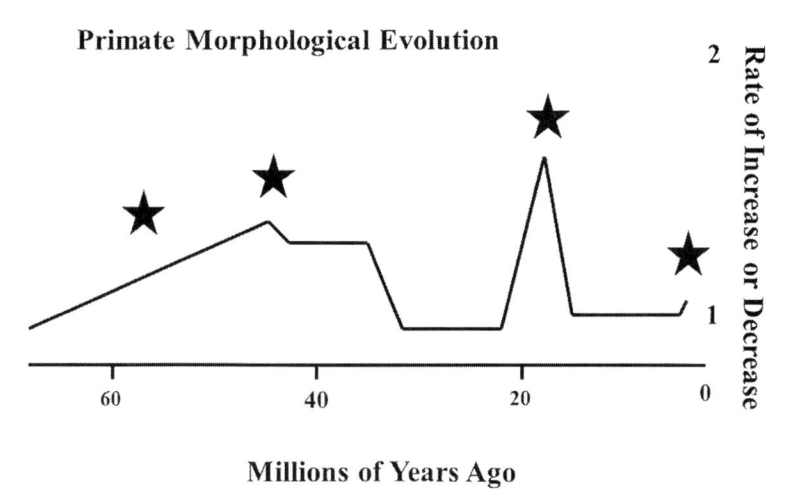

Figure 8.1. Rates of primate morphological evolution through time. The stars indicate major radiation events. In primates, these are associated with significant changes in morphology; in other mammalian orders (e.g. bats, rodents), rapid speciation can occur without major changes in morphology. From early to late, the stars refer to the Paleocene radiation, the Eocene radiation, Miocene radiations associated with the emergence of basal catarrhines, crown catarrhines, hominoids, and platyrrhine primates, and a Plio-Pleistocene radiation associated with the cercopithecoid monkeys, hominids, and the subfossil Malagasy prosimians. Modified from Venditti et al., 2011:Fig. 2b.

increase in evolutionary rates during the Eocene; after a significant decline that levels out and persists through the Oligocene, primates then experience a very brief pulse of high evolutionary rates during the Miocene (Venditti et al., 2011: Fig. 2b). In primates, evolutionary radiations appear to be associated with morphological change. Figure 8.1 illustrates primate morphological evolution through time, but also shows two modifications from Venditti et al. First, I introduce the plesiadapoid primates into the analysis. Primates then demonstrate a very strong burst of high morphological evolution during the Paleocene. This is the Paleocene radiation. It is followed by the Eocene radiation. The second modification from Venditti et al. is that I introduce the burst of primate morphological evolution that occurs with the Plio-Pleistocene radiations of the cercopithecoid monkeys, the hominids, and the subfossil Malagasy prosimians. The very high rates of primate morphological evolution that occur during the Miocene are caused by the appearance of basal catarrhines, crown catarrhines, the hominoids, and the platyrrhines.

In this chapter, I am including six families in the Superfamily Plesiadapoidea. Some researchers who have focused on this group recognize as many as 11 or 12 families (Boyer & Bloch, 2008; Silcox, 2008). The families that I include are shown in Table 8.1, which also includes major genera and some notes about adaptations. I use the term "plesiadapoid" to refer to members of this Superfamily, and I also use the term to refer to all members of the initial primate radiation (the Paleocene radiation) that took place at the beginning of the Cenozoic. Many specialists working with this group are convinced that convergent evolution is occurring–i.e.

Table 8.1. The Superfamily Plesiadapoidea.

Found only in North America (Canada and the USA) and Western Europe; a land bridge connects Western Europe with Greenland and North America; the North Atlantic does not open until after the Early Eocene.

Family Paromomyidae	(*Purgatorius* [earliest genus], *Ignacius*, *Paromomys*, *Phenacolemur*); *Purgatorius* is the earliest plesiadapoid; a single tooth of *Purgatorius ceratops* occurs in the latest Cretaceous of Montana, USA, which extends the primate order back into the terminal Cretaceous. The age of this tooth is disputed, because it occurs in a channel fill. The genus is also found in the earliest Paleocene of Montana and Saskatchewan, Canada; *Purgatorius janisae* has a lower dental formula of 3:1:4:3. Tarsal bones of *Purgatorius* indicate that the genus was arboreal. A colugo-like gliding membrane was thought to exist in some paromomyids, but recent study indicates that no plesiadapoids had postcranial gliding specializations.
Family Plesiadapidae	(*Pronothodectes* [earliest genus], *Plesiadapis*, *Platychoerops*, *Chiromyoides*, *Nannodectes*); the neurocranium is long and low; there is no post orbital bar; a single pair of enlarged, procumbent incisor teeth are followed by a diastema; *Plesiadapis* is the most well-known genus of plesiadapoids; a long, extremely bushy tail and clawed digits on both the hands and feet are seen in soft-tissue impressions of the cat-sized *Plesiadapis insignis* from Menat, France (Gingerich, 1976:Plate 12); *Plesiadapis* has six or seven species, and an age range of 29 my, from 63.3 mya to 33.9 mya, which is the beginning of the Oligocene; the genus is found in Canada, the USA, and Europe.
Family Carpolestidae	(*Elphidotarsius* [earliest genus], *Carpolestes*, *Carpodaptes*); found only in North America; lower P4 becomes extremely enlarged, with many bladelike cusps that extend in fine, vertical ridges from the crown to the cemento-enamel junction; these premolars resemble those that independently evolve in Family Saxonellidae, as well as in a suborder of the multituberculates (plagiaulacoids), and in living marsupials like the rat-kangaroos and the pygmy possum.
Family Saxonellidae	(*Saxonella* [the only known genus]); the genus is rare, but widely distributed, being found in Germany and Canada; the lower P3 becomes extremely enlarged and bladelike, versus enlargement of the lower P4 in carpolestids.
Family Microsyopidae	(*Palaechthon*, *Plesiolestes*, *Microsyops*); more than two dozen genera are recognized from North America and Europe; the bulla is not formed from the petrosal bone, and resembles that of insectivores—thus, there is controversy about including them in the Order Primates; microsyopids are widespread in North America, and they survive until the Late Eocene in the western USA.
Family Picrodontidae	(*Picrodus*, *Zanycteris*); found only in North America; the lower molar crowns become flat and relatively featureless, and resemble those of modern nectar-eating bats.

a number of lineages are independently acquiring plesiadapoid-like traits. However, details of evolutionary events are still uncertain. To reflect both this uncertainty, as well as the conviction that convergent evolution is taking place, many researchers use the term "plesiadapiform." The quotation marks refer to the probability that a number of lineages are independently entering a plesiadapoid-like adaptive zone, and traits that are associated with plesiadapoids are accreting differently in different lineages (Boyer & Bloch, 2008; Silcox, 2008). This situation would not be unique to early primates. It also occurred when mammal-like reptiles were evolving mammal-like traits. It took over 50 years to document how mammal-like traits were acquired in different lineages. It would take a similarly long time to document convergent evolution in plesiadapoid primates, and, of course, this documentation would be dependent on the discovery of well-preserved fossils.

Purgatorius is the earliest primate genus. It dates to 66–61.7 mya, and is found in northeastern Montana, USA, and in Saskatchewan, Canada. Its dentition is generalized: the lower molars exhibit a pronounced division between the trigonid and talonid, and the talonid basin lies well below the level of the trigonid. However, even at this early date, the upper molars are more quadrangular than in the basic therian molar (Figure 1.2). This molar pattern in *Purgatorius* indicates a shift toward herbivory that was rare in contemporary mammals, and is the reason why some researchers identified primate origins with specializations for herbivory (Van Valen & Sloan, 1965; Szalay, 1968). *Purgatorius* was arboreal, because characteristically primate-like tarsal bones have been discovered (Kaplan, 2012; Milius, 2012). The ankle joint was highly mobile, and allowed multidimensional movement. Yet, the tarsal bone was not elongated, as it is in specialized primate leapers. Hence, *Purgatorius* scampered and scrambled in the trees, engaging in a variety of movements that selected for ankle flexibility. The hallux and other foot bones are not known, and so one cannot argue that the hindlimb dominance and grasping hallux characteristic of primates had developed in *Purgatorius*. However, small arboreal marsupials like the living woolly opossums (*Caluromys* spp.) exhibit the diagonal sequence diagonal-couplets gait that is typical of primates (Cartmill et al., 2002). This is an example of convergent evolution. The presence of diagonal sequence diagonal-couplets gaits in primates and these marsupials, along with fruit- and insect-eating diets, has encouraged the use of woolly opossums and other small arboreal marsupials as models for the behavior of the earliest primates (Cartmill, 1972; Rasmussen, 1990).

The current consensus is that plesiadapoids are not primates, but close primate relatives ("primatomorphs"). The idea was first advanced by Martin (1968). However, the idea was later launched into popularity by the visual predation hypothesis of primate origins. Because plesiadapoids show no specializations for insectivory, they conflict with the visual predation hypothesis. Cartmill (1972, 1974b, 1975) thus suggested removing them from the order in his original publications on visual predation. Several researchers subsequently argued that some plesiadapoids possessed a gliding skinfold or patagium that stretched between forelimb and hindlimb (Beard, 1990; Kay et al., 1990). Although this idea was immediately contested on the basis of morphology (Krause, 1991), the notion of gliding plesiadapoids further distanced them from living primates, which have no gliding species, and appeared to ally them with colugos, members of the Order Dermoptera. And gliding specializations would be another trait linking some plesiadapoids to the living marsupial phalangers. Yet, gliding specializations have independently evolved many times over in the history of the vertebrates. For example, 15 living rodent genera exhibit gliding specializations. Twelve of these genera are in the squirrel family (Family Sciuridae), and three of these genera are in the scaly-tailed squirrel family (Family Anomaluridae). Most of these rodent gliders are mouse-sized, but the anomalurids are as large as cats. Struts of cartilage at the wrist (flying squirrels) or the elbows (anomalurids) may further extend the gliding membrane. Given the documented occurrence of convergent evolution, the presence of a gliding membrane in some plesiadapoids should not automatically exclude them from the primates. Yet, an exhaustive study of newly described and remarkably complete plesiadapoid skeletons finds no evidence of gliding specializations in any region of the postcranium (Boyer & Bloch, 2008). Thus, conclusions that strongly altered opinions about plesiadapoid affinities (Beard, 1990; Kay *et al.*, 1990) have since been falsified.

Major recent publications on fossil primates do not include any discussion of plesiadapoids (e.g. Hartwig, 2002). The genus *Plesiadapis* remains the most well-known form, and it is one of the largest plesiadapoids (Figures 8.2–8.4). Despite this consensus opinion, a shift is beginning to occur towards recognition of plesiadapoids as the first primates. This shift was initiated by the discovery and description of new plesiadapoid fossils (Bloch & Boyer, 2002, 2003; Bloch *et al.*, 2007; Boyer & Bloch, 2008). New preparation techniques that dissolve limestone matrix while allowing small and delicate fossils to be extracted without damage have opened a novel window into the world of plesiadapoid primates. In particular, three new taxa (*Carpolestes simpsoni*, *Ignacius clarkforkensis*, and

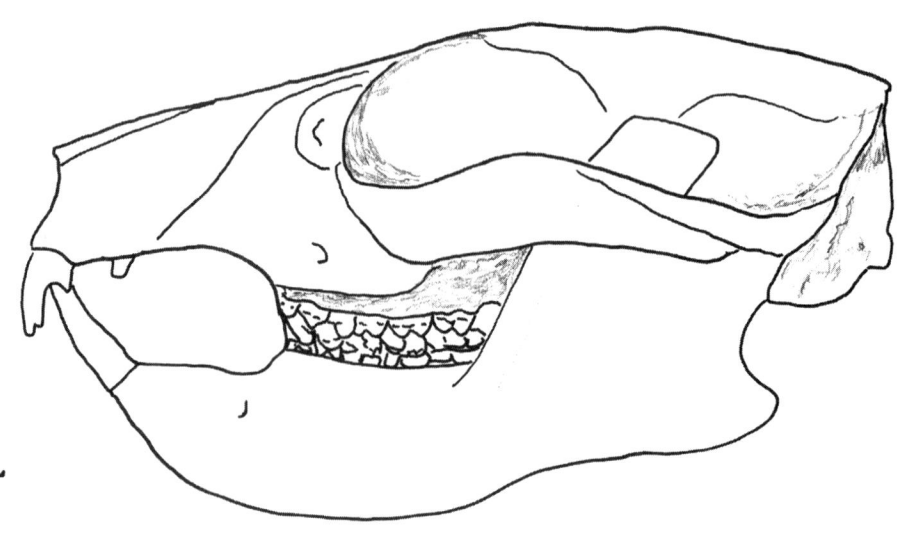

Figure 8.2. The skull of *Plesiadapis tricuspidens*, type species of the genus *Plesiadapis*. The neurocranium is low, and the relative brain size is small, compared to euprimates. The dental formula is 2:1:3:3/2:1:3:3. The incisors are enlarged, and there is a diastema in the upper and lower jaws. There is no postorbital bar, and the orbits are not convergent.

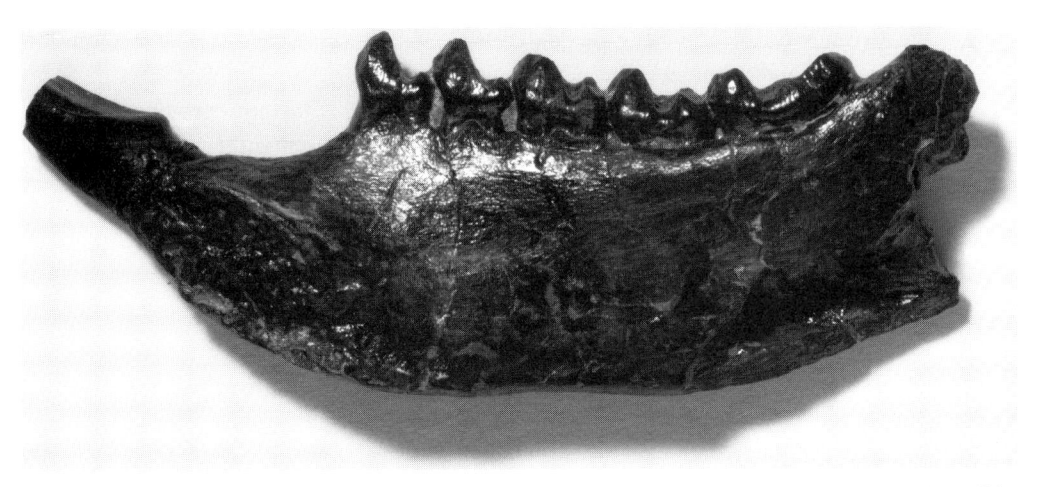

Figure 8.3. Mandibular fragment of *Plesiadapis fodinatus*, Late Paleocene, Wyoming, USA. Courtesy of the Mary Evans Picture Library/Natural History Museum.

Figure 8.4. A reconstruction of *Plesiadapis*, shown as a generalized arboreal quadruped. Illustration by Angela J. Tritz.

Dryomomys szalayi) have mandated a re-evaluation of the links between plesiadapoids and primates of modern aspect (euprimates).

A characteristic feature of plesiadapoid primates is the presence of a large pair of procumbent incisor teeth in the upper and lower jaws. These are followed by a diastema, and then by very generalized molar teeth. In living animals with a diastema, such as rodents, a fold of skin from the lips is drawn into the mouth as the animal uses its incisors to gnaw. This prevents fragments of inedible material from entering the mouth and being inadvertently swallowed.

Both carpolestids and saxonellids possessed a greatly enlarged, serrated, blade-shaped lower premolar, in which long, fine vertical ridges descend from the crown to the cemento-enamel junction of the tooth. The Paleocene primates evolve this morphology independently, because the P4 is enlarged in carpolestids and the P3 is enlarged in saxonellids. This premolar is termed a plagiaulacoid premolar, after the multituberculate suborder that classically evolves this trait. One might assume that this trait indicates bizarre specialization in the multituberculates and these Paleocene primates. However, the plagiaulacoid premolar evolves independently many times in mammal evolution, and is a good example of convergent evolution. It evolves twice in the multituberculates—in the North American plagiaulacoids and in the European kogaionids (Agustí & Antón, 2002:4). Besides its double origin in the multituberculates, the plagiaulacoid premolar is found in the carpolestid and saxonellid plesiadapoid primates, the polydolopids, an extinct group of South American marsupials, and in three forms of living marsupials: the potoroid rat-kangaroos, the potoroid bettongs (*Bettongia*), and the burramyid pygmy possum (*Burramys*). Thus, although the phylogenetic relationships of the living marsupials remain somewhat unclear, the plagiaulacoid premolar evolves independently in mammals at least seven times through convergent evolution. The plagiaulacoid premolar functioned in the same way in multituberculates and in plesiadapoid primates: food was ground against the upper tooth, and the shearing premolar therefore wears down from the apex (Dumont et al., 2000). In living marsupials, occlusion is different. The shearing premolar has a scissors-like occlusion with the upper tooth, and food is cut between the upper and lower teeth. What dietary items were processed with a plagiaulacoid premolar? In living marsupials, the blade-like lower premolar functions to slice through the tough husk of nuts and seeds, or the tough pericarp of fruit. Once the hard exterior is opened, the soft interior meat, kernel, or flesh is exposed. This soft textured material then requires very little processing. The rat-kangaroos and bettongs also exploit a unique food source, subterranean truffle-like fungi. The most reasonable explanation for the plagiaulacoid premolar of extinct plesiadapoid primates is that it functioned to crack open tough husks, seeds, or the pericarp of fruit.

New fossil material and ultra-high-resolution CT scans reveal that some plesiadapoids had several euprimate-like traits. Given the diversity of plesiadapoids and the accretion of euprimate-like traits in their lineages, it seems reasonable to anchor the origin of the euprimates within some generalized member of the Paleocene radiation. *Plesiadapis tricuspidens* and *Carpolestes simpsoni* clearly possess a petrosal bulla; *Ignacius* had a bony tube enclosing its internal carotid artery; and *Carpolestes* had a large internal carotid artery with promentory and stapedial branches (Boyer & Bloch, 2008). The limb length/body length ratios and vertebral and innominate morphology of *Ignacius* indicate that it was an agile, scansorial animal that emphasized use of its hindlimb for quick and powerful forward propulsion in the trees. It might have used its claws to cling to vertical supports like living tamarin monkeys (Boyer & Bloch, 2008).

Like all extant primates and Eocene euprimates, plesiadapoids had long digits relative to the metacarpals and metatarsals (Figure 8.5). When relative toe length is measured by the ratio of digit length/metatarsal length, *Plesiadapis*, *Dryomomys*, and *Carpolestes* exhibit typically long, primate-like toes (Hamrick, 2012). This trait is thought to have evolved for small-branch grasping. Plesiadapoids show variability in the morphology of their hands and feet. *Plesiadapis* and *Dryomomys* have elongated digits, but they are clawed, and the clawed hallux is not opposable or divergent. *Carpolestes*, however, had an opposable, divergent hallux that was nailed; claws appear on the other digits. This morphology is unique among the known plesiadapoids. Boyer *et al.* (2013: Tables 1 & 2) synthesize all available evidence on plesiadapoid hands, and summarize evidence for Eocene euprimates and two early fossil anthropoids (*Aegyptopithecus* and *Apidium*). They conclude that all known plesiadapoid hands resemble those of modern primates, both with regard to internal proportions of the hand, and with regard to hand size relative to body size. The hands of Eocene euprimates appear to emphasize more grasping and vertical clinging than plesiadapoid hands do.

The adaptive zone of plesiadapoid primates

Currently, about 120 plesiadapoid species are recognized, distributed across 11 or 12 families (Boyer & Bloch, 2008). This is significant biodiversity. Unlike modern primates, plesiadapoids are extremely abundant in Paleocene sites—as abundant as rodents are in modern sites. This alone indicates that the position of plesiadapoids in ancient communities was different from that of modern primates. The wealth of plesiadapoids in Paleocene sites puzzled paleontologists of the early twentieth century. Teilhard de Chardin (1922) first suggested that these species may have been terrestrial, because they occur in such profusion in Paleocene sites that they seemed to violate the dictum that arboreal species are mostly winnowed out of the fossil record through taphonomic processes. Arboreal species tend to be rare, because they occur away from sediment traps or sinks. These depositional areas are generally located in lake or river margins. Gingerich (1976) also posited that members of the Family Plesiadapidae were largely terrestrial, and had niches that resembled those of modern ground squirrels and marmots. The frontispiece of his 1976 monograph shows three plesiadapoid species (*Plesiadapis cookei*, *Plesiadapis dubius*, and *Chiromyoides major*) foraging on the ground in a Paleocene Wyoming palm forest.

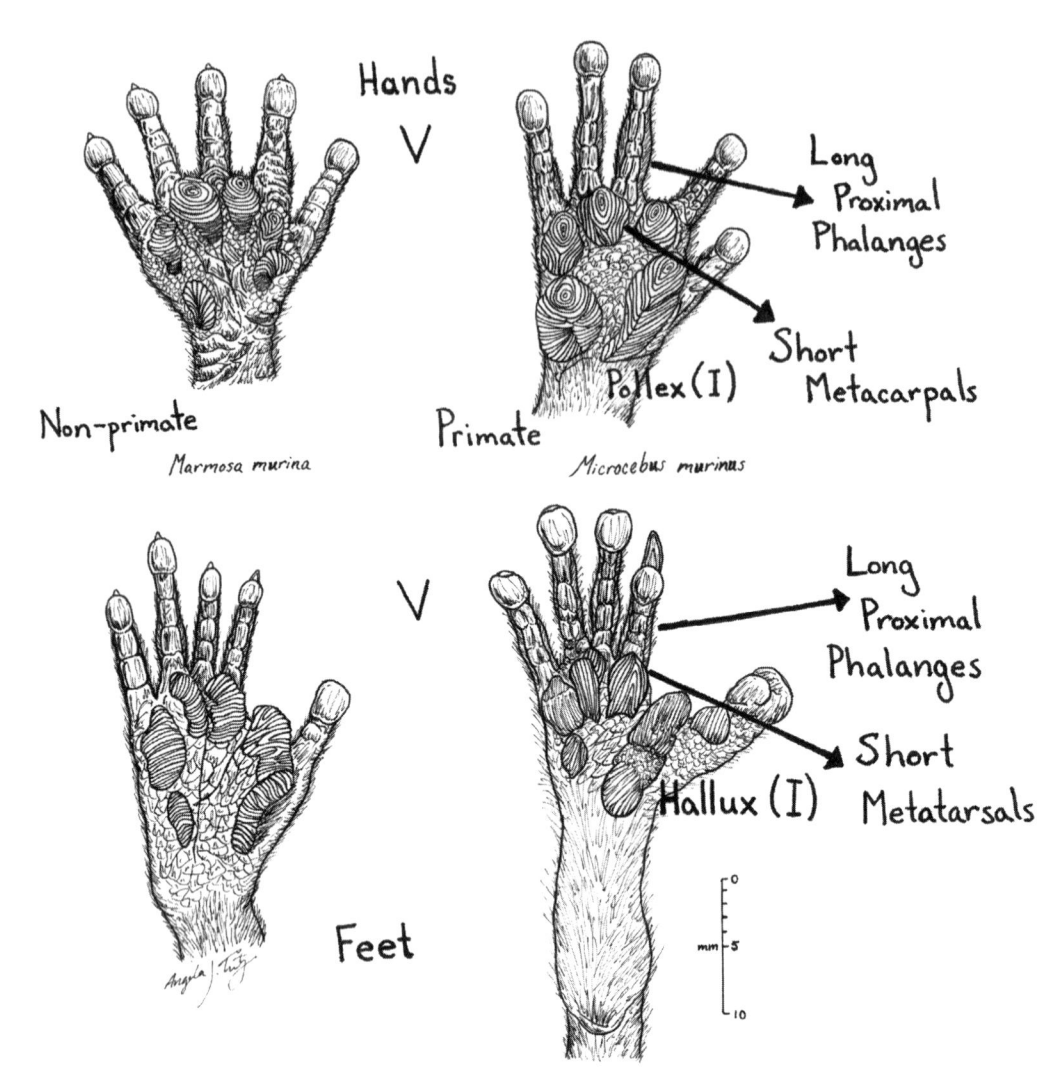

Figure 8.5. Primate hands and feet, represented by a mouse lemur (*Microcebus murinus*), are contrasted with those of a non-primate mammal, the mouse opossum (*Marmosa murina*). Note that the mouse opossum has a foot with a grasping hallux. The mouse opossum also has a prehensile tail, which aids in arboreal locomotion. Specializations for arboreality per se are thus not unique to primates. However, the grasping hallux is more enlarged and divergent in the mouse lemur. And primate hands and feet are both characterized by long proximal phalanges, relative to the length of the metacarpals or metatarsals. The high ratio of proximal phalanx length to metacarpal/metatarsal length that is typical for primates is also found in many plesiadapoid primates. The posterior extension of the foot in the mouse lemur reflects an elongation of the ankle bones that is associated with springing or leaping. See Figure 9.5. After Hershkovitz (1977:Fig. I.23). Illustration by Angela J. Tritz.

Sussman and Raven (1978) argued that living lemurs and galagos and small marsupials play an important role in the pollination of many plants, and that plesiadapoid primates had the same function in ancient ecosystems. In fact, Sussman and Raven (1978:735) believe that competition between nectar-feeding plesiadapoid primates and the earliest bats led directly to the extinction of the plesiadapoid primates and to the rise of powered flight in bats, as bats escaped competition by opening up a new adaptive zone that plesiadapoids could not fill. They also believe that euprimates of the Eocene were better able to exploit fruit, flowers, and insects available in small terminal branches than plesiadapoids were, because the euprimates develop grasping, nailed extremities and convergent orbits (Sussman & Raven, 1978:734).

However, although bats and some rodents are known to be important pollinators, the idea that lemurs and galagos, marsupials, and plesiadapoid primates were also pollinators and coevolving with flowering plants has been largely rejected (Lee & Cockburn, 1985). In order to be considered part of a plant/animal coevolutionary system, animals need to be able to transfer pollen from plant to plant without destroying the flower or plant, or eating the entire cargo of pollen. This leaves insects as the principal plant pollinators. Flying vertebrates like hummingbirds and bats can hover and sip nectar without destroying flowers. Several rodent species are known to be pollinators of sturdy *Protea* flowers (F. Proteaceae) in the unique fynbos habitat of the Cape of South Africa. The honey possum (*Tarsipes rostratus*) of Australia weighs between 7 and 20 g—one-third the size of mouse lemurs—and has a tubular mouth and specialized, extensible tongue to feed on nectar; it preens its fur clean of pollen between bouts of eating, and thus transfers pollen from plant to plant (Lee & Cockburn, 1985). But there is no evidence that living lemurs have a place in modern ecosystems as pollinators of flowering plants, and extinct fossil primates (plesiadapoids or later primates of the Eocene radiation) are also dubious plant pollinators. This point is not trivial, because it establishes whether primates have now or ever had an important role as pollinators in an ecosystem. Living primates (along with birds and bats) can function as seed dispersers, and thus serve a role in ecosystem structure. However, living primates are not keystone species in modern communities, and therefore tend not to be the focus of systematic conservation efforts. The rationale for mounting heroic conservation efforts for living primates is based solely on their relatedness to humans (Chapter 17).

How did the enlarged incisor teeth of plesiadapoid primates function? Based on dental microwear in *Plesiadapis*, Gingerich (1974) demonstrated that the occlusal surfaces of the incisors become worn and blunted with use—they are not used to gnaw through food like the chisel-like incisors of rodents do, and the occlusal edges of the incisors in *Plesiadapis* are not self-sharpening. The upper incisor has three cusps, and a narrow basin lies posterior to the two anterior cusps. The lower incisor occludes within the basin of the upper incisor; as the lower incisor entered this space, it may have sheared against the posterior cusp of the upper incisor. Gingerich suggests that this incisor activity would best function to shear through relatively soft plant stems. Because plesiadapoids were an order of magnitude larger than smaller Paleocene primates, they could have subsisted on a folivorous/frugivorous diet. Smaller Paleocene primates like the paromomyids would have a different diet—the teeth of some paromomyids like

Ignacius indicate exudate eating or the consumption of gums, saps, and resins from damaged bark (Boyer & Bloch, 2008).

Plesiadapoid primates existed in a world very different from the present. At the beginning of the plesiadapoid reign, extinct placental mammalian herbivores, especially the condylarths, which are ancestral to living ungulates, very rapidly radiated to fill niches that were left vacant by the removal of the dinosaurs. No obvious mammalian predators existed for the first million years of the Paleocene, which implies that competition between relatively small-bodied herbivores was driving community evolution (Van Valen, 1978). Understanding plesiadapoid biology mandates understanding the ancient communities in which they lived.

For the duration of their temporal span, plesiadapoids coexisted with multituberculates. Multituberculates were the longest-surviving order of mammals. Their fossil record extends from at least the Early Jurassic to the Early Oligocene. They were not therian mammals. Multituberculate molar crown morphology was absolutely distinctive, because a multitude of little uniform cusps formed the crown surface. They are very abundant in every well-sampled Cretaceous and Paleocene fossil mammal locality in North America. Having searched for mammal teeth by sieving sediments from sites in the American West, I can attest to the abundance and reassuring ease of identification of multituberculate molars. There are indications that multituberculates like *Ptilodus* of the Middle and Late Paleocene of North America were arboreal. It possessed a strong, prehensile tail, a mobile ankle joint, and a divergent hallux that could both abduct and adduct independently of the other digits—it may have been able to descend vertical tree trunks head first, as living squirrels do, because its toes could face backwards (Jenkins & Krause, 1983:Fig. 2). *Ptilodus*, like other plagiaulacoid multituberculates, had a serrated, bladelike lower P4, a specialization also seen in carpolestid and saxonellid (P3) plesiadapoid primates. Other multituberculates were terrestrial. Fossil taeniolabid multituberculates from the Cretaceous of Mongolia burrowed underground like modern woodchucks, and some were the size of living beavers. In general, multituberculates were vaguely rodent-like in their cranial and dental morphology, having a long, low neurocranium and enlarged, procumbent incisor teeth. This is also true for two other extinct orders of mammals, the Order Tillodontia and the Order Taeniodontia, which were never abundant, and existed from the Late Paleocene to the Late Eocene. The same morphology also exists in the plesiadapoids, and it is not accidental that tillodonts, taeniodonts, plesiadapoids, and multituberculates do not survive for long after the rise of the first true rodents, which appear at the end of the Paleocene.

In North America, true rodents suddenly appear about 56 mya, at sites that span the Paleocene/Eocene boundary. Their sudden appearance is associated with a simultaneous decline of multituberculates. Rodents originate in Asia, and immigrate into North America via a land corridor that connected Asia and Alaska in the region of the Bering Strait. Beard (1998b) argues that many mammal groups originate in Asia during the Paleocene—an idea called the East of Eden hypothesis.

That is, Asia was the major center of origin and diversity for later mammal groups that disperse and proliferate in the Eocene. Silcox (2008) strongly denies that Asia was the center of origin for primates or euprimates, and argues instead that North America, with its long, well-sampled record, probably has this honor.

Multituberculates immediately declined when encountering the first rodents. This is also true for the plesiadapiform primates (Krause, 1986:Table 1). The concomitant decline of multituberculates and rise of the rodents is one of the few unequivocal examples of taxonomic displacement or competitive exclusion in the mammalian fossil record (Krause, 1986). The rodents also deliver a death blow to the plesiadapoid primates, and for the same reason: competitive exclusion. Multituberculates, plesiadapoids, and rodents were all relatively small, arboreal, generalized herbivores. Rodents, however, had three herbivorous specializations that neither multituberculates nor plesiadapoids possessed (Figure 8.6). Their incisor teeth were ever-growing, and had a self-sharpening mechanism at the anterior, chisel-like occlusal surface of the incisors: a rim of hard enamel surrounds the softer dentine, and allows rodents to gnaw powerfully through tough material like wood. Rodent molar crowns have formidably complex shearing

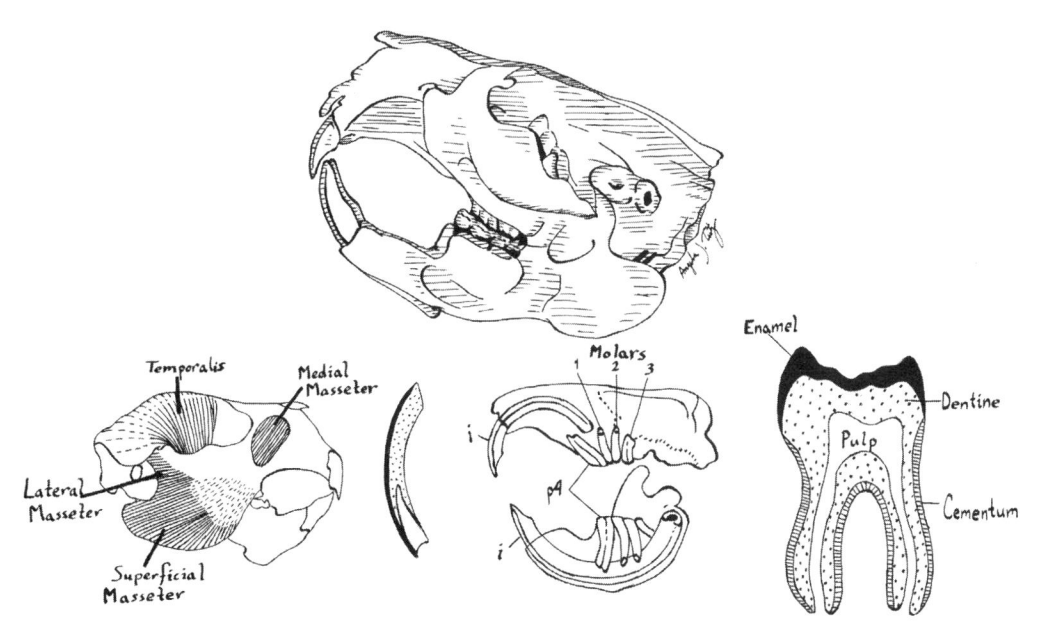

Figure 8.6. A beaver skull and teeth, illustrating some of the specializations that allowed rodents to out-compete the plesiadapoid primates. Both the incisors and posterior teeth (P4-M3) have open, ever-growing roots. The chisel-like incisors have a self-sharpening edge created by the tough enamel meeting the softer dentine. The enormous masseter muscle, largely responsible for sideways movements of the lower jaw, has an extra extension (the medial masseter) that originates on the face, in front of the orbit. Not shown is another rodent specialization: the intricate enamel and dentine patterns on the crowns of the posterior teeth. These act to shred plant food very finely, and thus allow the rodent to consume low-quality plant foods. Illustration by Angela J. Tritz.

surfaces, composed of alternating folds of enamel and dentine—the better to cut through vegetation. Lastly, rodents possess a very complex masseter muscle, with an extra slip that extends onto the anterior of the snout below the orbit. This allows rodents to move their lower jaws in a very complicated lateral fashion. They are therefore much better able to masticate vegetation than either multituberculates, plesiadapoids, or living primates.

After the rise and spread of the true rodents, primates decline, and never again achieve the abundance and diversity that they had during the Paleocene. No living primate has a rodent-like niche, because they cannot compete with rodents. Although it was originally identified as a rodent by French scholars of the seventeenth century, the aye-aye (*Daubentonia*) is a true primate, albeit a specialized one. The aye-aye actually occupies a woodpecker-like niche in Madagascar, where avian woodpeckers are absent. Its enlarged and procumbent incisor teeth allow it to bore through wood, and the thin, elongated third digit on the hand allows the aye-aye to extract grubs and larvae. The aye-aye's facial skeleton is shortened anteroposteriorly, but elongated dorsoventrally. The facial skeleton is bent down relative to the cranial base. These changes increase the vertical component of the hafting between neurocranium and upper jaw. This reduces deformation or strain in the face when the aye-aye uses its incisors to bore into wood (Cartmill, 1974a). The aye-aye is thus specialized for extractive foraging, and is not occupying a rodent-like niche. As peculiar as the morphology of the aye-aye appears, similar anatomy and behavior is seen in the New Guinea striped possum *Dactylopsila*, which has enlarged first incisors, an elongated, thin fourth hand digit, and taps on bark to detect hidden insect grubs. Woodpeckers are also absent in New Guinea. Extinct Early Cenozoic mammals called apatemyids (Family Apatemyidae, sometimes identified as a separate order, Order Apatotheria) also possessed two enlarged incisor teeth, and beautifully preserved Eocene specimens (*Heterohyus*) from the Messel demonstrate that the second and third hand digits were elongated. These three independent origins of dental and hand specializations for extractive foraging in mammals thus serve as another example of convergent evolution.

In summary, it is not an accident that interpreting the adaptations of plesiadapoid primates leads one to examine multituberculates and a host of small, living marsupials: sugar gliders, phalangers, pygmy possums, striped possums, and bettongs. Plesiadapoids arose in a world without rodents, like the Australasian marsupials east of Wallace's Line. Mouse-like murid rodents cross a water gap from Timor to enter Australia only late in time, at about 2.5 mya. Rats and mice are then accidentally introduced into oceanic islands over the last several thousand years by humans who are deliberately colonizing these islands using seaworthy boats. Australasian marsupials thus thrived in a world without rodents, and also filled niches otherwise occupied by prosimian and anthropoid primates elsewhere. Lands east of Wallace's Line are also devoid of non-human primates. Small phalangers and possums occupy mouse lemur and fat-tailed lemur niches; large, folivorous phalangers and tree kangaroos occupy monkey-like niches.

Considering plesiadapoid primates in this light, they no longer appear to be aberrant. A number of their traits (enlarged incisor teeth, a diastema, plagiaulacoid premolars) are developed through convergent evolution a number of times over in mammal evolution.

The Paleocene/Eocene Thermal Maximum

The paleogeography of the earth during the Middle to Late Paleocene is well known. The South Atlantic and Middle Atlantic basins were already extensive. Because sea-floor spreading was initiated in the south, the North Atlantic was still very narrow, and land connections existed between North America, Greenland, and Europe (Figures 8.7 and 8.8). This accounts for the wide dispersal of plesiadapoid primate taxa in North America and Europe.

Although the Paleocene itself was an epoch of global warmth, the transition from the Paleocene to the Eocene was a time of extraordinary warming. At about 55 mya, the temperature of the earth spiked upward very abruptly. In less than 10,000 yrs, mean annual global temperature rose by 5–8°C (Zachos *et al.*, 2008: Fig. 2). This temperature extreme is called the Paleocene/Eocene Thermal Maximum or PETM. This temperature spike has not been duplicated during the last 65 my, the entire span of the Cenozoic. Temperatures in the Arctic were warm and moist throughout the Paleocene and Eocene. Sediments taken from sea-floor

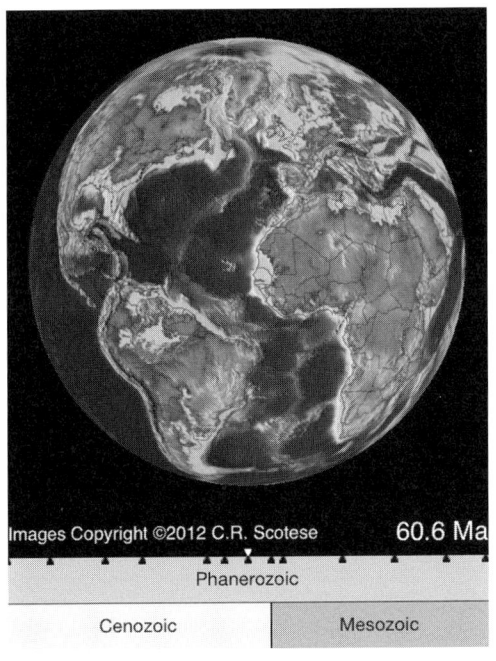

Figure 8.7. Paleogeography of the Atlantic Basin during the Paleocene, 60.6 mya. Note the extent to which the South, Middle, and North Atlantic Ocean have already opened, with plate movements initiated in the south. Courtesy of Dr. Christopher Scotese.

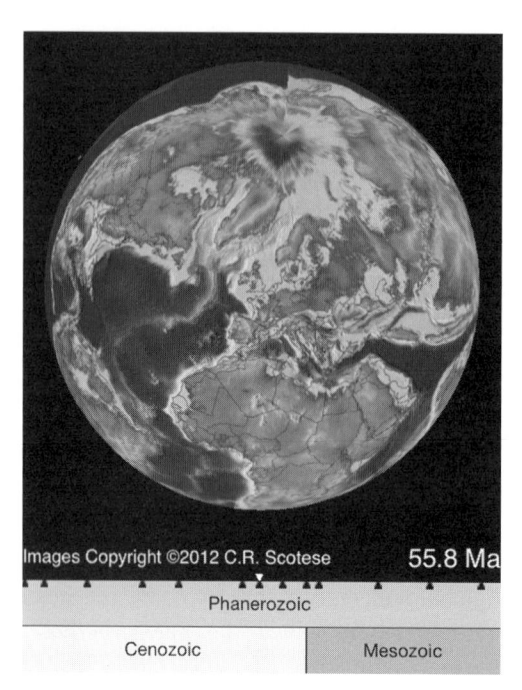

Figure 8.8. Paleogeography of the Arctic during the latest Paleocene at 55.8 mya, close to the Paleocene/Eocene Thermal Maximum. Courtesy of Dr. Christopher Scotese.

cores in the Central Arctic Ocean can be used to infer that mainland forests during the PETM had a relative humidity of over 70 percent, and soil moisture of over 60 percent (Pagani *et al.*, 2006). This would indicate tropical rainforest conditions in the High Arctic–an extraordinary situation–clearly unlike the modern world.

The PETM is thought to have been triggered by an influx of greenhouse gasses into the earth's atmosphere. This could have occurred in two ways. On the one hand, vast methane deposits were released from frozen hydrates encapsulated in ocean sediments lying on the continental shelves. This injection of methane bubbled up through ocean waters and was set free into the atmosphere.[1] The trigger for this release of methane from ocean sediments is debatable–it may have been initiated in a drastic way, either by earthquakes creating underwater landslides, or by an oceanic asteroid impact causing offshore landslides. Conversely, oceanic methane

---

[1] A small-scale example of such a methane release occurred on April 20, 2010, when The Deepwater Horizon, an offshore oil drilling rig in the Gulf of Mexico, experienced an explosion and fire that destroyed the rig and killed 11 workers. Methane gas emitted from the undersea BP well leaked into the well bore pipe, exploded, and blasted out the rig. Methane continued to leak out of the blown well, and rose to the ocean surface, along with light crude oil and natural gas. The amount of oil contaminating the US Gulf Coast became the worst environmental disaster in US history. There was also a continuous emission of methane gas, comprising about 40 percent of the mass spewing from the undersea well. Deep-water drilling had occurred in an area of volatile methane ice (methyl hydrate). Oil companies had previously avoided this danger by drilling in shallow, offshore sites. Continuing methane ice formation around the bore pipe hindered efforts to cap the oil flow. The well was finally sealed on July 15, 2010. See http://news.sciencemag.org/oilspill.

deposits may have been released by more gradual events. A 3,000 year warming of ocean bottom waters occurred off the coast of New Jersey—possibly caused by orbital forcing of oceanic and atmospheric circulation—and this may have warmed underlying sediments (Sluijs *et al.*, 2007). An alternative view for the origins of the PETM is that massive basalt flows in the North Atlantic were caused by the separating plate boundary between Greenland and Europe. Plate tectonic activity associated with the creation of the North Atlantic may have released greenhouse gasses (Storey *et al.*, 2007). Whatever the ultimate cause, there was a massive methane release at the Paleocene/Eocene boundary. A unique peatland deposit in England (the Cobham Lignite) demonstrates that terrestrial wetlands also generated methane during the PETM spike (Pancost *et al.*, 2007). Methane release created a global greenhouse gas effect, as measured by the carbon isotope excursion (CIE) retrieved from sediment cores. The carbon released during these events may have been twice as massive as previously estimated, because the carbon from Arctic terrestrial plants has a CIE excursion that is much larger than that from marine carbonates (Pagani *et al.*, 2006).

A number of mammalian lineages decrease their body size during the PETM. This response is especially well confirmed for the earliest horses, which first decrease their body size by about 30 percent in the first 130,000 yrs of the PETM, and then afterwards increase their body size by 76 percent (Secord *et al.*, 2012). Body size responses of two primate genera are known from the earliest Eocene (Secord *et al.*, 2012:Fig. 1). The 100 g microsyopid *Arctodontomys* shows a 25 percent increase in body size after the PETM; the 750 g adapid *Cantius* first appears in the PETM and shows a 10 percent increase in body size during that interval. Thus, primates, unlike many other mammals, increase body size during the PETM. The drastic spike upward in global temperature does not seem to have had a deleterious effect on primates.

9 The Eocene primate radiation

The Eocene is a critical turning-point in primate evolution. During this epoch, the archaic plesiadapoid primates begin to go extinct, and primates of modern aspect, the euprimates, first appear. The Eocene expansion of the euprimates results in the founding of the major extant primate lineages: lorisiforms, lemuriforms, tarsiiforms, and anthropoids. These modern lineages are all traceable to Eocene ancestors. And, at the end of the Eocene, primates suffer a major decline—something that is also true for other mammalian orders. What was the Eocene crucible of primate evolution like? Details of Eocene climate and geography are well known (Figure 9.1).

The first 30 million years of the Cenozoic were very different from the present climate. Global temperatures were much higher than they are now, and the poles had no or virtually no ice. After the extreme temperature excursion at the Paleocene/Eocene boundary—the Paleocene/Eocene Thermal Maximum (PETM)—Eocene temperatures remained high and stable for a long time. Besides the PETM (Chapter 8), the Eocene experienced a less extreme rise in temperature at 53 mya, and temperatures remained constantly very high until 51 mya. In fact, the very warm and steady global climate between 53 and 51 mya is known as the Early Eocene Climatic Optimum (Zachos et al., 2008). The temperature of the earth slowly declined after this, but tropical habitats were widespread. The Gulf Coast at Laredo, Texas, was ringed with tropical rainforest and mangrove swamp, and therefore resembled the modern mangrove-laden coasts of Southeast Asia (Westgate, 2009). Global temperature peaked again at 42 mya, during the Middle Eocene (Zachos et al., 2008:Fig. 2). By the Late Eocene, temperatures had fallen enough to remove much of the tropical forests from Southern California, and about 40 percent of the non-carnivorous mammals—including primates—go extinct (Tomiya, 2009). Genera over 100 g in size with crushing, bunodont teeth increase in number, and the rodents proliferate. The terminal Eocene witnessed a drastic plummeting of temperature that severely affected mammal evolution, including the evolution of the primates (Chapter 11).

Primates in the High Arctic

The warm climate of the Early Eocene enabled tropical mammals to exist even in what is now the High Canadian Arctic. Ellesmere Island in Hudson Bay supported alligators, primates, rodents, multituberculates, and terrestrial ungulates—tapirs,

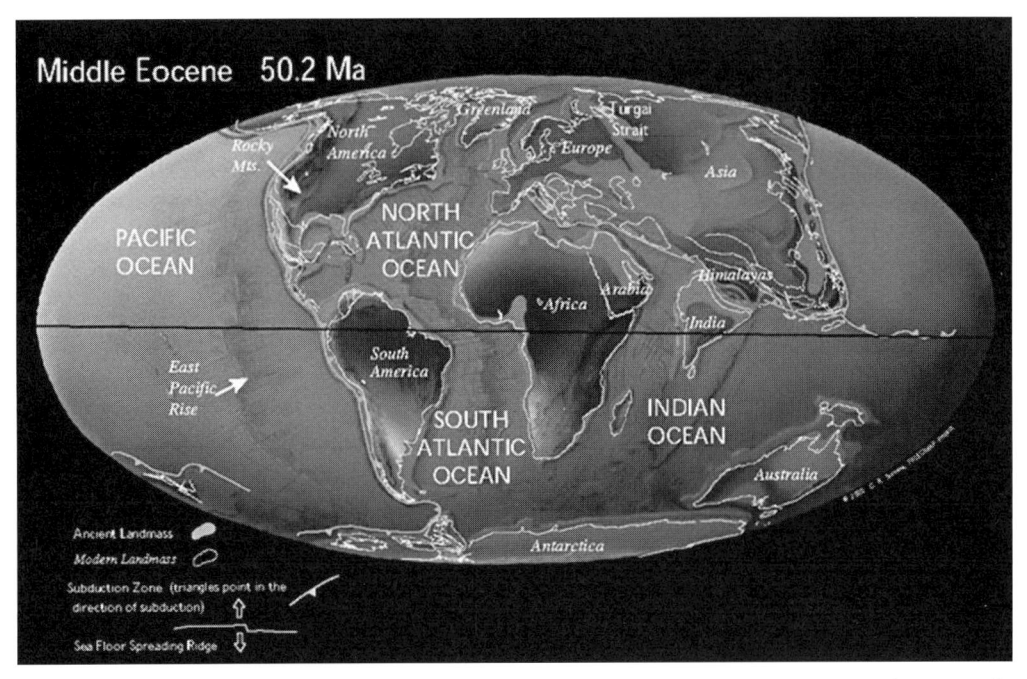

Figure 9.1. Paleogeography of the earth during the Middle Eocene, 50.2 mya. Courtesy of Dr. Christopher Scotese.

brontotheres, and *Coryphodon*, a hippopotamus-sized, swamp-dwelling pantodont with a widespread distribution. *Coryphodon* was the largest mammal of its day, and has the dubious distinction of having the smallest brain/body size ratio of any placental mammal. Islands of the Canadian Arctic are now at a very high latitude. The paleolatitude of Ellesmere Island was also high. At 78° North, and less than 400 km from the North Pole, it would have experienced some darkness for 6 months of the year (McKenna, 1980). The area would have been plunged into nearly continual winter darkness of more than 22 hours/day for at least 3 months, and an equally long wealth of sunlight of more than 22 hours/day during the 3 summer months (Schubert *et al.*, 2012). Thus, irrespective of a warm global temperature, animals inhabiting the area would have had to accommodate themselves to intense seasonal fluctuations in sunlight. Stable carbon and oxygen isotopes from the tooth enamel of *Coryphodon*, a tapir species, and a brontothere species reveal their ability to subsist yearlong on local vegetation without migrating south for the winter. *Coryphodon* was sampled the most, and individuals showed marked seasonal differences in diet. During the long summer sunlight, *Coryphodon* consumed swamp plants, flowers, and leaves; during the long winter darkness, *Coryphodon* consumed pine needles, twigs, leaf litter, and fungi (Eberle & Humphrey, 2009). A forest of dawn redwoods (*Metasequoia*) covered the Eocene High Arctic landscape (Figure 9.2). These trees were able to cope with seasonal lack of sunlight, and could also tolerate swamp conditions, because their roots were inundated. Masses of lotus flowers covered the stagnant waters. The forests

(a)

Figure 9.2. (a) A "dawn redwood" (*Metasequoia glyptostroboides*), typical of high-latitude forests in the Canadian Arctic of Eocene North America. This species was known only from fossils until rediscovered in China in 1941. (b) Dawn redwood leaf structure. Botanical Gardens, University of Bonn.

resembled the modern cypress swamps of the southeastern USA. The North Atlantic had not yet opened, and so land mammals could move freely back and forth between Europe, Greenland, and North America (McKenna, 1983). In addition, temporary polar land bridges also allowed land animals to disperse between North America and Asia in the region of the Bering Strait and across Arctic Canada.

The Eocene forests of Ellesmere Island and other sites in the Canadian Arctic can be reconstructed in detail. Leaf and other botanical remains are preserved in the fossil record. However, wood from Eocene trees is actually preserved in

(b)

Figure 9.2. (*cont.*)

modern permafrost. This wood is not fossilized; it has been freeze-dried in the permanently frozen soil of the Arctic. As a result, carbon isotope analysis within individual tree rings reveals the ancient climate. Summer rainfall was three times more abundant than winter rainfall. About 76 percent of the annual rainfall occurred during the summer months of nearly unbroken sunlight (Schubert *et al.*, 2012). This is the opposite from what occurs in the modern rainforests of the Pacific Northwest, where summer rainfall is only one-half to one-sixth that of winter rainfall. The closest analog in the modern world is coniferous forests of East Asia, like those in South Korea. In the Eocene Canadian Arctic, winter temperatures remained above freezing, despite the lack of sunlight; summer temperatures reached 20°C or higher. Note that the Eocene forests are above the Arctic Circle—i.e. above the modern treeline. Only sparse vegetation now exists in areas once covered by vast forests.

The flora of the Eocene Canadian Arctic has been studied in detail. Plants are preserved both as macrobotanical fossils and as freeze-dried wood buried in the modern permafrost. Besides dawn redwoods and other conifers, there are many angiosperm tree species, such as alder, birch, elm, and walnut. Fossils of very large, broad leaves indicate warm temperatures and heavy rainfall. Swamp forest conditions prevailed. Freshwater debouching into the enclosed Arctic Basin sometimes allowed mats of the floating fern *Azolla* to build up on the ocean waters. There are some marine vertebrates: three genera of sharks resembling sand tiger sharks and a warm-water ray have been found (Eberle & Greenwood, 2012). Alligators, 10 families of turtles, and bowfin, pike and gar fish inhabited the freshwater swamps that were overarched by forests. Among the mammals were the following forms: a multituberculate, a rodent family, *Coryphodon*, a pangolin, two families of creodont carnivores, two families of true carnivores, perissodactyl herbivores (tapirs and brontotheres), four genera of colugos (three of which are undescribed), and the paromomyid plesiadapoid primate genus *Ignacius* with several still undescribed species (Eberle & Greenwood, 2012:Table 2). About two-thirds of the Early Eocene mammal genera in the High Arctic are also found in middle-latitude states in the USA, such as Wyoming and Colorado. Artiodactyls are not present in the Eocene High Arctic, probably because their chambered stomachs, specialized for fermenting plant foods, did not allow them to cope with the low-quality, high-fiber food (e.g. pine needles, twigs, leaf litter, fungi) that was available during the winter.

The undescribed species of the primate genus *Ignacius* coexisted with a startling diversity of colugos. The High Arctic array of colugos indicates that this was a major center for colugo evolution. The coexistence of these two mammal groups when colugos were undergoing an evolutionary radiation must be significant. Colugos are currently recognized as the closest living relatives of primates. Phenotypic similarities caused by relatedness would be accentuated at this remote period, and the arboreal niches of colugos and these plesiadapoid primates must have been similar. Yet, how were colugos and primates able to cope with the prolonged seasonal darkness? There is no evidence that they migrated south during the winter. They could not subsist on low-quality vegetation like twigs or bark the way that giant sympatric herbivores like *Coryphodon* could. Besides, many of these resources (leaf litter, pine needles) occur on the ground, and would have been unavailable to arboreal animals. Both colugos and primates may have turned to insectivory during the winter darkness. It is possible that the procumbent incisors of *Ignacius* would have allowed it to gouge tree bark, eliciting a flow of gum, sap, or resin (Figure 9.3). Alternatively, colugos and primates may have undergone periods of semi-hibernation or torpor as winter darkness fell.

During the Early Eocene, an integrated North American/Europe faunal province existed, which includes areas now high in the Canadian Arctic, like Ellesmere Island. Primates were part of this integrated faunal province. Simpson's Coefficient demonstrates that Asia was a separate island-like faunal province, being separated from Europe by the Tugai Sea, and from North America by the Bering Strait (Flynn, 1986). However, during the early Eocene, euprimates suddenly appear in the fossil records of Asia, Western Europe, and North America. Two suggestions have been made about the trigger for this virtually simultaneous dispersal: either high global temperatures caused the spread of tropical rainforest habitats even in high latitudes, or barriers to euprimate dispersal were removed at the end of the Paleocene. Analysis of local faunas in Europe before and after the

Figure 9.3. Ellesmere Island in the High Canadian Arctic during the Paleocene/Eocene Thermal Maximum, 55 mya. The extinct pantodont *Coryphodon* feasts on lotus plants and other swamp vegetation in the background. Two specimens of the paromomyid primate *Ignacius* cling to the vertical trunk of a dawn redwood. This reconstruction of *Ignacius* is based on Boyer and Bloch (2008). One of these animals gouges the trunk to elicit the flow of edible resin. Small-bodied primates would be unable to subsist on dead vegetation and other low-quality food during several months of complete Arctic winter darkness, and are unlikely to have migrated south before seasonal darkness fell. Insectivory and exudate eating are possible responses to surviving the seasonal darkness. Illustration by Irene V. Hort.

PETM does not show much ecological turnover (Soligo, 2007). Thus, in spite of the fact that a greater number of arboreal niches may have been present in the Early Eocene of Europe—conducive to euprimate spread—new taxa (including euprimates) were occupying the same ecological niches. The euprimate invasion of Western Europe was therefore probably caused by the elimination of physical barriers to dispersal at the Tugai Sea (Soligo, 2007). The spread of euprimates from

Western Europe into North America is explained by the fact that a land bridge connected northern Europe, Greenland, and North America during the Early Eocene: the North Atlantic Ocean had not yet opened up.

Fissure-fill deposits from the karstic landscape of southeastern China reveal a hitherto unsuspected diversity of primates dating to about 45 mya (Beard *et al.*, 1994). Both adapoid and omomyid primates are found here. The genus *Eosimias*, represented by four species and sorted into a new family, Family Eosimiidae, is the most diverse taxon found in these fissure-fill deposits. *Eosimias* is postulated to be the earliest anthropoid primate, based on dental characteristics. However, the mandibular symphysis is unfused, and body size is extremely small, ranging from 67 to 136 g (Beard *et al.*, 1994). There are no cranial remains. Tali and a calcaneus also demonstrate a small body size, although the morphology of the calcaneus appears to resemble that of the living squirrel monkey (*Saimiri*)—an anthropoid (Beard, 2004). Interpretation of the Eocene fossils from China is thus intertwined with questions about where and when the first anthropoids are found.

The first euprimates

Teilhardina asiatica is currently recognized as the oldest euprimate (Figure 9.4). It is dated to the latest Paleocene, at about 56 mya, and is identified as a basal omomyid. Members of the genus *Teilhardina* spread very rapidly from Asia to Europe and then to North America during the Paleocene-Eocene Thermal Maximum (Smith *et al.*, 2006). Size of the orbital diameter relative to skull length indicates that *Teilhardina asiatica* was probably diurnal—in fact, its orbital diameter falls directly on the regression line measuring orbital diameter to skull length in living diurnal primates (Ni *et al.*, 2004:Fig. 4). The body size was very small, far smaller than other Eocene euprimate taxa, such as *Tetonius* and *Shoshonius*. Body weight is estimated to be about 28 g. Shearing crests on the teeth indicate an insectivorous diet, which might also be inferred from the small body size.

The oldest known fairly complete euprimate skeleton is slightly younger than *Teilhardina asiatica*, and it also comes from China. This is the type specimen of *Archicebus achilles*, dating to about 55 mya (Ni *et al.*, 2013). The body weight is very small. It is estimated to be between 20 and 30 g, approximately the size of a pygmy mouse lemur. The orbits are small, and there is a postorbital bar. The preserved distal phalanges of the foot indicate that all the digits were nailed. The tail is very long. The thigh and lower leg are also long relative to the forelimb. The feet are very big, especially the metatarsals and phalanges. *Archicebus* may therefore frequently have engaged in leaping behaviors. The taxon is classified as a tarsiiform primate, and may be the basal member of this group. Ni *et al.* (2013) argue that the shape of the calcaneus and the short tarsus in *Archicebus* resemble the anthropoid condition, and that this specimen therefore contains a mixture of prosimian and anthropoid-like traits. This taxon is probably better classified as an Asian member of the Eocene omomyid radiation. In any case, *Archicebus* indicates that the first euprimates may have been tiny, diurnal insectivores.

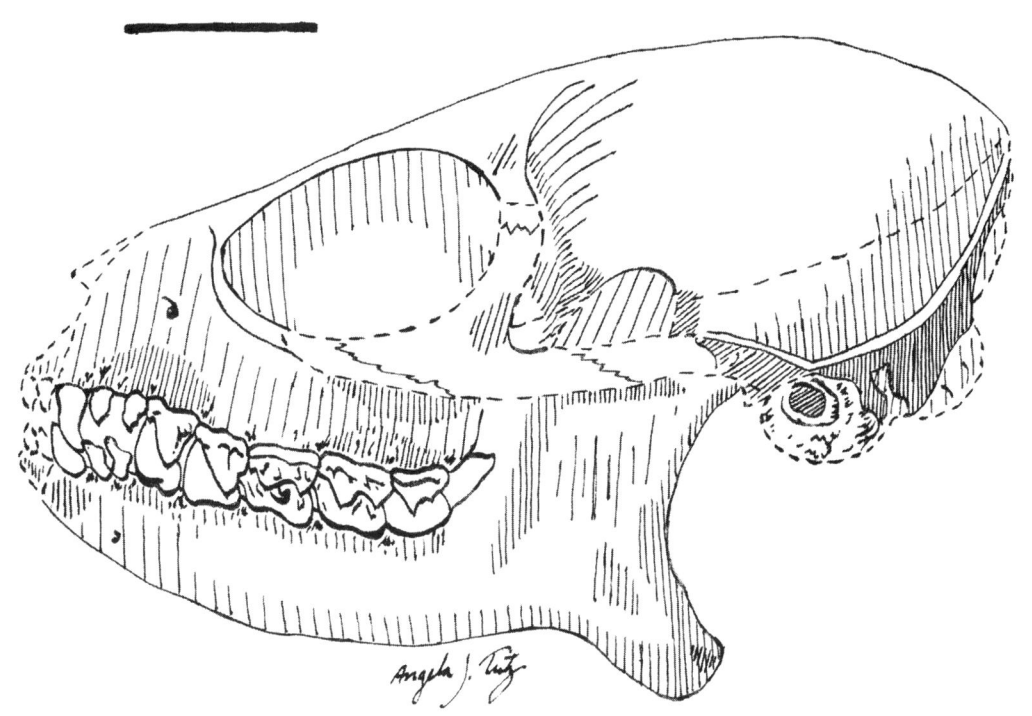

Figure 9.4. Skull of *Teilhardina asiatica*, possibly the earliest euprimate, dating to 56 mya. After Ni *et al.* (2004:Fig. 1b). Illustration by Angela J. Tritz.

Because of fairly abundant postcranial remains, the locomotion of Eocene euprimates has been extensively investigated. In general, the locomotion of Eocene taxa frequently resembles that of living lemurs that are generalized arboreal quadrupeds. Fossil ankle bones are especially well preserved. These often show some degree of elongation, especially in the omomyid primates. However, this need not indicate extraordinary leaping specializations in the omomyids. Their ankle bones often resemble those of the living mouse lemur (*Microcebus*) in their elongation (Figure 9.5). Because of its small size, the quadrupedal mouse lemur must frequently scramble and leap as it moves over discontinuous and swaying arboreal supports. One can infer a similar type of locomotion in many extinct omomyid prosimians.

Adapoids and omomyids

Franzen *et al.* (2009) ultimately credit William King Gregory with reifying a lemuroid versus tarsioid distinction in Eocene primates by virtue of his superb monograph on *Notharctus* (Gregory, 1920). G. G. Simpson (1937, 1940b), who later worked with Gregory at the American Museum, clearly articulated the idea that Eocene primates could be conveniently differentiated into a "lemur-like"

Microcebus   *Galago*   *Tarsius*

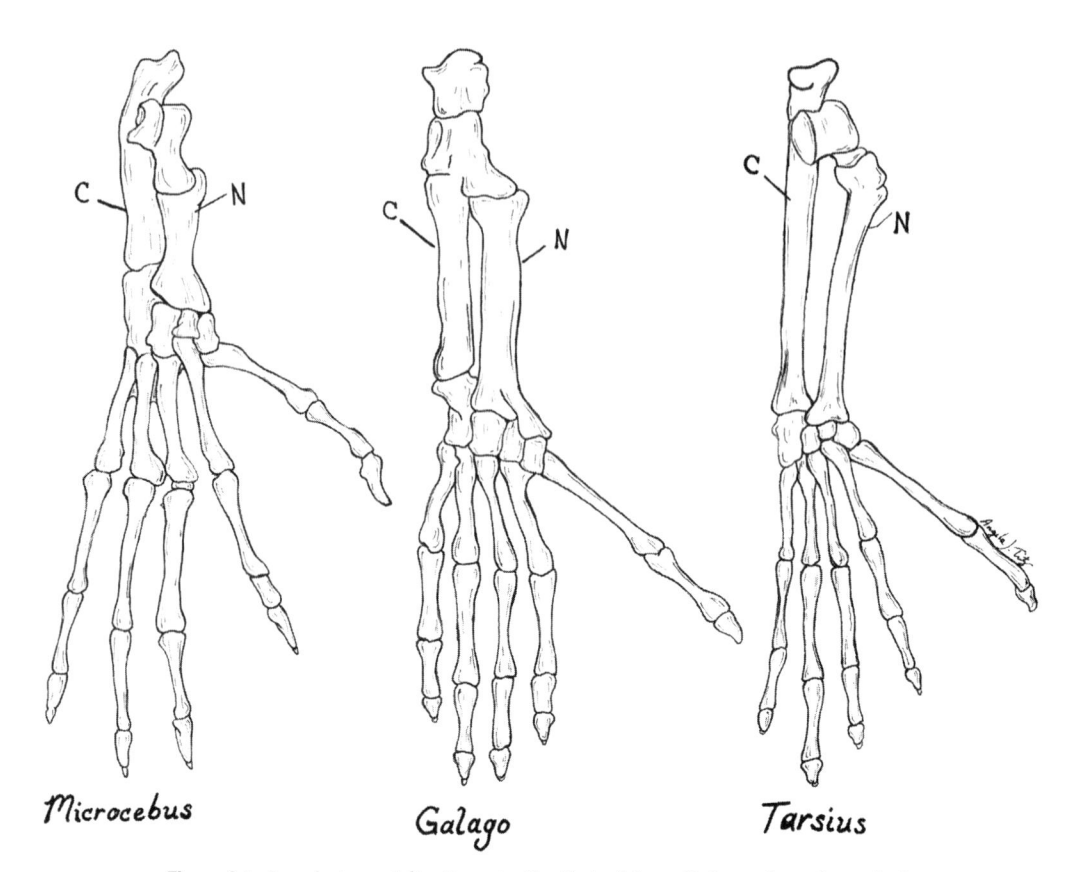

Figure 9.5. Leaping specializations in the feet of three living arboreal prosimians, *Microcebus* (the mouse lemur), *Galago* (the bush baby), and *Tarsius* (the tarsier). The illustrations are scaled to the same size. Note the elongation of the ankle bones (calcaneum = C, navicular = N) in these species. The bush baby and the tarsier are highly specialized leapers. However, the mouse lemur, because of its small size, frequently must leap and scramble as it moves quadrupedally over an irregular and discontinuous arboreal surface. Eocene omomyids often resemble the mouse lemur in the elongation of their ankle bones, and one can infer a similar kind of arboreal locomotion in them. Illustration by Angela J. Tritz.

group and a "tarsier-like" group. When cladistic taxonomy became prevalent in paleontology during the 1970s, this separation morphed into a fundamental dichotomy. This is because cladistics demands that strict monophyletic ancestor/descendant relationships trump other taxonomic considerations. The "lemur-like" group will be discussed here as adapoids; the "tarsier-like" group will be discussed here as omomyids. In general, the adapoids are more generalized than the omomyids, and have the primitive primate dental formula of 2:1:4:3. The adapoid *Adapis* was the first fossil primate ever described. It was studied by Baron Cuvier in 1822, although Cuvier did not realize that it was a primate.

William King Gregory used the functional morphology of living lemurs to discuss the adaptations of the extinct North American adapoid genus *Notharctus*

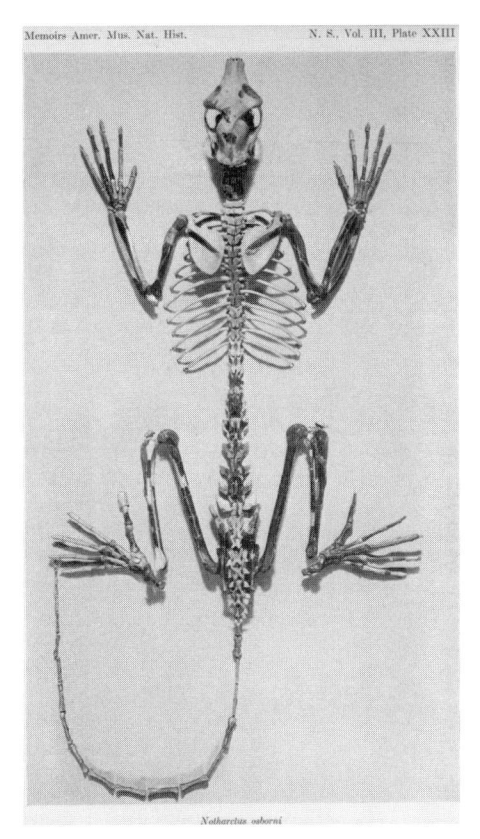

Figure 9.6. The skeleton of *Notharctus osborni*, an Eocene adapoid prosimian. From Gregory (1920:Plate XXIII).

(Gregory, 1920). This was an amazing accomplishment. The major portion of the monograph is taken up with descriptions of the functional morphology of living lemurs, and these are then used to elucidate the diet and locomotion of ancient *Notharctus* (Figure 9.6). Species of *Notharctus* emerge as clearly resembling modern lemurs. Thus, Gregory forever changed vertebrate paleontology by bringing fossil organisms to life (Rainger, 1989). In addition, Gregory organized a 1936 symposium on evolution above the species level, and invited his colleague G. G. Simpson to contribute a paper on rates and types of evolution. Although Simpson was uncertain about the value of this topic, and had an unusually difficult time producing the paper (Laporte, 2000:23–24), it later morphed into the chief accomplishment of Simpson's career: *Tempo and Mode in Evolution* (Simpson, 1944). Thus, Gregory was responsible for introducing both functional morphology and the study of evolutionary processes into vertebrate paleontology, and fossil primates figured heavily in this introduction.

Both adapoid and omomyid primate fossils are abundant and well dated in a number of sites in the American West. These include the remains of well-known adapoid genera, such as *Cantius* and *Smilodectes*, along with well-known

Figure 9.7. Reconstruction of *Cantius*, an Eocene North American and European adapoid prosimian dating to 55–50 mya. Illustration by Angela J. Tritz.

omomyid genera, such as *Tetonius* (Figures 9.7–9.9). So dense and well dated is the American primate record that researchers are able to study and quantify selection intensity on these organisms. The fossil record documents morphological change in lineages of both adapoid and omomyid primates (Gingerich, 1979, 1981; Bown & Rose, 1987). Morphological change is gradualistic, with no sign of abrupt punctuational change. For example, Bown and Rose (1987:Fig. 67) document change in the lower dentition of the *Tetonius–Pseudotetonius* lineage.

Figure 9.8. Reconstruction of *Smilodectes gracilis*, an Eocene North American adapoid prosimian. Illustration by Angela J. Tritz.

Although change between fossil material in adjacent strata is minor, major changes accumulate through time: the P2 is lost, I2-P3 are reduced in size, and I1 becomes hypertrophied in size.

Relative brain size in Eocene euprimates is larger than in plesiadapoid primates. For example, a natural endocast of *Tetonius homunculus* has a relatively expanded neocortex, when compared to primitive living insectivores (Radinsky, 1979). In general, relative brain size in the Eocene euprimates enters the lower range of values found in living prosimians. The size of the neocortex increases enough to cover the cerebellum. Olfactory bulbs have the same relative size as those of living prosimians, and they are generally distinctly visible in natural or artificial endocasts, or in computer images of the endocranium based on high-resolution CT scans. Eocene omomyids have shortened facial skeletons and smaller nasal cavities when compared to plesiadapoids or adapoids (Smith *et al.*, 2007). Based on the volume of the nasal cavity, adapoids appear to have a level of nasal complexity comparable to living lemurs. In some omomyid genera

Figure 9.9. Reconstruction of *Tetonius homunculus*, an Eocene omomyid prosimian from Wyoming. Based on the frontispiece in Bown and Rose (1987). Illustration by Angela J. Tritz.

(e.g. *Necrolemur*, *Tetonius*, *Shoshonius*), the orbits are sufficiently close together to form an interorbital septum. The distinctly large orbits of the Early Eocene *Shoshonius cooperi* impinge in the midline on the area otherwise occupied by the olfactory recess (Smith *et al.*, 2007).

The most complete non-human primate fossil ever discovered is the type specimen of *Darwinius masillae*, found in the Middle Eocene site of Messel, Germany (Franzen *et al.*, 2009). Only the left hindlimb is missing. The specimen is dated to about 47 mya. *Darwinius* appears to be a very generalized adapoid, and was placed in the Family Notharctidae by its describers. Although strikingly lemur-like, it did not have a dental comb or a grooming claw on its surviving right foot. Each digit ends in a nail. The orbits are large, relative to the size of the skull, which might indicate that *Darwinius* was nocturnal. Because there is no penis bone (baculum) present—a trait found in nearly all primate species—the type specimen was a female.[1] The fine-grained sediments surrounding the fossil preserve the outlines of soft tissue and dense fur. The contents of its gut have also been preserved. The last meal consisted of leaves and fruit. No insects were found. Because other Messel fossils have insects preserved in their gut contents, it appears that *Darwinius* did not eat insects. This is confirmed by the morphology of the teeth. Based on tooth eruption patterns, the specimen is thought to have completed about 60 percent of its growth, and had reached a body weight of 650–900 g. *Darwinius* is a juvenile whose age is estimated at 9–10 months.

The type specimen of *Darwinius* had broken its right wrist, probably in a long fall from the forest canopy. Enough time had passed since the injury for a large bone callus to form at the wrist—the wrist was therefore stiffened, and the right forearm was slightly shorter than the left (Franzen et al., 2012). This would have hindered the animal's mobility. As a result, the juvenile seems to have fallen again, this time from branches directly above the ancient Messel Lake, where its body settled into fine, muddy sediments at the lake bottom. Only eight primate specimens have ever been recovered from the Messel site. Except for *Darwinius*, all of these specimens come from two species of the adapoid genus *Europolemur*. This is a remarkably small number of primate fossils, considering that the Messel Lake was surrounded by a tropical rainforest that would have been first-rate habitat for primates. Is there a taphonomic explanation? Except for *Darwinius*, all of these specimens are fragmentary. Many show small carnivore tooth marks—a mandible fragment of *Europolemur* was found in a coprolite from an otter-like carnivore—and the broken tip of a crocodile tooth was embedded in a tail vertebra from another *Europolemur* specimen (Franzen et al., 2012). *Darwinius*, however, seems to have plummeted directly into the lake, thus accounting for its remarkable preservation. The other primates died, fell from the forest canopy, and decomposed on the floor of the rainforest until they were scavenged by carnivores.

Darwinius is identified as a cercomoniine notharctid (Family Notharctidae). The describers of the type specimen of *Darwinius* argue that it demonstrates that adapoid primates were the ancestors of anthropoids (Franzen et al., 2009). The consensus opinion holds that this is not true. Most researchers identify an unknown omomyid as the ancestor of anthropoids. One of the co-describers of *Darwinius*, P. D. Gingerich, has long argued that the most fundamental dichotomy in primate phylogeny occurs not between prosimians and anthropoids, but

---

[1] Only humans, spider monkeys (*Ateles, Lagothrix, Brachyteles*), black sakis (*Chiropotes satanas*), and tarsiers (*Tarsius*) lack a baculum (Osman Hill, 1972; Hershkovitz, 1977). The presence of a baculum is the norm for placental mammals.

between plesiadapiforms and tarsiiforms, on the one hand, and lemuriforms and anthropoids, on the other. He has therefore advocated the division of the Order Primates into two suborders: Suborder Plesitarsiiformes and Suborder Simiolemuriformes (Gingerich, 1976). He believes that higher primates evolve from adapoids. That is, anthropoid primates evolve from a generalized, lemur-like ancestor, and never passed through a tarsioid-like stage of morphology (Gingerich, 1973, 1975b). They may have evolved from cercomoniine adapoids (Gingerich, 1975a). It is therefore not an accident that the description of the cercomoniine notharctid *Darwinius* emphasizes its generalized, but characteristically lemur-like morphology. *Darwinius* is posited as a possible ancestor for higher primates because it does not possess specialized traits found in living lemurs, such as a tooth-comb or a grooming claw on the foot. The describers of *Darwinius* (Franzen *et al.*, 2009) argue that Eocene adapoids (lemuroids) and omomyids (tarsioids) are often very similar in terms of their postcanine teeth, but that omomyids often exhibit specialized anterior teeth, such as enlarged, procumbent incisors. On dental traits, adapoids are generalized enough to serve as possible anthropoid ancestors, but omomyids are too specialized. Franzen *et al.* credit William King Gregory with reifying the lemuroid versus tarsioid distinction by virtue of his superb monograph on *Notharctus* (Gregory, 1920). They believe that Gregory, by reconstructing *Notharctus* as being strikingly like a member of the living Malagasy lemur Family Lemuridae, poisoned the well for a lemuroid origin of anthropoids, and therefore launched the search for the origins of anthropoids among tarsioid primates. Note that Franzen *et al.* (2009) place *Darwinius* in the Family Notharctidae, a position obviously also held by *Notharctus*. Yet, Gregory's work on *Notharctus* demonstrates that it truly resembled living Malagasy lemurs in the Family Lemuridae. With regard to the absence of a grooming claw in *Darwinius*, it is now becoming clear that homoplasy may be occurring. Foot bones of *Notharctus tenebrosus* from the Early Eocene of Wyoming were found in association with dental material that unequivocally established their identity. Pedal digit II has a distal phalanx whose apical tuft is intermediate in form between a grooming claw and a nail (Maiolano *et al.*, 2012). Thus, grooming claws may have evolved and disappeared independently in primates. The lack of a grooming claw in *Darwinius* does not indicate its ancestral anthropoid status.

Omomyid primates have not been found at the Messel site. Given the rarity of primates at the Messel site, the absence of omomyids is not mysterious. Yet, the omomyids have the distinction of being the latest survivors of the Eocene primate radiation. They survive in many regions long after the extinction of the adapoids. In North America, omomyids are the last surviving prosimians, and the last primates in North America until the continent is colonized by ancestors of the Amerindians. *Ekgmowechashala philotau* from the Late Oligocene and Early Miocene is the last surviving prosimian in North America. It is also one of the largest of the omomyids, being approximately cat-sized. Another late-surviving omomyid, *Rooneyia viejaensis* from the Oligocene of West Texas, illustrates the morphology of an advanced omomyid (Figure 9.10). Bone in the posterior

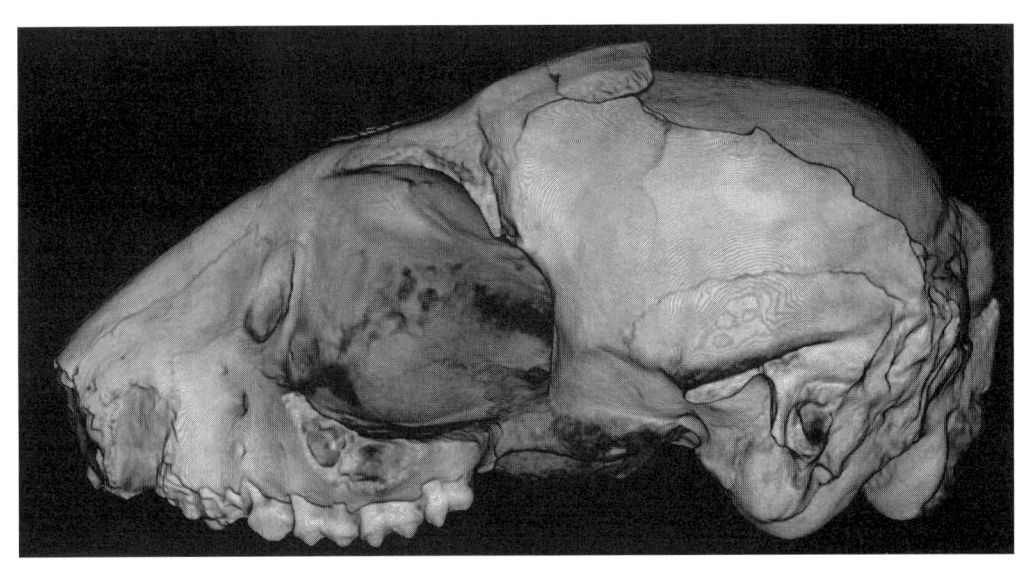

Figure 9.10. The cranium of *Rooneyia viejaensis*, an omomyid prosimian from the Oligocene of West Texas from the Digimorph Website. This is the type specimen. Funding for the CT scan provided by a National Science Foundation grant to the University of Texas at Austin. Copyright © Digital Morphology Group, in conjunction with The University of Texas High-Resolution X-ray Computed Tomography Facility. http://digimorph.org/.

neurocranium is absent, but a partial natural endocast is preserved in this region. The size of the neocortex has increased enough to cover over the cerebellum, which is not visible in the endocast. Details of the endocast are fine enough to allow the two cerebral hemispheres to be distinctly separated along the sagittal plane.

Body size increases through time in the omomyids, probably as a response to declining global temperatures, but early species are very small. *Altanius orlovi*, an anaptomorphine omomyid from the Early Eocene of Mongolia, was approximately 60 percent the size of the living mouse lemur (Dashzeveg & McKenna, 1977). Because it is clearly related to North American anaptomorphines, the climate and habitats of a Bering Strait corridor could allow for the presence of an ancient primate that weighed about 40 g. Although it is identified as a "tarsioid," it has no relation to the living genus *Tarsius*.

Study of the diet and locomotion of Eocene fossil primates reveals that substantial specialization occurs, and it rivals the degree of diet and locomotor specialization observed in living primates. For example, three prosimian species occur between 39 and 37 mya in southern France: the adapoid *Leptadapis magnus*, and the omomyids *Necrolemur antiquus* and *Pseudoloris parvulus* (Ramdarshan *et al.*, 2012). Microwear on the teeth of the Eocene fossils was investigated to reveal the texture of dietary items, using microwear on the teeth of living prosimians as a comparative set. *Leptadapis* was highly folivorous, although it ate both leaves and fruit when it lived in tropical rainforest habitats.

Necrolemur had a variable diet that included both fruit and insects, and the predominant food type depended on the habitat. *Pseudoloris* was an insectivore.

Prosimian descendants of the Eocene radiation

Living and subfossil Malagasy prosimians are the most obvious descendants of the Eocene primate radiation. The radiation of Malagasy prosimians is so complex that it will be presented in a separate chapter (Chapter 10). Even in the early twentieth century, it was clear that unknown fossil adapoids in the Infraorder Lemuriformes were the ancestors of the lemurs of Madagascar (Gregory, 1920). The Infraorder Lorisiformes is first represented by the earliest galagos (Family Galagidae), found in the Late Eocene of the Fayum (Chapter 12). Two genera are present—*Wadilemur* and *Saharagalago*. Three other galago genera (*Mioeuoticus*, *Progalago*, and *Komba*) are found in the Early and Middle Miocene of Kenya and Uganda (Harrison, 2010b). These date to 20–15 mya. The history of the Infraorder Tarsiiformes is not so straightforward. As discussed above, a substantial radiation occurred during the Eocene, with many lineages going extinct. The ancestry of the living tarsiers can be traced back to the Eocene (Chapter 12).

Much interest has centered on the first appearance of the prosimian tooth-comb or tooth-scraper (Figure 7.3). The tooth-comb is formed by lower incisors and canines that are elongated and slender, and that form a procumbent unit in the anterior mandible. Upper incisors are lost, reduced, or moved to accommodate the tooth-comb. The tooth-comb first unequivocally appears in Miocene primates, although an isolated, elongated canine that resembles the morphology of a canine that is part of a tooth-comb has been attributed to the Late Eocene Fayum prosimian *Karanisia clarki* (Godinot, 2010:Fig. 19.7F). The principal function of the tooth-comb is not to groom fur, but, rather, to collect oozing gum, sap, and resin from damaged vegetation. In living marmosets (*Callithrix* and *Cebuella*), long incisors in a V-shaped mandible approximate the size of the canines. These anterior lower teeth function like a prosimian tooth-comb, because the "short-tusked" callitrichids use their anterior lower teeth to collect oozing gum. Because gums, saps, and resins are carbohydrates rich in calories, the ability to collect these food sources is a major dietary adaptation for many living and fossil primates. Among living primates, the tooth-comb is found in all prosimians, except for the aye-aye and the tarsiers. It occurs in all living Malagasy lemurs, but was secondarily lost in the derived, terrestrial fossil archaeolemurids of Madagascar, which converge remarkably on living baboons (Chapter 10).

Lorises are found in several localities in the Middle to Late Miocene (13–7 mya) of Pakistan. One of these taxa is *Nycticeboides simpsoni*, which has a partial skeleton. Preserved elements of the orbital region demonstrate that *Nycticeboides* had the same degree of orbital convergence and frontation that living lorises have—the greatest in the primate world (MacPhee & Jacobs, 1986:Fig. 5). This orbital topography was one of the features that led to the development of the visual predation hypothesis of primate origins (Chapter 7). The most remarkable

Figure 9.11. Reconstruction of the lorisid *Nycticeboides simpsoni*, from the Miocene of Pakistan. *Nycticeboides* enters a characteristic defensive posture while fending off attack by an arboreal predator. The expanded ribs of this taxon, similar to those of the living golden potto (*Arctocebus*), implies this defensive mode. After golden potto behavior illustrated in Charles-Dominique (1977:Fig. 40). Illustration by Irene V. Hort.

feature of *Nycticeboides*, however, is its rib structure. In spite of being a small animal (500 g or less), its ribs are wide and thick (MacPhee & Jacobs, 1986: Fig. 10). The cross-sectional area of the dorsal one-third of the largest ribs rivals the cross-section of the humerus at mid-shaft. This resembles the condition in some living lorises, like *Nycticebus* (the slow lorises) and *Arctocebus* (the golden potto). This trait is not merely a signifier of phylogeny. It indicates an astonishing anti-predator functional adaptation in *Nycticeboides* (Figure 9.11). When pottos and golden pottos are threatened by an arboreal predator like a civet or another small cat, their first response is to use their pliers-like hands and feet to remain powerfully attached to their arboreal support. They subsequently draw their head down and roll their body up into a ball (Charles-Dominique, 1977). Then they strongly butt or lunge against the predator, which might dislodge the startled animal from the tree, because the predator clings to its arboreal support with claws alone, and is not as powerfully attached as its would-be primate prey. Alternatively, pottos and golden pottos bite the predator. The golden potto hides its head in an armpit but keeps its mouth open, and bites by raising its arm if it is seized by the predator from above or behind. The rolled up defensive posture appears early in life, and is clearly an innate behavior. Both of these living prosimians have skeletal adaptations that protect the neck and spinal column when they roll up into their defensive posture: the potto has a scapular shield with vertebral spines and thick skin (Charles-Dominique, 1977:Fig. 35); the golden potto has expanded ribs. One can imagine Miocene *Nycticeboides* behaving in a similar fashion if attacked. One can refine the reconstruction even further. The golden potto moves through heavy vegetation near the ground, and often descends to the ground. When threatened, the rolled up golden potto elicits attack from the rear by raising a circle of black hair around the pale tuft of its tail. Its defensive bite then comes from an unexpected direction. It is possible that Miocene *Nycticeboides* exploited a similar habitat, and attracted an adversary into biting range with a special clump of hair at its rear. Obviously this reconstruction of behavior is speculative.

10 The Malagasy primate radiation

Lost Lemuria

In 1864, the English zoologist Philip Lutley Sclater published a paper called "The Mammals of Madagascar." In this paper, he proposed that a mighty continent occupying most of the Pacific and Indian Oceans had once connected South America, Africa, and India in the southern hemisphere. This was approximately 100 years before the universal acceptance of plate tectonic theory. The present geography of the world was thought to have been in existence since the origin of continents and oceans, and only the rise and fall of mountain chains and sea levels could affect the surface of the earth. Given the view that geography had been static throughout earth history, the distribution of living and fossil organisms was sometimes a mystery to biologists. They often invoked the rapid appearance and disappearance of land bridges to account for otherwise inexplicable distributions. Given this mindset, Sclater argued for a previously unknown land connection on a grand scale–in fact, he argued for the existence of an unknown giant continent. Because he thought that this continent accounted for the distribution of lemurs and other prosimian primates, he named it "Lemuria." Similarities between South American and African monkeys were also explained by this land bridge connection. Sclater also noted the peculiar nature of the animals and plants of Madagascar–many species were endemic to Madagascar– and the sharp rift between African and Malagasy species even though Madagascar lies off the coast of Africa. He argued that the homeland of the prosimian primates would appear to be Madagascar, given their modern distribution. He believed that prosimians were different enough from anthropoid primates to be put in a different order of mammals, although he retained the traditional zoological classification that sorted prosimians and anthropoids together in the Order Primates (Sclater & Sclater, 1899). Yet, Sclater considered the prosimians to be very ancient, and argued that their original homeland, in remote geological time, had once been the now lost continent of Lemuria.

The existence of Lemuria was widely accepted by biologists. Ernst Haeckel, a biologist with profound influence in the German-speaking world, not only endorsed the existence of Lemuria in the Indian Ocean, but also announced in his 1870 book that Lemuria had been the "probable cradle of the human race." An elaborate chart in this book depicted how various branches of humanity had dispersed from their point of origin in Lemuria. Delisle (2006) describes how important this kind of conjectural evolutionary theorizing was to late nineteenth-century science, given the paucity of human fossils. At this time, conjectural theorizing was the purlieu of respectable scientists, and not idiosyncratic eccentrics. However, the eccentrics soon arrived. Lemuria became the focus of intense occult speculation and conjectural history. The influential occultist Madame H. P. Blavatsky, in particular, placed enormous emphasis on how human evolution and transformation were linked to the history of Lemuria. Blavatsky detailed in *Isis Unveiled* (1877) and *The Secret Doctrine* (1888) how Lemuria had been the source of the Third Root Race of Humanity, which at first resembled a giant ape standing 16 feet tall, before it evolved a smaller and more human-like aspect. The continent was eventually destroyed by catastrophic earthquakes and volcanoes—a southern hemisphere analog to the more fabled Atlantis—with Madagascar, Australia, and the Pacific Islands remaining as isolated remnants. Delisle (2006) recounts how important this type of revelatory, speculative history was to ideas about human evolution in the late nineteenth century, and how a multitude of human and non-human primate phylogenies were generated mainly by conjecture.

Although one might be quick to deride the occultist's notions of life on Lemuria, Lost Lemuria is still being highly touted by South Indians promoting and embracing specific political agendas.[1] For example, modern Tamil nationalists use the existence of a lost continent (called Kumarinatu) in the Indian Ocean to protest their perceived disenfranchisement and cultural domination by North Indians (Ramaswamy, 2004). Until 1981, Tamil schools taught that their ancestors had migrated north from this continent now lost beneath the sea. This hypothetical continent stretched over 11,000 km south from India and Sri Lanka. This sunken continent in the Indian Ocean had once been the fount and source of all humanity, and thus some modern South Indians believe that they at least now deserve their own country. Lost Lemuria is thus used to invoke nationalistic Tamil endeavors, and to endorse and justify the establishment of a separate Tamil state. The phenomenon of scientific endeavor being hijacked to serve political agendas occurs frequently today in archaeology, where sites are used to affirm an ancient presence on the landscape. At least these archaeological sites once existed, and were not merely the dream of first

---

[1] Many modern New Agers are convinced of the existence of a lost, hidden colony of Lemurians dwelling inside the volcanic Mt. Shasta, towering over 14,000 feet tall in northernmost California. The Lemurians will reveal their presence to the outside world when ultimate global catastrophe threatens, at which point they will save a corrupt and unenlightened humanity with their ancient, secret knowledge. Colonies of New Agers live in towns at the foot of the mountain, and beleaguered forest rangers often find and dispose of symbols, altars, and occult paraphernalia used in rituals performed on the mountainside. Thus, some of the nineteenth-century occult history of Lemuria is still being invoked in twenty-first-century America.

biogeographers and then occultists, as Lost Lemuria was. Yet, living Madagascar itself was bizarre enough to invoke a dream-like state in the first explorers of its terrain.

Malagasy natural history

The history of Madagascar is strange whether its native flora and fauna are considered, or whether its human discovery and settlement are considered. The island is very large, virtually a mini-continent—only Greenland, Borneo, or New Guinea are larger. Madagascar is 1,600 km long, and the axis of the island is aligned from north to south. It thus crosses latitudinal gradients, but the rugged topography also ensures the presence of many diverse habitats. In this extensive and varied landscape, native species could radiate to fill a broad array of ecological niches. A peculiarity of river hydrology guaranteed further speciation. Rivers whose sources lay at low altitudes were isolated geographically, thus becoming sources of extremely localized endemic species (Wilme et al., 2006). The original complement of species was an extraordinary collection of ancient organisms that once inhabited ancient Gondwana, because Madagascar was a remnant of this super continent that remained isolated from other land masses when Gondwana began to break up in the middle of the Cretaceous. Madagascar separated from Africa in the Jurassic. The western part of the island was under water by 160 mya, although Madagascar maintained a connection with India until about 88 mya (Wells, 2003). At this point, an isolated India very quickly traveled northward. Madagascar established its current position with Africa by about 120 mya, but Africa lay far to the south of its modern location. The two land masses subsequently drifted north until they achieved their present position in the Late Cretaceous. Because Madagascar lay south of the tropical regime and close to Antarctica for long ages of time, many tropical Gondwana species were lost, and modern tropical species evolve from temperate ancestors.

Many Malagasy species are endemic, and the floral composition of the island shows close ties to the ancient species of Gondwana (Figures 10.1–10.3). Students of Malagasy natural history thus encounter a vivid world unlike that found anywhere else (Tyson, 2000; Goodman & Benstead, 2003). Eighteen national parks currently exist, and there are plans to triple the amount of protected habitat, although this will not be enough to preserve the unique biodiversity of the island (Conniff, 2006).

Madagascar has two major types of tropical forest. The dry tropical forest is composed of spiny thorn species, 95 percent of which are unique to Madagascar. The Malagasy rainforests are among some of the oldest rainforests preserved alive today. This is because plate tectonic reconstructions show that Madagascar has remained within a tropical climatic zone since the Late Cretaceous. However, humans have apparently interfered with the natural rainforests on Madagascar

Figure 10.1. *Euphorbia milii*, a Crown of Thorns shrub, endemic to Madagascar. Botanical Gardens, University of Bonn.

since they colonized the island. Forests have been extensively affected by slash and burn swidden agriculture, as well as by logging, and fragmentation by roads and villages.

Colonizing Madagascar

Madagascar's terrestrial mammals are depauperate. Madagascar has one endemic order of terrestrial mammals known only from meager fossils—Order Bibymalagasia, apparently convergent with aardvarks. Hippopotamuses arrive very recently, apparently swimming the Mozambique Channel. Other endemic terrestrial mammals occur in four orders: primates, insectivores, carnivores, and rodents (Figure 10.4). Lemur-like primates and tenrecid insectivores are highly diverse. Both of these groups are monophyletic, and molecular divergence dates establish that they arrive in Madagascar during the early Tertiary (Douady *et al.*, 2002; Yoder & Yang, 2004). Endemic herpestid carnivores and nesomyine rodents arrive much later, during the Miocene.

Figure 10.2. *Pachypodium geayi*, endemic to Madagascar. This taxon is a member of the angiosperm Family Apocynaceae, and is one of the largest tree species in Madagascar. Botanical Gardens, University of Bonn.

Malagasy prosimians are all closely related. They are the descendants of probably only a single founding species. This is true for all Madagascar colonizers—if a radiation occurs, it is initiated by a single founding species, because of the improbable nature of the first colonization event. Ancient DNA extracted from

Figure 10.3. *Alluaudia ascendens*, endemic to Madagascar. This species is a member of the angiosperm Family Didiereaceae, typical of the dry, spiny forest habitats of lowland Madagascar. Botanical Gardens, University of Bonn.

the remains of subfossil lemurs confirms that the Malagasy lemurs derive from a solitary founding species, rather than evolving from ancestors from multiple colonizing events (Karanth et al., 2005).

Land mammals are notoriously poor at crossing water barriers (Lawlor, 1986). Most researchers think that these solitary founding species reach Madagascar via

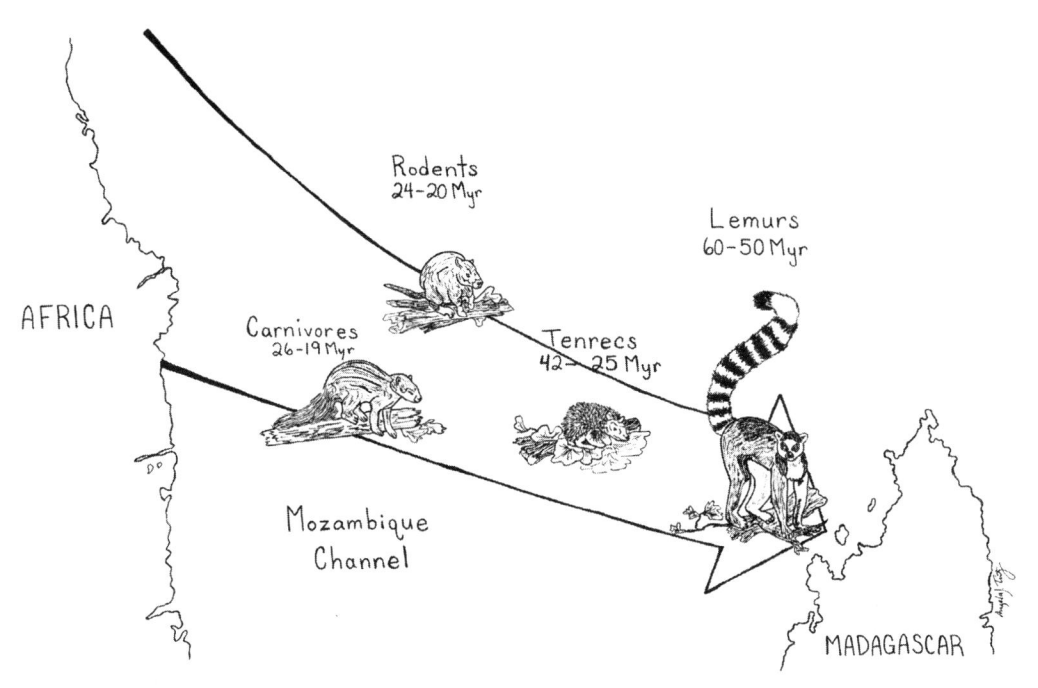

Figure 10.4. Four land mammal groups disperse to Madagascar from Africa via sweepstakes dispersal rafting over a water gap: primates, tenrecid insectivores, rodents, and viverrid and herpestid carnivores. Primates appear to have been the earliest colonizers. DNA evidence indicates that they would have closely resembled the living fat-tailed lemurs (genus *Cheirogaleus*). If they had possessed the ability to hibernate and live off stored fat, as the living fat-tailed lemurs do, surviving a long ocean passage without food or water would have been possible. Illustration by Angela J. Tritz.

sweepstakes dispersal (Simpson, 1940a; Chapter 13). That is, they cross a water gap on a raft of floating vegetation. However, some researchers are convinced that the Malagasy colonizers island-hop (Figure 10.5). They suggest that the sea-floor was intermittently exposed in the Mozambique Channel along the Davie Ridge from 45 to 26 mya (de Wit & Masters, 2004; Tattersall, 2008). However, no continuous land bridge was formed, because the African and Malagasy faunas are too dissimilar—the Mozambique Channel remained a strong barrier to land mammals from Africa throughout the Tertiary. Yet, using plate tectonics to reconstruct the paleogeography of Africa and Madagascar during the early Tertiary demonstrates that Madagascar and Africa were approximately 1,650 km south of their modern position at 60 mya. This would place Madagascar in a different ocean gyre, with a strong surface ocean current moving from northeastern Mozambique and Tanzania east towards Madagascar (Ali & Huber, 2010; Krause, 2010). Simulation of ancient ocean currents 60 mya therefore reveals a window of time during which both lemur-like primates and tenrecid insectivores could raft from Africa to Madagascar. A fuller discussion of the biological properties of land mammals that cross water gaps to colonize new lands is found in Chapter 13.

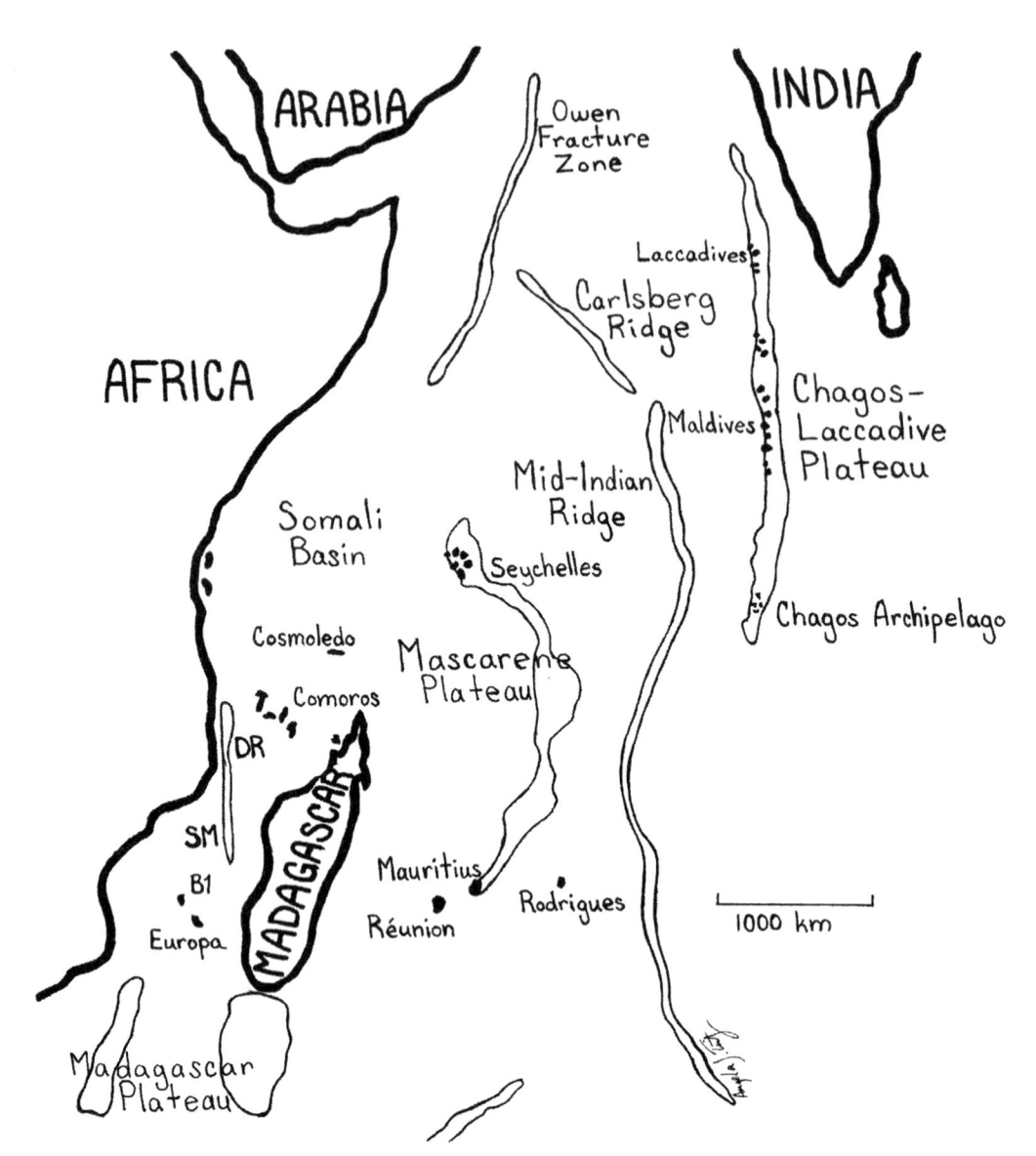

Figure 10.5. Bathymetry of the ocean basin surrounding Madagascar, showing ocean rises and plateaus. Some researchers suggest that land mammals reach Madagascar by island-hopping during periods of relatively shallow ocean depths, although this seems less likely than sweepstakes rafting. DR = Davie Ridge, SM = Sakalava Seamounts. Illustration by Angela J. Tritz.

Taxonomic inflation in the living Malagasy lemurs

Some of the species diversity of the living Malagasy lemurs is caused by taxonomic inflation (Chapter 2). In an effort to explore lemur species, especially nocturnal species, in the highly threatened landscapes of Madagascar, surveys of genetic variability are carried out in national parks or reserves. Mitochondrial DNA is especially easy to examine, because only 37 genetic loci exist on a single ring chromosome, which limits the amount of genetic sequencing that needs to be done. Using three mtDNA regions—the D-loop, 12S rRNA, and PAST fragments—two new cryptic species of Malagasy prosimian (*Avahi mooreorum* and *Lepilemur scottorum*) have been recognized at Masoala National Park in Madagascar (Lei et al. 2008).

The subfossil lemurs

The extinct lemurs of Madagascar are known from subfossil remains. The bone is not fossilized, and the preservation is excellent. Nor are the bones ancient—all of the material dates from the Holocene, with the exception of the genus *Megaladapis*, which dates back to the terminal Pleistocene (Figure 10.6). Living lemur species are found along with the extinct species. This intermixture indicates that the extant lemurs were obviously part of the ecosystem that also contained the subfossil species. This underscores the fact that the natural history of Madagascar, as it is now understood, reflects a relict world that has lost most of its past diversity, and this diversity was present until very recent time ranges. A virtue of

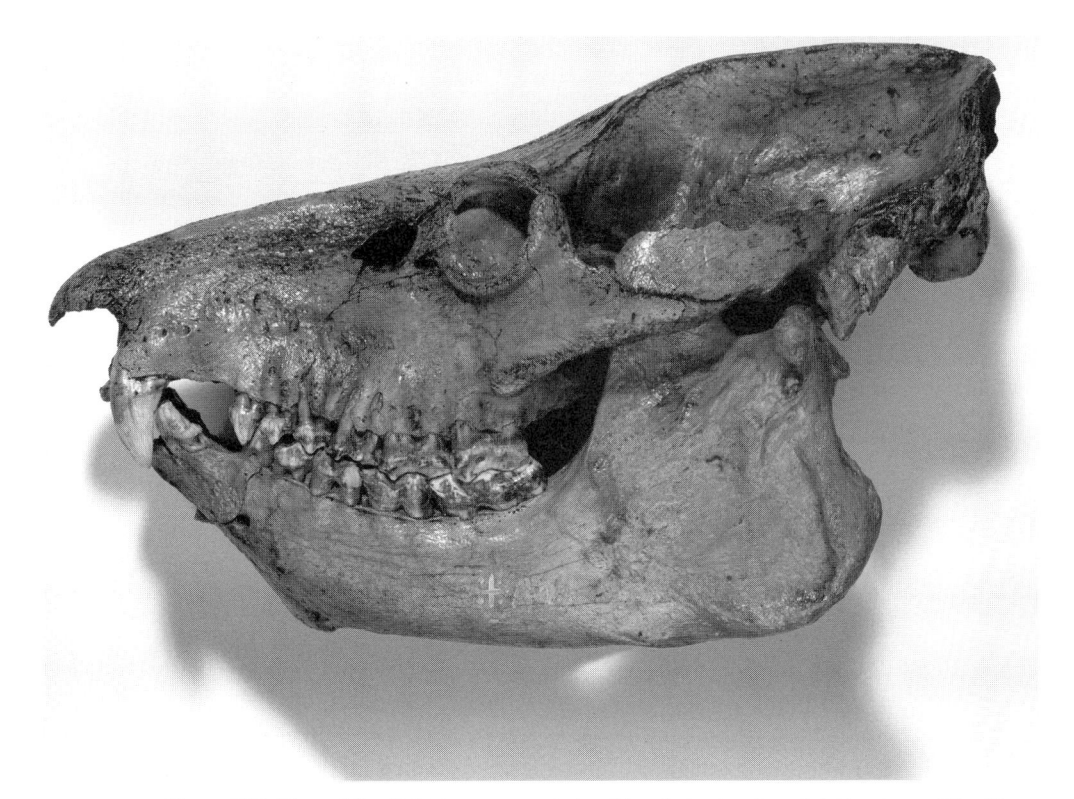

Figure 10.6. The subfossil Malagasy prosimian *Megaladapis edwardsi*. Courtesy of the Mary Evans Picture Library/Natural History Museum.

Table 10.1. The subfossil lemurs of Madagascar.[1]

Family Palaeopropithecidae (the sloth-lemurs)
Palaeopropithecus maximus
Paleopropithecus ingens
Paleopropithecus kelyus
Archaeoindris fontoynontii
Babokotia radofilai
Mesopropithecus glopiceps
Mesopropithecus pithecoides
Mesopropithecus dolichobrachion

Family Archaeolemuridae (the monkey-lemurs–convergent to baboons)
Archaeolemur majori
Archaeolemur edwardsi
Hadropithecus stenognathus

Family Megaladapidae (the koala-lemurs)
Megaladapis edwardsi
Megaladapis madagascariensis
Megaladapis grandidieri

Family Lemuridae (related to the living ruffed lemurs, *Varecia* spp., but 3–4 times larger)
Pachylemur insignis
Pachylemur julli

Family Daubentoniidae (five times larger than the living aye-aye)
Daubentonia robusta

[1] Godfrey *et al.* (2010).

this recent decimation is that Holocene habitats were equivalent to those now. Because plants and animals are the same as they are when the subfossil lemurs roamed, there is no difficulty in reconstructing habitats, as there would be if one were reconstructing habitats from 14 or 40 mya.

Table 10.1 lists the subfossil lemur taxa of Madagascar. In general, the now extinct subfossil lemurs were larger than the surviving species. Some genera (e.g. *Megaladapis* or *Archaeoindris*) were among the largest primates that ever lived. This difference in body size is accounted for by two things: the extinct species were occupying niches that no longer occur on Madagascar; and there is a general tendency for Pleistocene mammals to exceed their modern relatives in body size. Living coyotes, wolves, deer, cougars, and beavers are smaller than their Ice Age relatives were, and the living aye-aye of Madagascar is only a fifth the size of its extinct subfossil relative (*Daubentonia robusta*).

It is possible that some survivors of the now extinct subfossil radiation lingered on until recent times. The first comprehensive book on the island was titled *Histoire de la Grande Isle de Madagascar*. It was published in 1658 by the French Admiral Étienne de Flacourt, after a long sojourn on the island. De Flacourt was the first European to document aspects of Malagasy natural history, and his name is preserved in the genus name of a flowering plant (*Flacourtia*), as well as in the name of an entire family of flowering plants (Flacourtiaceae). What fame! De Flacourt had served in the French East India Company, and had been the governor of the first French colony at Fort Dauphin from 1648 to 1655. He diligently sent plant specimens back to King Louis XIV for the royal garden at Versailles. De Flacourt was finally killed by pirates in 1660, upon his return to the island to form his own trading company. His death epitomizes the violent history of Madagascar.

But important for purposes of this chapter is the fact that *Histoire de la Grande Isle de Madagascar* included an eyewitness description of what appears to be one of the now extinct indriids. It is not clear whether this description was part of a long oral tradition about a now-extinct animal, or whether the description reflects relatively recent sightings.

Trétrétrétré or *Tratratratra* is an animal as big as a two-year-old calf, with a round head and a man's face; the forefeet are like an ape's, and so are the hindfeet. It has frizzy hair, a short tail and ears like a man's. It is like the *tanacht* described by Ambroise Paré. One has been seen near the Lipomani lagoon in the neighborhood of which it lives. It is a very solitary animal, the people of the country are very frightened of it and run from it as it does from them. (Heuvelmans, 1995:610–611).

I believe that the description refers to a recently sighted animal, not an animal remembered from dim folklore. After all, the sighting is placed in a precise locale. I further believe that the description refers to a real species, rather than a mythological creature. The telling point is that the hands and feet of this animal are described as being ape-like or human-like. This identifies it as a primate, rather than some other kind of unknown mammal or mythological beast. Note also the round head, man's face, and ears like a man. This was not something with an elongated muzzle, or the upstanding ears typical of most mammals. The *trétrétrétré* also clearly possessed a large body size. How large was a 2-year-old calf in mid-seventeenth century Madagascar? The local zebu cattle breed does not attain the size of modern American beef cattle, and animals 350 years ago were significantly smaller than modern cattle (Ajmone-Marsan *et al.*, 2010:Fig. 2). I estimate that the *trétrétrétré* was approximately 300–350 kg in weight—clearly a large mammal.[2] Was de Flacourt's description of the *trétrétrétré* based on rare sightings of *Megaladapis* or *Archaeoindris*—genera that were possibly teetering on the brink of complete extinction in the mid seventeenth century? One might wonder why observers did not then mention the long arms and short legs that exist in these genera. However, if both the animals and human observers were fleeing in opposite directions, one should probably not expect that fine morphological details would be noted. A body weight of 300–350 kg is well above the weight estimated for the largest species of *Megaladapis*, so it is possible that *Archaeoindris* was observed.

---

[2] The slaughter weight of 2-year-old heifers and steers in modern America ranges between 360 and 550 kg.

And what was the *tanacht* mentioned by de Flacourt in the description of the *trétrétrétré*? Ambroise Paré quotes the French traveler and priest Father André Thévet, a Renaissance natural historian. Writing in *Cosmographie Universelle* in 1575, Thévet described the *tanacht*, an animal observed near the port of Calicut (Kozhikode), in India. The *tanacht* was an animal the size and shape of a tiger, although it lacked a tail. It had the head, face, and ears of a man, as well as a man's hands. Its nose was snubbed. The possession of human-like hands is, again, the unequivocal indication that the creature was a primate. It may have been an Asian colobine monkey of the genus *Rhinopithecus*. Monkeys of this genus are large, with red/orange or golden-toned fur. Their noses are extraordinarily snubbed, so that the nostrils point straight forward, giving the nasal region a gruesome, skeletal-like aspect. The tailless status of the *tanacht* poses a problem in this identification. Yet, members of the species *Rhinopithecus roxellanae* have relatively short tails. A *Rhinopithecus* monkey could well have been brought to Calicut, on the Malabar Coast, especially if the genus were more widespread in the sixteenth century. Yet, it is unlikely that a large Malagasy lemur, either known or not known, could have been transported alive to the southwestern coast of India in the late sixteenth century. This would be true in spite of extensive trade across the Indian Ocean.

Although the extinct sloth-lemurs (Family Palaeopropithecidae) are related to the living indris and sifakas, and the extinct koala-lemurs (Family Megaladapidae) are related to the living sportive lemurs, the subfossil lemurs are, for the most part, clearly occupying niches no longer found among the living Malagasy primates. Although the indris and sifakas are folivorous, the extinct sloth-lemurs are hyperfolivorous, as indicated by the degree of cresting on their postcanine teeth. The sloth-lemurs also evolve exaggerated postcranial specializations for prolonged suspensory posture and slow arboreal locomotion while hanging under tree limbs. In fact, they were the primate equivalent of the sloths found in New World tropical forests. And, like the true sloths, the subfossil sloth-lemurs probably reached astonishing population densities, because of the ubiquitous nature of their food. The metabolic rate of true sloths is astonishingly low, however, because of both the poor quality of their food and their slow locomotion. It is possible that the extinct sloth-lemurs had a similarly low metabolic rate. It is noteworthy that the sloth-lemurs achieved the greatest biodiversity of all of the subfossil lemurs, because they are represented by four genera. This might indicate some specialization for different forest habitats. However, because of the peculiarities of Malagasy colonization, no other mammalian arboreal folivores were present, leaving these niches open for primates.

Archaeoindris fontoynontii was the largest of the subfossil lemurs—far larger than any living Malagasy lemur—and was also a member of the sloth-lemur family. Because of its size, it was surely terrestrial. This would have made *Archaeoindris* convergent to the ground sloths of the New World. In both cases, animals that are highly specialized for arboreal suspension and slow, hanging locomotion become terrestrial. Extremities that were evolved for hanging must then function in terrestrial quadrupedalism. *Archaeoindris* must have engaged in a kind of fist-walking on the ground, similar to what living orangutans do when they are terrestrial. The diet was folivorous, but *Archaeoindris*, because of its size and the length of its forelimbs, could have easily reached up and pulled down low-hanging branches. Because of the virtual absence of large mammalian carnivores—the endemic viverrids are the largest Malagasy carnivores—body size in *Archaeoindris* and other subfossil lemurs did not increase as a defense against predation.

The koala-lemurs (Family Megaladapidae) are closely related to the living sportive lemurs (Family Lepilemuridae), and are sometimes sorted into this family. All koala-lemur species are members of the genus *Megaladapis*. Like the living sportive lemurs, which have the smallest brain to body size ratios of living primates, *Megaladapis* also had a small relative brain size. The cranial anatomy of *Megaladapis* demonstrates that the orbits are strongly angled upward (Figure 10.6). This is a trait that is seen in other primates that are giants of their lineage (Figure 5.4). It does not, as argued by some early researchers, imply that *Megaladapis* led an aquatic existence, in which its upward facing eyes would allow it to survey the world above the water while the rest of its body lay submerged. Extensive study of the postcranial anatomy of *Megaladapis* indicates that it was an arboreal genus that engaged in vertical clinging and slow vertical climbing, as living koalas do (Jungers, 1977, 1978).

The monkey-lemurs (Family Archaeolemuridae) highly resemble living macaques and baboons, although they are related to the living indriid prosimians. Resemblances to macaques and baboons indicate that similar forces of natural selection are independently producing similar morphology in different lineages. The extinction of this family thus removed a group that, given more time, might have evolved even more anthropoid-like morphology. Skeletal remains are so fresh that detailed reconstruction of the muscles of mastication can be made. The central maxillary incisors are very broad, and there is no tooth-comb. Dental and craniofacial traits relating to mastication clearly approach the anthropoid condition. Some doubt whether the brain and special senses showed any advance over the modern indriid condition (Tattersall, 1973), but others assert that the archaeolemurids are the only subfossil lemurs that approach anthropoid-like brain to body size ratios (Godfrey *et al.*, 2006). However, there is no doubt that the two genera in this family (*Archaeolemur* and *Hadropithecus*) were large terrestrial quadrupeds, feeding and foraging in open-country habitats (Tattersall, 1973). Species of *Archaeolemur* had diets similar to that of the common baboon (*Papio*), while *Hadropithecus stenognathus* had a diet similar to that of the gelada baboon (*Theropithecus*). The craniofacial skeleton of *Hadropithecus stenognathus* has been examined in detail (Figure 10.7). Besides comparing *Hadropithecus* to the gelada baboon, one researcher also compares the craniofacial anatomy of this genus to the fossil hominid *Australopithecus* [*Paranthropus*] *boisei* (Rak, 1983). Based on biomechanical analyses, *Hadropithecus* may have specialized in eating small, hard objects, such as grass seeds.

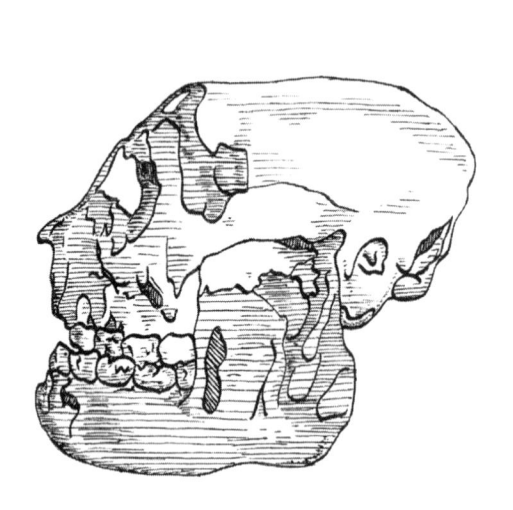

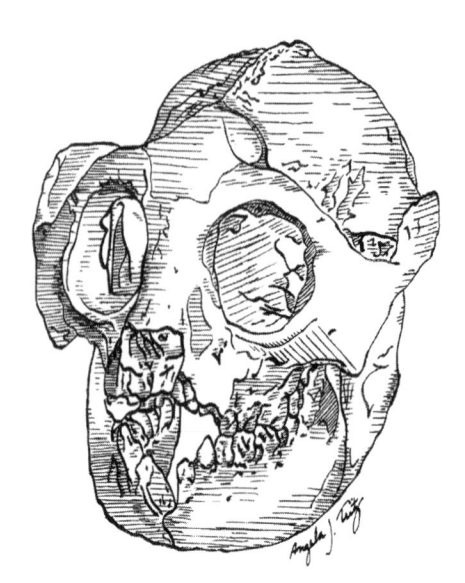

Figure 10.7. Two views of the skull of *Hadropithecus stenognathus*, a subfossil Malagasy prosimian. Illustration by Angela J. Tritz.

But *Hadropithecus* is unique among the subfossil lemurs because of another trait: although the other subfossil lemur genera (*Paleopropithecus, Megaladapis, Archaeolemur*) show an extremely accelerated rate of molar crown formation—indicating dental development unique among primates—*Hadropithecus* has the slowest rate of dental development among primates (Godfrey *et al.*, 2006). In fact, the subfossil lemurs demonstrate the alpha and omega among primates, from the fastest rate of dental development (*Palaeopropithecus*) to the slowest (*Hadropithecus*). The M1 crown formation process of *Palaeopropithecus* is essentially complete at birth, that of *Megaladapis* is slightly more prolonged, and that of *Archaeolemur* resembles that of *Papio*, although the M2 and M3 crown formation time in *Archaeolemur* is much quicker than in *Papio* (Godfrey *et al.*, 2006:Fig. 1). It is noteworthy that these rates of dental development are decoupled from body size, although they are linked together in anthropoid primates and other mammals. One expects slower growth in large-bodied species, but *Palaeopropithecus* and *Hadropithecus* are approximately the same size (35 kg). What is happening? Folivorous genera (*Palaeopropithecus* and *Megaladapis*) have accelerated rates of tooth crown formation; genera that consume harder foods (*Archaeolemur* and *Hadropithecus*) have slower rates of tooth crown formation. The most specialized hard-object feeder among the Malagasy subfossil lemurs (*Hadropithecus*) has the slowest rate of tooth crown formation among primates.

Another revelation about subfossil lemur life histories discerned from dental development is that, when judged against the rate of tooth growth, archaeolemurids have a fast rate of both craniofacial and brain growth (Godfrey *et al.*, 2006). This results in a larger brain to body size ratio than in other subfossil lemurs.

In fact, the relative size of the archaeolemurid brain approaches the anthropoid level—another indication of the monkey-like grade of this extinct group.

A peculiarity of Malagasy existence – extreme seasonality

Because rainforest constituted much of the natural habitat of Madagascar, one might expect that seasonality would not be a marked feature of Malagasy climate. Nevertheless, the land is often parched. The effects of natural aridity on the landscape are exacerbated by human farmers who cut and burn the forests. The impact of the dry season is profound, and harsh, dry forests with unique xerophilous species originally covered a significant portion of the island. There is no shade in these spiny thorn forests. Western fat-tailed lemurs (*Cheirogaleus medius*) are unique among tropical mammals, because they respond to heat and drought by undergoing true hibernation for 7 months a year within sleeping nests in tree hollows. They store fat at the base of their tail when food is abundant. Eastern fat-tailed lemurs (*Cheirogaleus* spp.) have recently been observed hibernating underground, in shallow depressions that they dig out just under the leaf litter on the forest floor (Blanco et al., 2013). This behavior is remarkable, and could only have evolved in a Malagasy ecosystem that is depauperate in terms of native carnivores. Otherwise, the hibernating lemurs would be easily located and scooped out from such shallow refuges. The falanouc (*Eupleres goudotii*), one of the endemic Malagasy viverrid carnivores, also stores fat at the base of its tail. This represents convergent evolution, because primates and carnivores are remote from each other in terms of phylogeny. The appearance of this food-storage trait through convergent evolution thus illustrates the harshness and severity of the Malagasy climate. This contradicts the superficial appearance of a tropical paradise.

In addition to energetic stress generated by striking seasonality, Malagasy lemurs have low-quality diets and suffer food competition if they live in large social groups. Conservation of energy becomes paramount. Lemurs respond by developing a low metabolic rate, seasonal hibernation or torpor, cathemeral ranging and foraging, seasonal breeding, seasonal changes in body growth and fat deposition, female priority access to food, and a unique female social dominance over males. Heat preservation behaviors include basking in the sun and huddling together in groups. Female lemurs produce dilute or low-quality milk, and lemur neonates have a body composition that is strikingly high in water, rather than protein or fat (Tilden, 2008).

Milne-Edwards' sifaka (*Propithecus edwardsi*) does not drink water, and thus obtains all of its water requirements from food. Nursing infants obtain all of their water, as well as nutrients, from their mother's milk. Years of very low rainfall affect milk production in this sifaka species. Infants born to older females have a significantly higher mortality rate than infants born to younger females (King et al., 2005). The worn teeth of older mothers affect neither their survivorship nor fertility; however, their reproductive success is affected. Under especially harsh conditions, when rainfall is ≤ 5 mm per day during lactation, the infants of older females do not live beyond 1 year of age (King *et al.*, 2005). This appears to be caused by the mother's dental attrition, and not hormonal changes, or loss of the unique female dominance over males with senescence (Taylor, 2008). Note that these rainforest species are affected by relatively subtle climatic fluctuations of temperature and rainfall.

The human colonization of Madagascar

Despite the fact that Madagascar lies only 400 km off the southeast coast of Africa, the original human inhabitants did not come from Africa, and they did not arrive early. Not until before 4,000 yrs B.P. is there archaeological evidence that humans were present on the northeast coast, and were foraging in the forests and coasts. These people came long before farmers and herders arrived, and exercised no discernible impact on the environment (Dewar *et al.*, 2013). This is important, because it indicates that human presence per se does not alter the environment, but farming and herding does. The Late Holocene extinctions that took place in Madagascar occur after the introduction of slash and burn agriculture and the introduction of cattle and other domesticated animals. Humans entered Madagascar in numbers only during Iron Age times at about 1,800 yrs B.P. They crossed the entire length of the Indian Ocean to find Madagascar, setting out on a deliberate colonizing endeavor from somewhere near the island of Borneo. The islands of Indonesia were already suffering from over-population. The distinctive Malagasy culture was subsequently encountered by Arab slave-traders, who first introduced sub-Saharan African people to the island as commodities in the slave trade. The native Malagasy today are thus an admixture of humans from three major geographic centers—Indonesia, the Afro-Arabian Peninsula, and sub-Saharan Africa. European explorers finally arrived, and systematic scientific exploration of the island began during the seventeenth century.

Madagascar endured a colorful and violent history, in part because of diverse colonial contacts, in part because of the vagaries of local life as the island remained a bastion of the international slave trade. Piracy was rampant. The mad queen Ranavalona reigned for 33 years. Given the status of a goddess, she managed to kill at least one-third (possibly one-half) of the population of Madagascar through murder, starvation, and forced labor before her undeservedly natural and peaceful death in 1861 (Laidler, 2005). She abhorred Europeans, and persecuted Christians, both missionaries and Malagasy converts, slaughtering and burning them in a fashion not seen since the days of the Emperor Nero. Slavery existed throughout Madagascar, and had done so for ages. The Merina people of Indonesian descent settled in the central plateau region, and enslaved people from the coast, as well as buying slaves from Africa. Yet, Britain abolished the slave trade in 1807, and finally abolished slavery itself in 1833. Constant battles between local despots, slave-traders, and the British fleet (attempting to bring a halt to the slave trade) ensued in the early nineteenth century before the slave trade was finally quelled in Madagascar, and thus halted worldwide (Jolly, 2004).

Climate change is often invoked as a cause of habitat loss. However, cultural or anthropogenic causes may also operate. What is the relative contribution of climate versus culture in causing habitat loss in tropical areas? This has been studied in Africa. Pollen and charcoal records from the lowland Congo rainforest have been examined in the Goualougo area of the Nouabalé-Ndoki National Park, Republic of the Congo. Both climate fluctuations and human burning of the rainforest influence plant species composition in the Congo, but the importance of fire increases at about 1,000 yrs B.P. (Brncic et al., 2007). Under natural conditions, rainforest tree species composition fluctuates to reflect oscillations between relatively arid and relatively humid precipitation cycles. With the introduction of a humanly generated burning regime, tree species that can survive artificial burning events begin to predominate in an area.

In Madagascar, human burning of the forest has denuded the thin tropical soils to such an extent that vistas of barren red soils are a common feature of the landscape. This has generated one of Madagascar's soubriquets: The Red Island. Satellite photographs reveal that these barren, red vistas are the principal feature of the island, when viewed from space. One could almost be viewing a mini-Mars. Another soubriquet is The Isle of Fire, for good reason. A Malagasy proverb says "Where there is fire, there are people."[3] About 2,000–7,000 km^2 of rainforest and secondary bush are deliberately set afire every year by Madagascar's farmers; pastoralists set afire a quarter to a half of the island's grasslands (Kull, 2004). The government enacts laws, and international conservation groups promote anti-fire propaganda, but to no avail. Violence and disputes have occurred for more than 100 years about the appropriate use of natural resources (Kull, 2004). Slash and burn swidden agriculture and pastoralists in Madagascar have thus removed vast amounts of rainforest cover. Currently, rainforest remains the most debatable habitat—a true No Man's Land. Yet, government edicts for over 70 years have not succeeded in controlling slash and burn agriculture; this is the necessary mode of subsistence for many Malagasy farmers (Kull, 2004:145). As a consequence, the eastern Malagasy rainforest (last haven of many lemur species) is fast disappearing. And forest species composition is altered by persistent burning, even if cover returns.

Although it is often thought that the existence of local taboos or *fadys* can protect endemic Malagasy species, this is not always true, particularly in modern times. Animal *fadys* may invoke instant spiritual retaliation if taboos are violated, but younger Malagasy are not concerned about this. Elders are more culturally conservative, and anxious about animal *fadys*. Large lemurs like the indriids (the indri and the sifakas) are usually considered to enjoy the most protection from local *fadys*. Because of their large body size and round, flat faces, the indriids most resemble humans, and are therefore supposed to be especially protected from human interference. They appear to resemble ancestral forms, and ancestors are

---

[3] *Tany misy afo misy olona* (Kull, 2004:145).

greatly revered by all Malagasy, irrespective of present-day religion. Nevertheless, individual Malagasy may kill and eat sifakas entering a village as they flee forest burning, despite the existence of *fadys* that prohibit this behavior. Villagers may kill and eat sifakas with both local and international conservation officers in the vicinity (Sodikoff, 2007).

The elaborate Malagasy ancestor cults have long been famed among social anthropologists. The elaborate burials of the dead, secondary burials, and exhumation of entombed corpses in elaborate multi-day festivals and ceremonies (*famadihana*) are practiced even today, and even by Christian converts (Mack, 1986; Jolly, 2004). These rituals appear to establish important continuing relationships with the ancestors, and they are thought to generate the awe necessary for maintaining the *fadys* that protect local animals. While it is true that tombs and cenotaphs are everywhere, lemurs were a common food in a number of Malagasy communities, even 50 years ago, when local traditions were stronger (Rudd, 1960). Snares, traps, blow-pipes, and arrows were used to kill these animals. Some species were killed for magical reasons. Pregnant women were enjoined to eat fat-tailed lemurs (*Cheirogaleus* spp.) because their children would be born with the round, beautiful eyes seen in these lemurs (Rudd, 1960). Sifakas were generally revered, because they were thought to resemble humans. Nevertheless, sifakas were sometimes killed and tortured for magical reasons— the animals being first skinned alive, sprinkled with cayenne pepper, and then roasted (Rudd, 1960).

It is therefore clear that *fadys* and a reverence for ancient traditions alone will not ensure lemur species survivorship. Yet, current conservation efforts in Madagascar center on lemurs, because they are considered "charismatic" species. The Malagasy conservation group "Association Aiza Biby" (Where are the Animals?) uses a lemur mascot to generate local interest and support (Sodikoff, 2007).

Lemur extinctions and loss of disparity

Dating of fossil material indicates that the subfossil lemurs, as well as other Malagasy megafauna (e.g. elephant-birds, pygmy hippopotamuses, giant tortoises), were present and thriving on the island before human colonization (Burney *et al.*, 2004). A great decrease in the megafauna occurs at 1,720 yrs B.P. This is followed by a large increase in charcoal retrieved from sediment cores, indicating human presence and alteration of the landscape by fire. Humans appear first in the southwestern part of Madagascar, but then spread throughout the island and build up dense populations over the following millennium.

Because the subfossil lemurs in Madagascar occupied a unique series of niches not present today, their extinction completely removed major trophic levels within the community. Lemurs are the principal mammalian herbivores among endemic Malagasy species. The extinct elephant-birds were also herbivorous, but, as flightless terrestrial species, they occupied a series of niches distinct from those of lemurs. The subfossil lemur extinctions affected every species over 9 kg in weight. Current Malagasy ecosystems are therefore ruined in terms of community structure and energy flow.

Biodiversity is often thought of in terms of species diversity. However, there is another way of examining biodiversity: focusing on phylogenetic diversity or disparity, which emphasizes the length of evolutionary time that separates various lineages. Disparity, in fact, may be a better measurement of the evolutionary potential of a habitat. The Zoological Society of London maintains a list of 100 mammal species on a special website[4] devoted not only to species in danger of extinction, but species that also are survivors of distinctive lineages found nowhere else. They thus represent not only endangered species, but the rarest of the rare. These species are called EDGE species—Evolutionarily Distinct and Globally Endangered. Extinction is not the only risk—disparity is threatened. Nine primate species appear on this list; six of these nine species are Malagasy prosimians. They are as follows, in their order on the EDGE list: the aye-aye (*Daubentonia madagascariensis*), the golden bamboo lemur (*Hapalemur aureus*), the greater bamboo lemur (*Hapalemur simus*), the golden-crowned sifaka (*Propithecus tattersalli*), the indri (*Indri indri*), and the hairy-eared dwarf lemur (*Allocebus trichotis*). In fact, predictions about imminent primate extinctions assess that the Malagasy prosimians will disappear first. This will affect the entire ecosystem, because prosimians constitute 44 percent of the terrestrial mammal fauna, and there can be no prospect of a recovery after an ecosystem collapse of such magnitude (Jernvall & Wright, 1998). Furthermore, the prosimian niches are unique, and so there can be no simple replacement by invading species that occupy vacant niches. Consider the aye-aye, which occupies a woodpecker-like niche, or lemurs that act as plant pollinators. The morphology and behavior of lemurs in these niches are not evolved easily. The extinction of the diverse sloth-lemurs has probably opened niches for large arboreal folivores in Madagascar, but no local mammals appear to be exploiting these niches. True sloths attain an incredible biomass in Neotropical forests, and one can presume the same of the extinct sloth-lemurs. Their removal must have affected Malagasy forest structure in a significant fashion. What would happen to Neotropical forests today if they were cleared of true sloths?

The problems of species loss on Madagascar are compounded by the fact that it is an island, and islands are much more threatened by extinctions than continental areas are. It is clear that Madagascar contains degraded habitats, but how does it compare to other tropical environments? Tropical forest cover in eastern Madagascar is highly threatened. It represents a global deforestation hotspot, as documented by the TREES II project, which utilizes remote sensing techniques to quantify worldwide loss of tropical forest cover (Mayaux *et al.*, 2005:Fig. 2). The effect of dense human settlement and deforestation occurring against the background effect of natural erosion in rugged terrain is easily discernible from satellite data.

---

[4] The URL of this website is the following: http://www.edgeofexistence.org/home.asp.

11 The Oligocene bottleneck

The Oligocene epoch begins at 33.9 mya. Plate tectonic reconstructions demonstrate that, for the first time, the geography of the earth largely resembled that seen today. In particular, the Atlantic and Mediterranean basins had finally achieved a modern aspect (Figures 11.1 and 11.2). Discussions of anthropoid primate origins and dispersal (Chapter 12) center on these two ocean basins and sweepstakes or rafting events that cross these basins.

During two episodes that straddled the Eocene/Oligocene transition, the world abruptly shifted from a warm to a cold phase. These two episodes lasted a total of 300,000 years. The cold phase that began after this shift has persisted ever since. Antarctica, which had previously been ice-free, except for some areas in the high interior mountains, acquired a permanent continental ice sheet. The Oligocene itself was an epoch of global aridity. So harsh were conditions that global sea level was drastically lowered, and signs of blowing aeolian dust are found in ocean sediment cores worldwide.

Mammals have never recovered from the harshness of the Oligocene. For the first time since the beginning of the Cenozoic, the extinction rate of mammalian families exceeded their origination rate. At the beginning of the Oligocene, 95 families of mammals exist. These fall to 77 families by the end of the epoch. Mammal diversity was thus reduced by 19 percent in only 11 million years—a severe reduction from which the class has never recovered. In Europe, the mammal turnover during the Eocene/Oligocene transition was so marked that it was noted over a century ago by the paleontologist Hans Stehlin (1909), who labeled it "*La Grande Coupure*" ("The Big Cut"). There are also significant mammal extinctions during the Eocene/Oligocene boundary in Asia. Here the extinction event is labeled "The Mongolian Remodeling" (Meng & McKenna, 1998; Hartenberger, 1998).

Major transient glaciation events occurred immediately after the Eocene/Oligocene boundary and across the Oligocene/Miocene boundary. Large ice sheets appear on Antarctica. The Oligocene is considered to be an early "icehouse" epoch—the first appearance of an "icehouse" earth that later culminated in the dramatic global temperature fluctuations of the late Cenozoic. Global temperatures appear to have plummeted by 6°C exactly at the Eocene/Oligocene boundary (Zachos *et al.*, 2008:Fig. 2). Thus, during the Oligocene, the earth is transformed from a warm planet, with virtually no high-latitude ice, into a colder planet with enduring Antarctic ice. Carbon and oxygen isotopes from sediment cores in the tropical Pacific reveal that the Antarctic ice sheet grew in two rapid, stepwise

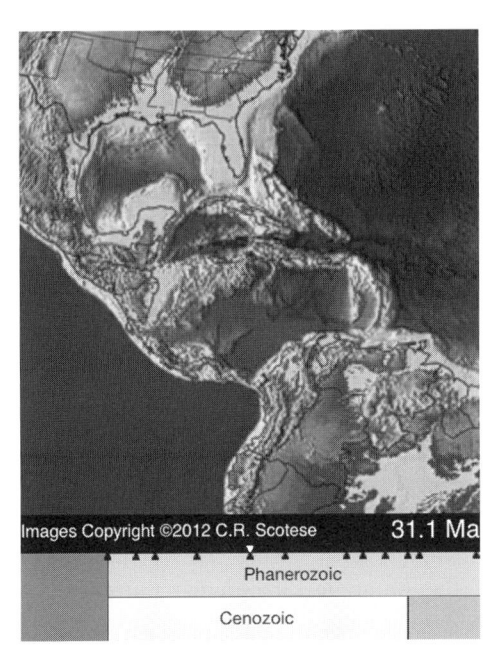

Figure 11.1. Paleogeography of the Oligocene earth, dated to 31.1 mya. The map is centered on the Atlantic Basin to highlight the geography of the Caribbean and the degree of separation of Africa and South America. Courtesy of Dr. Christopher Scotese.

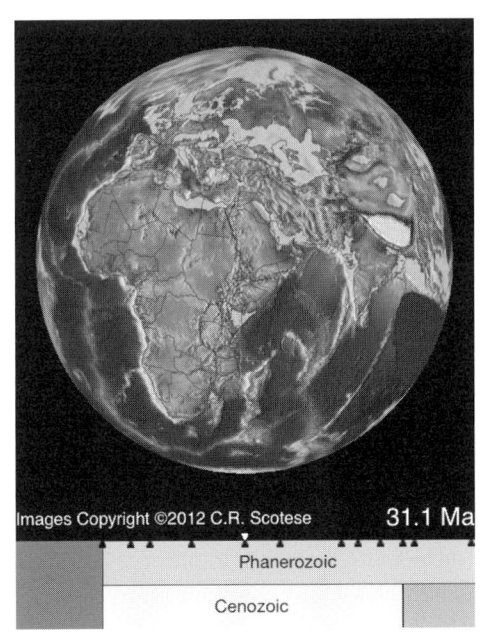

Figure 11.2. Paleogeography of the Oligocene earth, dated to 31.1 mya. The map is centered on the Mediterranean (ancient Tethys Ocean) Basin to highlight the geography of North Africa, where the earliest undoubted anthropoids are found. Courtesy of Dr. Christopher Scotese.

increments at the Eocene/Oligocene boundary (Coxall *et al.*, 2005). Additional cooling factors must have occurred to account for isotope changes across this boundary. Either worldwide cooling was occurring, or a combination of global cooling and northern-hemisphere glaciation was taking place (Coxall *et al.*, 2005).

Plate tectonic movements were the ultimate cause of these global climate changes. Final rifting between South America and Antarctica created the modern Drake Passage and Tasmanian Passage. The Tasmanian–Antarctic Passage opened first, and the Drake Passage opened between 33.8 and 33.5 mya (Zachos *et al.*, 2001b). Sea water flowed through this gap, and the Antarctic Circumpolar cold ocean current was thereby established. The Antarctic Circumpolar Current reduced ocean circulation between high and low latitudes, and allowed the ocean around Antarctica to remain cold. A wide disparity in global temperature between high and low latitudes was created, and, for the first time during the Cenozoic, seasonal climates, rather than equable ones, became possible at high latitudes. Antarctic ice began to grow, although it became reduced in the late Oligocene. The Antarctic ice sheet again achieved significant growth during the Middle Miocene at about 15 mya. The launching of the Antarctic Circumpolar Current also established modern ocean layering. Modern ocean structure contains four layers: surface, intermediate, deep, and bottom waters. Hence, the Early Oligocene witnesses the advent of both modern climatic seasonality and modern ocean structure.

Variations in the earth's orbit magnified climatic changes caused by tectonics. A continuous deep-sea sediment core from the equatorial Pacific that spans the Oligocene reveals clear oscillations that relate to orbital variations of the earth (Pälike *et al.*, 2006). Modeling of the Oligocene carbon cycle demonstrates how it promoted the effect of long-term orbital cycles and reduced the effect of short-term cycles. Decreasing atmospheric CO_2 levels occur across the Eocene/Oligocene boundary (Pearson *et al.*, 2009), and orbital forcing initiates glaciation once these levels reach a critical threshold level (Pälike *et al.*, 2006). At the end of the Oligocene, orbital variations create cooler summer temperatures and an increasing volume of Antarctic ice during the Miocene transition (Zachos *et al.*, 2001a).

Because much of this global climate record has been extracted from sea-floor sediments, it is important to note that supporting evidence comes from terrestrial records. This terrestrial evidence indicates cooling and aridity that is even more severe than reconstructed from marine records. Playa lakes on the northeastern part of the Tibetan Plateau disappear completely, as continental Asia experiences aridification (Dupont-Nivet *et al.*, 2007). This disappearance is not caused by tectonic uplift, but by global aridification. As the sediments record abrupt aridification, fossil pollen concomitantly documents cooling and drying, and mammal species experience a drastic turnover. Even more impressive is the climate record preserved in the tooth enamel and bone of fossil vertebrates from three localities in Central North America—the American West of Wyoming, South Dakota, and Nebraska (Zanazzi *et al.*, 2007). The authors first use oxygen isotopes in tooth enamel to estimate oxygen isotope composition of a fossil animal's drinking water. Once this is established as a baseline, the oxygen isotope composition of fossil bone, which picks up oxygen from surrounding sediments after death, can be used to reconstruct ambient temperatures. This yields a detailed record of terrestrial temperature during the Eocene/Oligocene transition. The mean annual temperature appears to have fallen by a startling 8°C over 400,000 years (Zanazzi et al., 2007). This is far larger than estimates based on oceanic records. Because only a small possible increase in seasonality is indicated, and no discernible increase in aridity, the drastic drop in mean annual temperature per se may account for the recorded extinctions or turnovers in ectothermic animals (gastropods, amphibians, and reptiles). Because they are endothermic, and their body temperatures are also stabilized by size, nearly all of the large mammals in these localities in the American West survive this Early Oligocene faunal crash.

Finally, there seems to have been an extraterrestrial factor in the global cooling that occurred during the terminal Eocene. An impact cluster consisting of at least three asteroids has been dated to around 35 mya. Giant impact craters formed by asteroids that were larger than 5 km in diameter have been found at this date in Popigai, Siberia, Chesapeake Bay, off the coast of Virginia, USA, and in the western Timor Sea (Glikson et al., 2010). Chesapeake Bay itself was formed from this impact. The impact also created the North American Strewn Tektite Field, where small, glassy rocks from molten terrestrial sediments (tektites) were ejected and catapulted hundreds of kilometers away from the impact site. There are immediate climatic effects from the impact of a large asteroid or a cluster of large asteroids. Shattered rock, molten ejecta particles, and sediments expelled into the atmosphere create a high-altitude pall of dust. This mantle of dust and ejecta particles was a direct and abrupt trigger for global cooling.

The Oligocene was also a period of low global sea levels. Low Oligocene sea levels are associated with a major decline in the rate of sea-floor spreading, which ultimately reflects a major change in plate tectonic dynamics (Zachos et al., 2001b:Fig. 2). As sea-floor spreading declines, less light, hot crust is created, and deeper ocean basins are formed. This lowers global sea level. Petroleum geologists in the late 1970s initiated research into long-term fluctuations in sea level (Vail & Hardenbol, 1979). This research was refined a decade later (Haq et al., 1987), and has now been exhaustively investigated for the last 100 my of earth history (Miller et al., 2005). At the beginning of the Late Eocene, global sea level was 25 m above its present level. By the beginning of the Oligocene, it had fallen to 25 m below the present level—a total excursion of 50 m (Miller et al., 2005: Fig. 3). Global sea level stayed low through the Oligocene until the earliest Miocene, when it reached its present level.

The Oligocene bottleneck was certainly real for primates. Figure 8.1 documents a startling plummet in the rate of primate morphological evolution and a subsequent long plateau in morphological change. This plateau is lower than at any time after the origin of primates. Since the time of Stehlin (1909), the *Grande Coupure* has been thought to signal the extinction of primates in Europe. Prior to this event, adapid and omomyid primates had been abundant in Europe. Yet, for a long time, only a single adapid molar from the earliest Oligocene of the Isle of

Wight, England, documented a primate presence in Europe. However, a mouse-sized omomyid (*Pseudoloris godinoti*), found in two sites in the Ebro Basin of Spain, apparently survived for at least 2 million years during the Early Oligocene (Köhler & Moyá-Solá, 1999). Thus, the *Grande Coupure* marks a major primate extinction event in Europe. In North America, prosimians of the late Middle Eocene (e.g. *Diablomomys dalquesti*, dated to 44–43 mya, from southwestern Texas) are found in relict rainforests of the Gulf Coast and California. These serve as refuge areas for primates going extinct in the north. Prosimian primates are virtually extinct in North America by the end of the Oligocene.

There is a marked trend for an increase in body size among the surviving omomyid prosimians. The largest of the North American omomyids (*Ourayia uintensis*) achieves the size of a large living owl monkey. Finally, in the Early Miocene, only *Ekgmowechashala* is left, of all the great North American Eocene primate radiation. It is also the size of an owl monkey. A larger body size would allow the last prosimian survivors to stabilize their internal body temperature in spite of colder and more seasonal ambient conditions (Chapter 5). However, a larger body size would mandate additional food to support this increase in size. The quest for more food would itself generate new selection pressures. In any case, after the Middle Miocene, North America—which had the earliest primate fossils and which remained a major center of primate evolution through the Paleocene and Eocene—is devoid of primates until the ancestors of the Amerindians cross the Bering Land Bridge between 18,000 and 16,000 yrs ago.

12 Rise of the anthropoids

Because anthropoids are higher primates, much research has focused on their origins. However, there are a number of factors that make anthropoid origins problematic. To begin with, there are a number of purported "first" specimens. Fossils from Burma (Myanmar), North Africa, and Germany have been advanced as the first anthropoids. This diverse material indicates that there are differences in opinion about the proto-anthropoids. Were they tarsiiform or adapiform primates? If they were tarsiiforms, were they close relatives of the living tarsiers or an extinct member of the tarsiiform omomyids? A concomitant consideration is this: what is the position of the tarsier lineage? Is it a sister group to the anthropoids, which would be reflected by sorting both tarsiers and anthropoids into the Suborder Haplorhini? Alternatively, did anthropoids arise from adapiform primates?

Resolving the phyletic position of the living tarsiers (Infraorder Tarsiiformes) is crucial before these questions can be answered. During the Eocene, tarsiiform primates had a widespread distribution throughout the northern hemisphere. Living members of this Infraorder (tarsiers) are now found only in some Indonesian islands (Borneo, Sumatra, and Sulawesi) and in the Philippines. Genomic sequences from living primates, spread across about 90 percent of the living genera, demonstrate that tarsiers are an ancient relict lineage distantly related to anthropoids (Perelman *et al.*, 2011). The living tarsiers have the longest lineage documented by fossils of all primates. The earliest member of the Family Tarsiidae is *Xanthorhysis*, found in the Middle Eocene of Shanxi Province, China, at about 45 mya (Beard, 1998a). But a number of phenotypic traits found in living tarsiers occur in fossil primates with no close relationships to the family. For example, 50 mya crania of *Shoshonius cooperi* from Wyoming superficially resemble the cranium of *Tarsius* (Beard *et al.*, 1991). Although the orbits of *Shoshonius* are enlarged, and fall well above a regression line comparing orbital diameter to cranial length (Beard *et al.*, 1991:Fig. 4), they are certainly not as large as those of the living tarsiers. And, unlike the living tarsiers, which are highly specialized for vertical clinging and leaping, the postcranial remains of *Shoshonius cooperi* demonstrate that it was a much more generalized arboreal quadruped with some leaping behavior (Dagosto *et al.*, 1999). It resembled living cheirogaleid prosimians or fat-tailed lemurs identified as leaper/quadrupeds–i.e. arboreal quadrupedalism, climbing, and leaping are all equally developed. *Shoshonius cooperi* possessed a short, robust femur, a relatively long humerus (and thus a high humerofemoral index), and a spherical femoral head. Thus, like fat-tailed lemurs, *Shoshonius* was an arboreal walker, climber, and scrambler. Any leaping that it did was of a short, squirrel-like type. It was not the advanced vertical clinger and leaper that tarsiers are.

Many researchers now favor the idea that the living tarsiers are the sister group of higher primates. This idea is not new. It was advocated first by several scholars in the early twentieth century, especially by the English scholar Frederick Wood Jones (1916), who argued that the first members of the human lineage diverged from all other primates early in the Cenozoic, and remained highly arboreal until their descent to the ground. These earliest humans closely resembled living tarsiers, although they did not leap like tarsiers (Figure 12.1 and 12.2). Wood Jones was impressed by the upright posture of tarsiers, their large, forward facing eyes, and what he perceived to be a large, globular braincase. He considered other

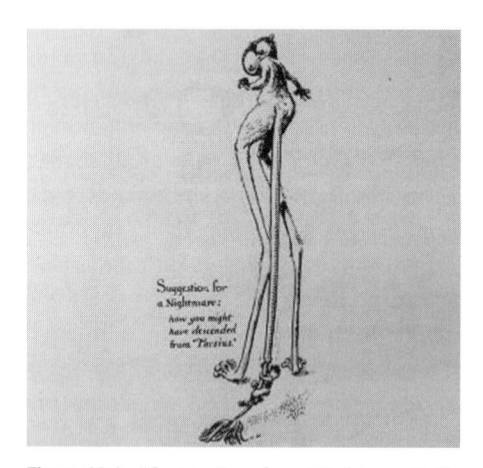

Figure 12.1. "Suggestion for a Nightmare": Descent from *Tarsius*. From Reed (1930:265).

Figure 12.2. *Tarsius* as ancestor. From Reed (1930:266).

living primates, especially the great apes, to be much too specialized to have given rise to humans. Wood Jones's strange ideas were gently mocked by the American anatomist and paleontologist William King Gregory (1949), who identified Wood Jones as a member of the "homunculist" school of human evolution, characterized by the search for a tiny, elfin very early ancestor for genus *Homo*.

The latest taxonomic revision of the primates links the six living species of tarsier with higher primates in the suborder Haplorhini (Groves, 2001). Like higher primates, the tarsier lacks a rhinarium, and has a free upper lip, not bound to the underlying bone—hence the designation "Haplorhini." Mothers give birth to single youngsters, although females have two or three pairs of nipples. This indicates that the ancestral state was to bear a litter. The extraordinary size of the tarsier eye—each eye is equal to the size of the entire brain and cannot be moved within the orbit—necessitates that bony support evolves to shore up the eyeball. Because the eyeballs are immobile, tarsiers must move the entire head to peer in different directions. As a consequence, tarsiers can rotate their heads nearly 180° away from the forward position. This owl-like ability to peer backwards is based on a unique and relatively lax position of the articular processes of the cervical vertebrae. Tarsier orbits are so enlarged that the tips of the upper molar roots (M_1–M_3) are visible within the base of the orbital floor. The extraordinary hypertrophy of the eyes occurs because tarsiers are highly carnivorous nocturnal hunters that rely on vision to locate their prey (Gursky-Doyen, 2010). The nocturnal specialization of tarsiers appears to be secondary, because they lack a tapetum lucidum, indicating that they evolved from a diurnal ancestor.

Tarsiers have a postorbital septum, but this appears to be convergent on the anthropoid condition. That is, the postorbital septum of tarsiers is a homoplasy. Adult tarsiers have an opening between the maxilla and the zygomatic bone in the posterior floor of the orbit. Newborn tarsiers lack both a postorbital bar and septum (Smith *et al.*, 2013). Instead, a membrane connects the frontal bone and the orbital process of the zygomatic bone (Figure 12.3).

One of the principal traits that links tarsiers to higher primates is the possession of a hemochorial placenta. This is the type of placenta possessed by humans, and it was long thought to represent a derived condition. It has now been shown that, not only is the hemochorial placental the ancestral state for primates, but it also is possibly the ancestral state for placental mammals (Wildman *et al.*, 2006). In short, the reproductive specialization that links the tarsiers to higher primates has disappeared. It is the primitive or ancestral primate condition.

An unknown omomyid primate, labeled "tarsiiform" because it is not a "lemuriform" primate (Chapter 9), is the probable ancestor of the anthropoids. An alternative viewpoint of anthropoid origins argues that they arise from Eocene adapiform primates. This was the opinion of Gingerich (1979), who arrived at it through developing the school of taxonomy called stratophenetics. When describing the Eocene taxon *Darwinius masillae*, Gingerich and his colleages later identified this taxon as a proto-anthropoid, but received considerable criticism for this identification (Chapter 9).

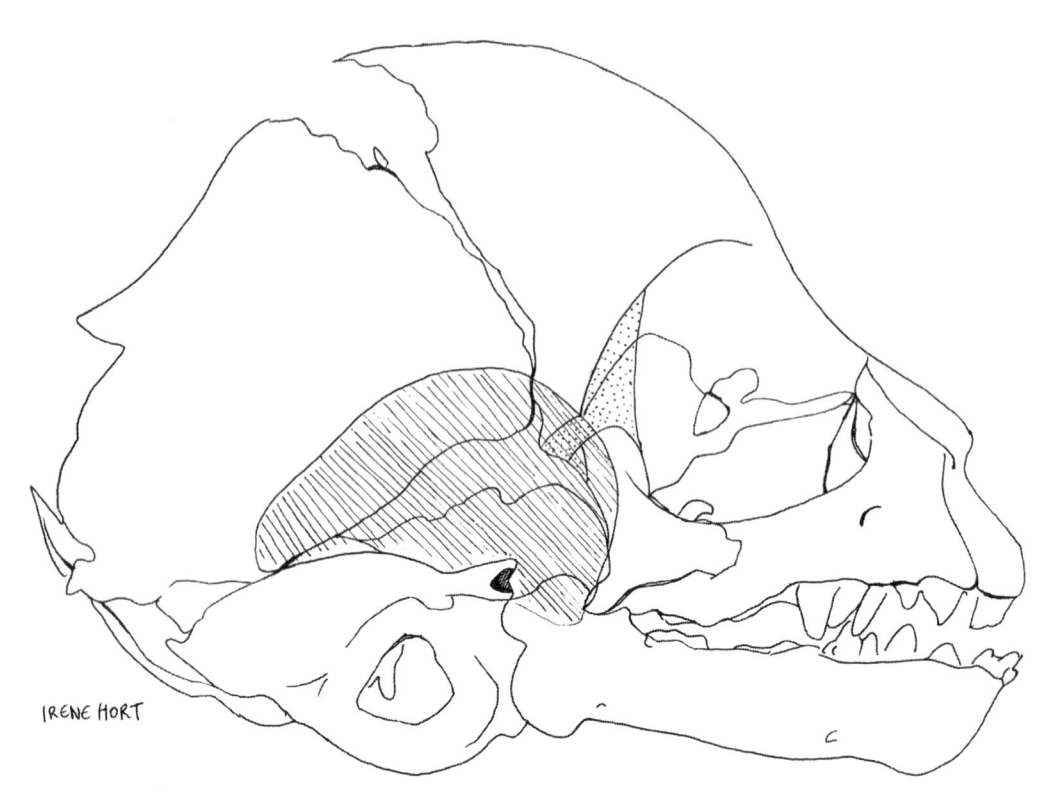

Figure 12.3. The skull of a newborn tarsier (*Tarsius syricta*), illustrating the incomplete postorbital septum. The postorbital membrane (dotted surface) connects the frontal bone and the orbital process of the zygomatic bone. The temporalis muscle (straight lines) covers the posterior part of the postorbital membrane. After Smith *et al.* (2013:Fig. 4). Illustration by Irene V. Hort.

Convergent origin of the anthropoids?

Figure 12.4 presents a model for anthropoid origins, based on known forces of natural selection. If anthropoid primates develop independently in the Old and New Worlds, they are a polyphyletic group. Such groups are rejected by the cladistic system of classification. However, anthropoid primates represent a distinctive grade or adaptive zone in primate evolution. I utilize the traditional system of primate taxonomy, which recognizes this grade by a Suborder Anthropoidea. If one were to insist on a monophyletic classification, Old and New World higher primates would be sorted into different suborders. The initial trigger for the cascade of morphological changes in Figure 12.4 is an increase in body size. This occurs as a response to deterioration in global temperature at the end of the Eocene. An increase in body size increases body volume relative to skin surface. This helps to stabilize the internal core temperature of the body. Fossil omomyids demonstrate a trend for increasing body size through time (Szalay, 1976:Fig. 177). The latest omomyid genera such as *Ourayia*,

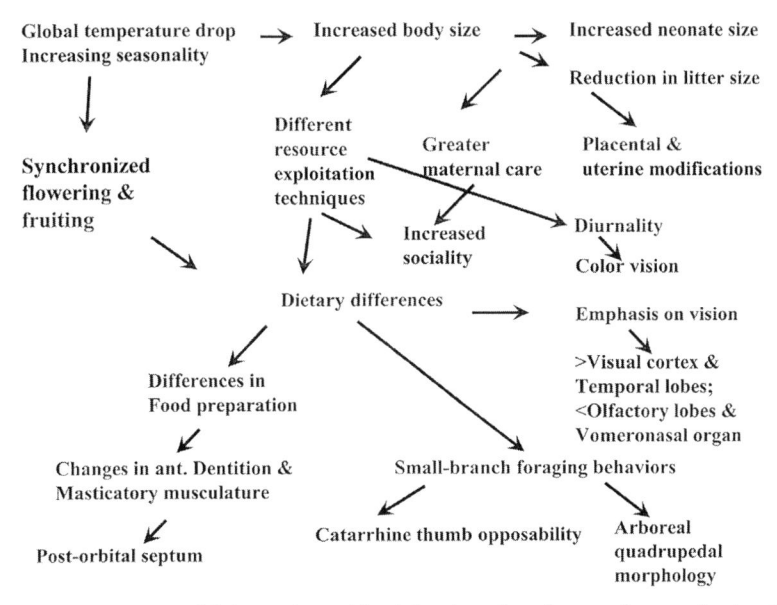

Figure 12.4. A model for anthropoid origins, based on known forces of natural selection.

Ekgmowechashala, *Rooneyia*, and *Macrotarsius* are within the size range of the living platyrrhine genera *Aotus* and *Pithecia*.

Note that a dietary shift triggers a number of morphological changes in Figure 12.4. I suggest that, in addition to eating the range of foods that their prosimian ancestors did, the earliest anthropoids broadened their diet by feeding on fruit with a hard pericarp. An examination of skull shape in New World leaf-nosed bats, which exhibit the greatest dietary diversity of any group of mammals, demonstrates that species that eat harder foods have skulls that are adapted for a more efficient production of bite force (Santana et al., 2012). This greater efficiency results from a larger biomechanical advantage when the mouth is gaping open at an intermediate angle. The morphology that is created by selection pressure for eating hard objects is a short skull with a larger temporalis muscle. This shape is seen in anthropoid primates, which have shorter skulls, and which possess larger temporalis muscles (especially anterior temporalis muscles) than prosimian primates do (Cachel, 1979a).

Unlike the mandible of prosimian primates, which separates into two hemi-mandibles after death, the mandible of anthropoid primates is a complete, single bone. The symphysis between the two halves of the mandible is fused into bone. This condition is not unique to higher primates, but also occurs in other mammal groups. When diet and mandibular fusion are investigated across mammals, a bony symphysis is associated with strong biting and chewing along the postcanine tooth row (Scott et al., 2012). Electromyographic study of living primates demonstrates that a bony symphysis serves to integrate the activity of the chewing muscles on both sides of the head, even if a bolus of food is being chewed primarily on one side of the mouth.

Figure 12.5. Lake Turkana, looking west from the Koobi Fora Spit, east side of the Lake Turkana Basin. This lake is land-locked, has no outlet, and is fed almost entirely by the Omo River. The North Island lies directly ahead. The delta of the Omo River lies to the northwest.

The recognition of plate tectonics in the geological sciences initiated speculation that a narrower South Atlantic might have facilitated rafting of early anthropoids from Africa to South America or vice versa. Derived omomyid prosimians might have rafted from North America to South America via volcanic Caribbean island arcs. The Oligocene palaeogeography of both the South Atlantic and the Caribbean basins is well known (Figures 11.1 and 11.2). Sweepstakes dispersal is indicated in either case—animals traverse a substantial water gap by traveling on floating mats of vegetation. Are there actual examples of such vegetation mats? Figures 12.5 and 12.6 illustrate a modern large, land-locked lake (Lake Turkana in northern Kenya) and a sizeable floating mat of vegetation that wafted as flotsam across this lake from the Omo Delta. The straight-line distance from the Omo Delta to the spit of land on which this mat rests is 55 km.[1] This vegetation mat is the kind that would make sweepstakes dispersal across a substantial water gap possible (Simpson, 1940a). A more detailed discussion of sweepstakes dispersal appears in Chapter 13.

---

[1] My colleague, Professor Craig Feibel of Rutgers University, has seen even larger mats of vegetation washed up on the Koobi Fora Spit after strong flooding of the Omo River. In addition, he notes that rafts of floating vegetation regularly blow from one side of Lake Naivasha (central Kenya) to the other side. This distance is only about 12 km, but the drifting mechanism could operate across much larger distances.

Figure 12.6. A vegetation mat, washed up on the Koobi Fora Spit, east side of the Lake Turkana Basin. The mat is approximately 6.75 cubic meters in volume (3 m high, 1.5 m wide, and 1.5 m long). It has floated across Lake Turkana from the Omo Delta, about 55 km to the northwest.

Detailed morphological investigation of higher primates in the Old and New Worlds demonstrates that key anthropoid features show either homoplasy or a different morphological state. An example of homoplasy is the sutural pattern of the bones in the postorbital region. These are different in platyrrhines and catarrhines (Figure 12.7).

Other examples exist of morphological differences between platyrrhines and catarrhines. The grasping pollex is pseudo-opposable in platyrrhines, but truly opposable in catarrhines. True opposability is a function of a saddle-shaped joint surface between the trapezium and metacarpal I (Napier, 1993). A saddle-shaped joint ensures that the thumb also rotates medially when it is flexed. The fleshy pad of the thumb can therefore be pressed against the fleshy pad of every other digit. The joint surface between the trapezium and metacarpal I is cylindrical in platyrrhines. In addition, platyrrhine hands often have a different functional axis. Rather than the functional axis falling between digits I and II, as in catarrhines, platyrrhines often have a zygodactylous (schizodactylous) hand structure, in which the functional axis falls between digits II and III (Hershkovitz, 1977:Fig. I.32B, Fig. I.33). Color vision evolves differently in platyrrhines and catarrhines (Jacobs, 2008; Jacobs & Nathans, 2009). Living platyrrhines have dichromatic vision, including all males and about one-third of the females. Their retinas have no pigment responsive to red-wavelength light. Two-thirds of the females might inherit allele variants for color vision on each of their two X chromosomes. But the normal condition in living catarrhines is trichromatic vision, dependent upon

Saguinus   *Macaca*

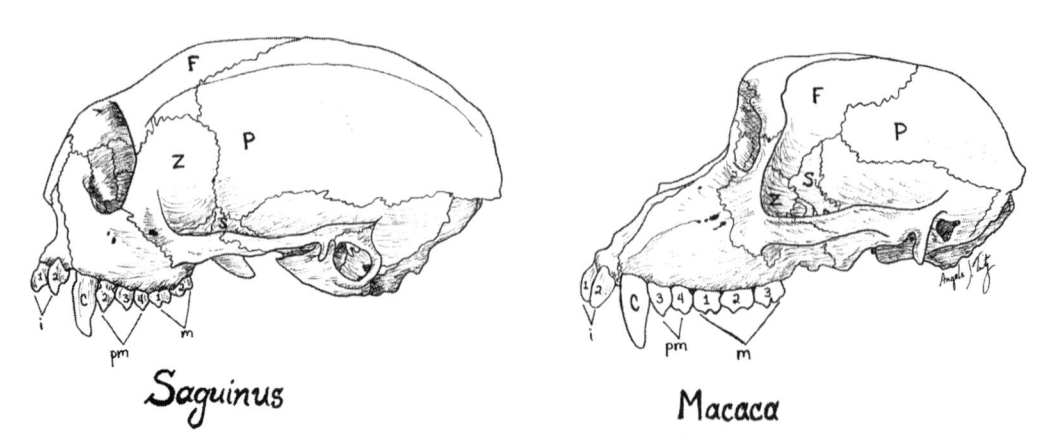

Figure 12.7. The sutural pattern of the bones in the postorbital region of anthropoid primates. Note the contrast between the platyrrhine condition, seen in the tamarin *Saguinus*, where the zygoma (Z) contacts the parietal (P), and the catarrhine condition, seen in the cercopithecoid monkey *Macaca*, where the frontal (F) and the squamosal (S) contact each other and separate the zygoma and parietal. After Hershkovitz (1977:Fig. IV.84). Illustration by Angela J. Tritz.

three types of retinal pigment that are responsive to light. This condition arose through an error in recombination during meiosis in a catarrhine female. Two pigment alleles then were located on the same X chromosome. This mutation then spread because of its selective advantage.

The Fayum primates

The Fayum Depression of Egypt is located in the Western Desert of Egypt, about 80 km southwest of Cairo. The Fayum has been yielding remarkable vertebrate fossils since the late nineteenth century. During the Late Eocene, this sedimentary basin represented an area where the Tethys Sea (predecessor of the Mediterranean) was regressing, and it therefore is a transitional area between alluvial environments and nearshore marine environments. A very large river, sometimes called the Ur-Nile, although it was not hydrologically affiliated with the modern Nile, was building up huge deltaic deposits as it debouched into the Tethys. The Late Eocene Birket Qarun Formation is the earliest formation in the Fayum sequence. It is dated to 37 mya, and it preserves both freshwater and marine vertebrate fossils (Murray *et al.*, 2010). Eocene and Oligocene fossils from the Fayum have contributed to our understanding of the evolution of many mammal orders, including primates. Our knowledge of early whales, sirenians, hyraxes, elephants, and elephant-shrews has been shaped by the discovery of Fayum fossils. Twelve orders of mammals are found here (Simons, 2008). Besides primates, these orders are chiropterans (bats), cetaceans (whales), artiodactyls (even-toed ungulates), ptolemaiids (an endemic insectivore-like order), sirenians (manatees and dugongs), embrithopods (a rhinoceros-like order found elsewhere only in Ethiopia, Turkey, and Romania), proboscidians (elephants), hyracoids (hyraxes), rodentia (rodents), creodonts (an order resembling carnivores), and peradectines (opossum-like marsupials). The hyraxes—as is true of other African Oligocene sites—were the dominant herbivores, being represented by four extinct families. They ranged from the size of a rabbit to the size of a rhinoceros, and were a major component in community structure. Fayum fossil sites have declined in productivity. Simons (2008) notes that the pace of fossil collecting at the Fayum has slowed, even for large vertebrates, such as whales and elephants. He attributes this to the fact that fossil discoveries are outrunning the rate of desert weathering that is taking place, which would reveal more fossils.

The Fayum is a major source of primate fossils, including the undoubted anthropoid specimens. These were long considered to be the earliest evidence of anthropoid primates, although slightly earlier material is found elsewhere in North Africa, and a number of sites in Asia have been championed as the font of the earliest anthropoids. Personnel from the American Museum of Natural History began collecting primate fossils here in 1907 (Osborn, 1908), and collection continues to the present. There are over 20 genera and about 30 species of fossil primates that are currently recognized from the Fayum deposits (Godinot, 2010; Seiffert *et al.*, 2010b). These fossils date from the late Eocene and Oligocene. With the exception of *Kamoyapithecus* (Chapter 14), *Nsungwepithecus*, *Rukwepithecus* and *Saadanius* (see section below), Oligocene catarrhine primates are found only in the Fayum (Table 12.1). There are 12 anthropoid genera, eight prosimian genera (including *Afrotarsius*, not related to tarsiers, and sometimes classified as *incertae sedis*), and one genus whose status is clearly uncertain. This is *Nosmips aenigmaticus*, dating to 37 mya, and described from 12 isolated teeth (Seiffert *et al.*, 2010a). These upper and lower premolars and molars are unique in their configuration, because the P_3 is very elongated, the P_4 is molariform, and the upper molars are very simple. Enamel pitting indicates that seeds or fruit pits could have been processed. Despite considerable analysis, the position of *Nosmips* within the anthropoids, modern prosimians, or adapiform prosimians remains undecided, although the majority of its traits identify it as a basal anthropoid.

The oldest anthropoid remains in Africa come from the late Middle Eocene site of Bir El Ater in Algeria, dating to about 40 mya. These fossils are represented by three species of the parapithecid *Biretia*. This genus also occurs in the BQ-2 site in the Late Eocene of the Fayum, and it will be discussed later. The site of Dur At-Talah, in South Central Libya, dates from the late Middle Eocene (39–38 mya), and has a greater diversity of primates: three anthropoid species distributed across three families, and one prosimian species (Jaeger *et al.*, 2010a). The authors describe this assemblage as evidence of the earliest radiation of anthropoids. Caution is warranted. All fossils are represented by isolated teeth, and they occur in a heavily bioturbated horizon, so that time-averaging is taking place. And Dur At-Talah may not represent the font of early anthropoids: local sedimentary conditions may simply be capturing more diversity than other sites

Table 12.1. Primate species from the Fayum.[1]

Taxon	Locality	Age
Incertae sedis		
Nosmips aenigmaticus	BQ-2	37 mya
Prosimians		
Family indeterminate		
Afrotarsius chatrathi	Quarry M	~30 mya
Family Adapidae		
Afromonium dieides	L-41	~34.2 mya
Afradapis longicristatus	BQ-2	37 mya
Family Djebelemuridae		
Anchomomys milleri	L-41	~34.2 mya
Family Plesiopithecidae		
Plesiopithecus teras	L-41	~34.2 mya
Family Galagidae		
Wadilemur elegans	L-41	~34.2 mya
Saharagalago misrensis	BQ-2	37 mya
Lemuriformes		
Family indeterminate		
Karanisia clarki	BQ-2	37 mya
Anthropoids		
Aegyptopithecus zeuxis	Quarries M & I	~30 mya
Apidium phiomense	Quarries M & I	~30 mya
Parapithecus grangeri	Quarries M & I	~30 mya
Propliopithecus chirobates	Quarries M & I	~30 mya
Qatrania fleaglei	Quarries M & I	~30 mya
Apidium bowni	Quarries G & V	~31.2 mya
Apidium moustafai	Quarries G & V	~31.2 mya
Propliopithecus ankeli	Quarries G & V	~31.2 mya
Proplipithecus haeckeli	Quarries G & V	~31.2 mya
Oligopithecus savagei	Quarry E	~33.1 mya
Qatrania wingi	Quarry E	~33.1 mya
Abuqatrania basiodontos	L-41	~34.2 mya
Arsinoea kallimos	L-41	~34.2 mya
Catopithecus browni	L-41	~34.2 mya
Proteopithecus sylviae	L-41	~34.2 mya
Serapia eocaena	L-41	~34.2 mya
Biretia fayumensis	BQ-2	37.5 mya
Biretia megalopsis	BQ-2	37.5 mya
(the genus *Biretia* is also found at Bir El Ater, Algeria [40 mya])		

[1] Godinot (2010), Seiffert et al. (2010a, 2010b).

in North Africa. But a major question triggered by this assemblage is whether anthropoids originate in Africa after a long period of *in situ* evolution, or whether they originate in Asia, and are seeded into Africa by colonizers. This is the opinion of Jaeger et al. (2010a). Major support for an Asian origin is given by the fossil rodents. The rodents from Dur At-Talah, like those of the Fayum, are members of the Family Phiomyidae. They are ancestors of the modern African cane rats and dassie rats, and are hystricognath rodents, one of the three rodent suborders. The rodents of Dur At-Talah have affinities with South Asian fossils, indicating that hystricognath rodents originate in South Asia, and disperse from there to North Africa (Jaeger et al., 2010b). More recently, *Afrasia djijidae*, a fossil primate from the late Middle Eocene of Burma (Myanmar), has been described as being startlingly similar to *Afrotarsius* from the Fayum, although the Burma fossil – known only from teeth – is dentally more primitive (Chaimanee et al., 2012). It is important to note that Burma was part of the Indian tectonic plate until the Late Oligocene (Morley, 2000), so its distance from North Africa was not as great as it is today. In a commentary on the paper announcing the discovery of *Afrasia*, Kay (2012) points out that catarrhines group together in a natural African clade, making their origin in Asia unlikely. The sheer taxonomic diversity of primates in the Fayum and other North African sites in Algeria and Libya appears within a limited time frame from the late Middle Eocene through the Oligocene. This indicates that a major evolutionary radiation was occurring in North Africa. Furthermore, the fossil record of mammals in South Asia and Africa is scant during the early Cenozoic, and many small mammal species crossed the Tethys Sea between South Asia and Africa during this time span. What was the vector of travel? Out of Africa, or into Africa?

Dur At-Talah has yielded early proboscidean fossils and five rodent taxa. The four primate taxa found at Dur At-Talah are one lorisiform prosimian species (*Karanisia arenula*) and three anthropoid genera distributed across three families: *Afrotarsius* (Family Afrotarsiidae), *Biretia* (Family Parapithecidae), and *Talahpithecus* (Family Oligopithecidae). Note that the authors here solve the problem of the uncertain status of *Afrotarsius* by placing it in a separate family. Considerable discussion involves the small body size of the four primate taxa. Adult body mass, estimated from the area of M_1, ranges from 120 to 470 g (Jaeger et al., 2010a). The Dur At-Talah primates are smaller than the species occurring later in the Fayum. However, it is important to note that size rapidly increases, and this increase continues throughout the Oligocene. This trend need not be mysterious, if one considers factors that affect changes in body size (Chapter 5). A trend of increasing body size through time has long been noted within the Fayum material (Simons, 1992). The smallest taxa (*Biretia*, *Arsinoea*) occur early in the sequence, and appear to be at or below Kay's Threshold of 500 g that separates insectivorous primates from herbivorous ones. Larger taxa occur later in the sequence, and range from 1,200 g to 6,000 g, the body weight estimated for *Aegyptopithecus zeuxis*, the giant among the Fayum taxa. It is noteworthy that the orbits of *Aegyptopithecus* angle upwards. This morphology appears to be associated with primates that are the giants of their lineage (Chapter 5), because it independently develops in several different primate genera (e.g. *Megaladapis* and extinct, large-bodied species of *Theropithecus*).

Position of the Parapithecoidea

The most abundant primates at the Fayum are parapithecids. They had a brief efflorescence, and quickly became extinct. The earliest Fayum species are members of the genus *Biretia*, which also occurs at the earlier site of Bir El Ater, Algeria. The genus is extremely important, not only because of its antiquity, but also because one of its species (*Biretia megalopsis*) may have been nocturnal. If this is so, it documents the transition from a nocturnal activity pattern to a diurnal activity pattern typical of anthropoids.

Remains of parapithecoids, especially the genus *Apidium phiomense*, are abundant in terrestrial Fayum sites (Figure 12.8). *Apidium* can lay claim to being the most abundant land mammal at the Fayum. Postcranial remains, especially a distal tibia and fibula that are partly fused, indicate leaping specializations. However, semicircular canal evidence from *Apidium* contradicts this (Ryan *et al.*, 2012). *Apidium phiomense* appears to be slow moving and lacking in agility. In this, it resembles another parapithecoid (*Parapithecus grangeri*), as well as other Fayum taxa. The semicircular canals of five Fayum primate taxa have been examined in order to discern their mode of locomotion—specifically, their rapidity of movement and degree of agility. These taxa are *Catopithecus browni*, *Proteopithecus sylviae*, *Apidium phiomense*, *Parapithecus grangeri*, and *Aegyptopithecus zeuxis* (Ryan *et al.*, 2012: Table 2). With the exception of *Proteopithecus*, whose acrobatic abilities seem to resemble those of the living Goeldi's marmoset of South America (*Callimico*

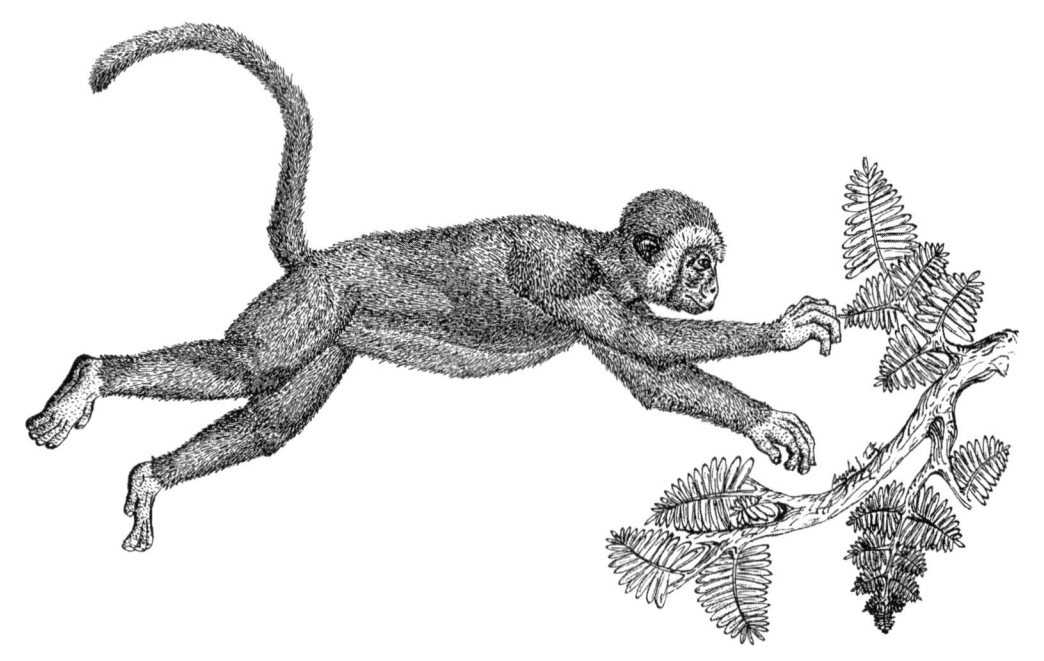

Figure 12.8. Reconstruction of *Apidium phiomense*, a parapithecoid anthropoid primate from the Fayum. Illustration by Angela J. Tritz.

goeldi), which engages in vertical clinging and leaping, all of the Fayum taxa appear to have been slow to only moderately agile in their locomotion. The semicircular canal evidence coincides with that of available postcranial evidence, except for *Apidium phiomense*, whose fused tibia and fibula entail leaping specializations. The totality of the locomotor evidence implies that the first undisputed anthropoids of the Fayum were engaging in slow, calculated arboreal locomotion. This would therefore appear to be both the basal anthropoid and the basal catarrhine condition.

High-resolution CT scans of the cranium of *Parapithecus grangeri* reveal endocranial volume, as well as volume of the olfactory bulbs and the optic foramina (Bush et al., 2004). The size of the olfactory bulbs relative to body size is larger than in living anthropoids, and the relative endocranial volume is far smaller. *Parapithecus grangeri* has a brain to body size ratio that is in the lower range for living prosimians, but this is also true for *Aegyptopithecus*. The olfactory bulb volume of *Parapithecus* is in the middle of the living primate range, and is perhaps midway between living anthropoids and prosimians. *Parapithecus* thus had a very small relative brain size, and a possible greater dependence upon olfaction than living anthropoids.

In summary, members of the Family Parapithecidae had an adaptive zone that is no longer found among living catarrhines. Although they were first described as being ancestral to Old World monkeys, it soon became clear that they had undergone a considerable separate adaptive radiation, and that they had no connection to any living catarrhine group (Simons & Kay, 1983). However, the parapithecoids offer insight into the adaptive zone of the earliest anthropoids. They existed at a period when prosimian primates were also undergoing a considerable radiation in the Fayum. These prosimians include the first galagos, adapids, and two other extinct prosimian families (Family Djebelemuridae and Family Plesiopithecidae). This last family contains one species, *Plesiopithecus teras*, whose skull superficially resembles that of the living aye-aye (*Daubentonia*). *Plesiopithecus* has enlarged upper canines and lower incisors, a rounded neurocranium, klinorhynchy, and large, anteriorly positioned orbits. Like the aye-aye, *Plesiopithecus* may have been able to gouge into bark to extract grubs and larvae (Godinot, 2010). In summary, there was clearly no competition for resources at the Fayum between prosimian and anthropoid. These adaptive zones were already well differentiated. Yet, like the Fayum prosimians, the parapithecoids disappear after the Oligocene climatic bottleneck. Was this a function of smaller body size yielding unstable internal temperatures during climatic decay? Was it a function of habitat restriction during the Oligocene? Recall that the once abundant prosimians of North America become isolated in relict rainforest distributions on the Pacific and Gulf Coasts, and then teeter into extinction during the Oligocene. Another important point to consider is that the parapithecoids flourish at a time before the separation of basal catarrhines into the crown catarrhine groups of cercopithecoid and hominoid. The parapithecoids thus offer an insight into the lifeways of the basal catarrhines.

Other Fayum taxa

Large body size signals better fossil preservation. *Aegyptopithecus zeuxis* is the largest of the Fayum taxa, and appears to be the giant of its lineage, given the slightly upward angulation of its orbits. This orbital position is determined by allometry, and can also be seen in other phyletic giants, such as *Theropithecus brumpti* (Figure 5.4A) and *Megaladapis edwardsi* (Figure 10.6). There is a long tail, and the limb morphology indicates generalized arboreal quadrupedalism. Based on evidence of the semicircular canals, *Aegyptopithecus* moved fairly slowly and deliberately (Ryan *et al.*, 2012). This rather slow and cautious movement appears to be the norm for basal anthropoids. The relatively abundant and intact crania of *Aegyptopithecus zeuxis* allow for a fair sample of endocranial volumes. The species resembled modern prosimians in having a low brain to body size ratio—one adult female specimen had a remarkably small endocranial volume (Simons *et al.*, 2007). This signals that the large brain to body size ratio found in modern anthropoids was not present in early anthropoids. A further implication is that catarrhine and platyrrhine anthropoids independently evolve a large brain size relative to body size.

Besides supplying data about anthropoid brain evolution, adult crania of *Aegyptopithecus* yield an additional significant fact. They are widely divergent in size (Figure 12.9). The bimodal distribution of these sizes can only mean that a

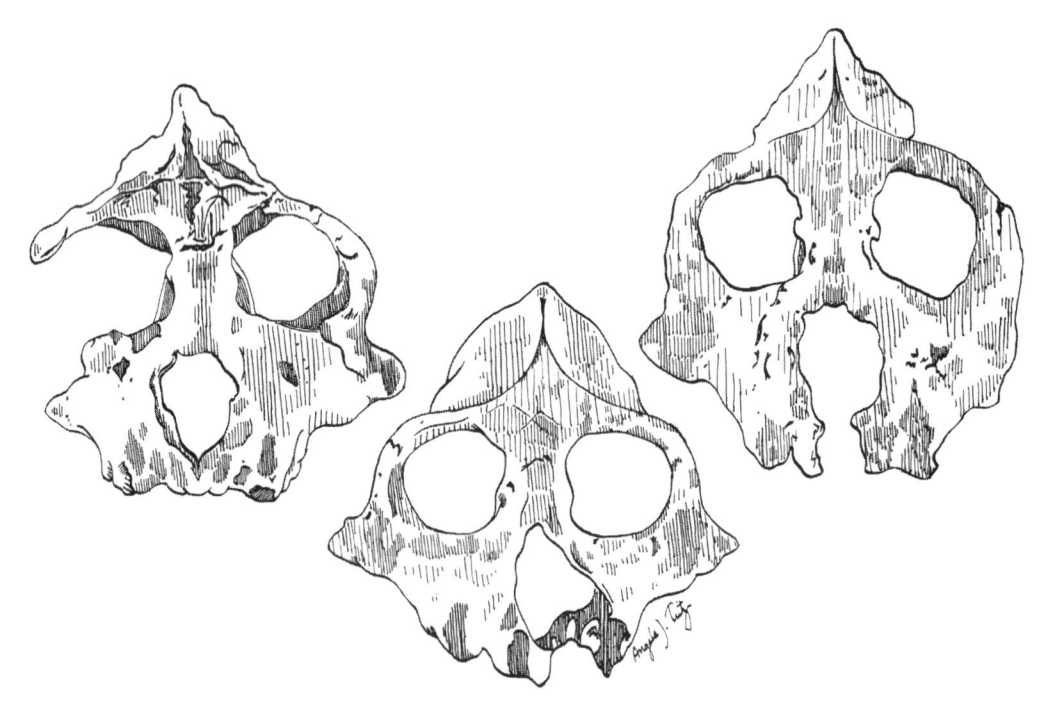

Figure 12.9. Three adult crania of *Aegyptopithecus zeuxis*, an anthropoid primate from the Fayum, illustrating the pronounced sexual dimorphism in this taxon. The cranium in the middle is a presumed female. Illustration by Angela J. Tritz.

significant degree of sexual dimorphism existed in this genus. In fact, in spite of the basal position of *Aegyptopithecus*, the degree of dimorphism in body size in this taxon is typical of that found in modern catarrhines. This has been used to infer the existence of a polygynous mating system in *Aegyptopithecus* (Simons *et al.*, 2007; Cachel, 2009). Multiple adult females and their young were associated with a single adult male. This type of mating system is widely distributed in living catarrhines, and largely accounts for the great disparity in body size between adult males and females. Adult males compete with other males for access to females. This sexual selection favors males with larger body sizes, and so the two sexes begin to diverge in size. The great sexual size dimorphism typical of living catarrhines occurs in *Aegyptopithecus*, and a polygynous mating system may be equally ancient in catarrhines.

In the earliest part of the Fayum sequence, at about 37 mya, prosimians are diverse. They are represented by djebelemurids, adapiforms like *Afradapis*, the earliest galagid, *Saharagalago*, and *Karanisia*. Anthropoids are represented by *Biretia*. In spite of the proliferation of prosimians in the earliest part of the Fayum sequence, they seriously decline after the early Late Eocene. Prosimians no longer occur in the Fayum after the Eocene/Oligocene boundary. Scaly-tailed, gliding anomalurid squirrels, a rodent family endemic to Africa, disappear from the upper levels of the Fayum sequence, during the latest Eocene and earliest Oligocene (Sallam *et al.*, 2010). This signals the disappearance of suitable rainforest habitats in northern Africa at this time. Even the gallery forest around the Ur-Nile must have been reduced. The decline of the Fayum prosimians was probably also caused by the loss and fragmentation of rainforest habitat at the end of the Eocene. This signals a profound reorganization of the Fayum paleocommunity. Even more important, it signals an evolutionary explanation for the origin of anthropoid primates.

Later catarrhine divergence

The importance of Afro-Arabia as the wellspring of the earliest catarrhines has been cemented by the discovery of *Saadanius hijazensis* in western Saudi Arabia (Zalmout *et al.*, 2010). This species is 29–28 my old, and thus postdates the Fayum fossils. It is siamang-like in size (15–20 kg), has a long snout, and possesses the tubular ectotympanic bone of later catarrhines. It is therefore more advanced than *Aegyptopithecus*, and has been put into a novel taxonomic family and superfamily (Family Saadaniidae, Superfamily Saadanioidea). The semicircular canals of *Saadanius* resemble those of the Fayum taxa *Catopithecus* and *Aegyptopithecus*, implying that it was slow and deliberate in its locomotion (Ryan *et al.*, 2012). It is important to note that *Saadanius* is neither an Old World monkey (cercopithecoid) nor an ape or human (hominoid), and so precedes the divergence of these two superfamilies. These two superfamilies, representing living animals, are called "crown catarrhines." *Saadanius* and the Fayum taxa are related to crown catarrhines, but do not fall within either of these groups. They are therefore called "stem catarrhines."

Pozzi *et al.* (2011) significantly point out that *Saadanius* does not indicate that the bifurcation between cercopithecoids and hominoids occurred at 29 mya. This genus demonstrates that the split between cercopithecoids and hominoids occurred after 29 mya, but only fossils with derived cercopithecoid or hominoid traits can indicate the presence of these groups. A 25.2 mya site in the Rukwa Rift, Tanzania, preserves such fossils. The material comes from a locality in the Oligocene Nsungwe Formation. A partial mandible with an M_3 is the type specimen of *Nsungwepithecus gunnelli*, the earliest known cercopithecoid; a partial mandible with ascending ramus and P_4-M_3 is the type specimen of *Rukwapithecus fleaglei*, the earliest known hominoid (Stevens *et al.*, 2013). The first well-known cercopithecoid is *Victoriapithecus macinnesi*, dating to 19 mya. The well-known hominoid genus *Proconsul* appears at 22.5–20 mya. The genus *Kamoyapithecus* is far older (27.8–23.9 mya), but its attribution is uncertain—it is known only from isolated teeth, and its hominoid status is controversial.

In addition to the earlier Fayum species, *Saadanius* further illustrates that early catarrhines were diverse, and exhibited a radiation of species with no living representatives. Early East African hominoids are more diverse than living hominoids. However, they are already less disparate than the Fayum species, because the parapithecoids and propliopithecoids were already winnowed out by the Early Miocene. Living catarrhines are either cercopithecoids (Old World monkeys) or hominoids (apes), yet they are all morphologically strikingly similar. Unlike higher primates in the New World, living catarrhines are all closely related, and they show little real evolutionary divergence, in spite of their formal separation into two superfamilies. The uniformity of living catarrhines in terms of anatomy has long been recognized (Schultz, 1936a, 1936b, 1950). Comparative genomics further supports this uniformity: the genome of the rhesus macaque monkey has a 93 percent similarity to humans (Rhesus Macaque Genome Consortium, 2007). Comparison of the rhesus macaque and human genomes also indicates a separation of the cercopithecoid and hominoid lineages at about 25 mya. This date agrees remarkably well with the 25.2 mya cercopithecoid and hominoid fossils from the Oligocene Nsungwe Formation, Tanzania (Stevens *et al.*, 2013). Thus, there is concordance between the fossil record and estimates of phylogenetic divergence based on DNA evidence.

13 The platyrrhine radiation

South America has had a strange biogeographic history. It was literally an island continent for most of the Tertiary. This is because the last land connection between Africa and South America was lost in the Middle Cretaceous (about 105 mya), and the South American connection with Antarctica—and hence the South American connection with Australia—was lost in the Late Cretaceous (about 80 mya). Central America did not yet exist, and only gradually came into being after a complicated succession of increasingly larger volcanic island arc systems finally culminated in the current land mass. The Isthmus of Panama did not form a complete land bridge between North and South America until 3.5 mya. The "splendid isolation" of South America yielded a unique fossil mammal history with peculiar ecosystems in which carnivorous marsupials and giant flightless birds preyed on a unique array of placental ungulates. Deciphering the fossil history of South America became a major research subject for vertebrate paleontologists investigating convergent evolution within long-separated lineages of marsupial and placental mammals (Scott, 1937; Simpson, 1980, 1984; Flynn, 2009).

Habitats in South America are also affected by strange types of forest vegetation, such as the narrow, tall *Araucaria* trees (Family Arucariaceae), whose leaves form tight spirals at the end of branches (Figure 13.1). These trees are otherwise found in Australia. Forests of southern beech (*Nothofagus*) are also characteristic of tropical and temperate forests in southern South America. There are almost 40 species in the genus, which is classified as a separate family (Family Nothofagaceae). Members of this genus are found throughout the South Pacific Rim, including Australia, Tasmania, New Guinea, and New Zealand. Fossils of southern beech are found in Antarctica. These *Nothofagus* forests were once found throughout the ancient landscape of the Gondwana continent, before plate tectonic movements tore the continent apart. The lowland tropical rainforests of Amazonia are now a major feature of the modern Neotropics, but their range has drastically expanded and contracted with Pleistocene climatic fluctuations. Archaeological evidence indicates that pre-contact Amerindians extensively altered the Neotropical rainforests through sophisticated agro-forestry, and so the current expanse of lowland tropical rainforests was apparently much different even as late as A.D. 1491.

Although it was long thought that Australasian and South American marsupials demonstrate convergent evolution, it is now known that all living Australasian marsupials originate from ancient South American ancestors about 80 mya, when

Figure 13.1. Araucaria trees like this species (*Araucaria aracana*) were widespread in the southern Supercontinent of Gondwana, and are typical of the ancient forests of South America. Note the intricate nature of leaf physiognomy, which demonstrates the ancient nature of this tree group. Botanical Gardens, University of Bonn.

the continents finally separated. This is shown by the position of nested or inserted retroposons ("jumping genes") in the DNA of living marsupials. Newer retroposons nested within older ones in identical DNA locations assure that homoplasy is not occurring. These nested retroposed elements are highly unlikely to arise in the same position within the genome of unrelated organisms (Nilsson *et al.*, 2010). In fact, all living marsupials are monophyletic, diverging from a single ancestor about 130 mya. Unlike any other mammals, their genomes have 10 nested retroposons in the same position. The four Australasian marsupial orders diverge from an ancestor that migrated from South America to Australia via Antarctica 80 mya. This ancestor resembled the rare South American "monito del monte" (*Dromiciops gliroides*), which is the only surviving species within the order Microbiotheria (Nilsson *et al.*, 2010). Although South America, Antarctica, and Australia were connected for a long time, only one successful marsupial migration ever penetrated Australia. The reason for this is unknown. After this single migration event occurred, the Australasian marsupials deviate widely from those in South America.

The "monito del monte" ("little mountain monkey") is important not only because its ancient relatives seeded Australia with marsupials, but also because members of its order (Order Microbiotheria) were once more speciose. Another marsupial order, the shrew opossums (Order Paucituberculata) were also species-rich during the Early Miocene (Abello *et al.*, 2012). These small, arboreal insectivores/frugivores would have occupied niches now occupied by small living New World monkeys like the marmosets and tamarins.

Colonizing South America

G. G. Simpson first argued that South American mammal evolution could be summarized by three event horizons (Simpson, 1980). The first horizon begins in the Cretaceous with monotremes, multituberculates, and other non-therian mammals. Marsupials appear between 70 and 60 mya (Flynn & Wyss, 1998). Archaic, indigenous species appear after 60 mya—specialized marsupial carnivores, edentates (sloths, anteaters, armadillos), and endemic ungulates (members of the Orders Litopterna and Notoungulata). The second event horizon has two new immigrant mammal groups (rodents and primates), and experiences a faunal modernization triggered by climatic cooling and aridity during the Eocene/Oligocene transition. A fossil assemblage (the Tinguiririca Fauna) dated to 36–31.5 mya from the Central Chilean Andes is the oldest mammal fauna in the world to be dominated by hypsodont herbivores (Flynn & Wyss, 1998). Because these were grazing animals, this implies that open-country grasslands first arose in South America, and preceded those found elsewhere in the world by 15 my.[1] The third event horizon begins with infrequent North American immigrants, and finally culminates in a great interchange of species across the newly formed Central American land bridge. This "Great American Biotic Interchange" began 3.5 mya, and continues today (Stehli & Webb, 1985).

Primates first appear in South America at the beginning of the second event horizon, arriving at the end of the Eocene. Primates appear suddenly, along with the first unique South American rodents (caviomorph rodents—i.e. agoutis, capybaras, guinea pigs, and their relatives). Caviomorph rodents are found only in South America, although a few species have dispersed to North America. These are the New World porcupines. Caviomorph rodents are one of the two divisions of the rodent Suborder Hystricognatha, the other being the phiomyids, dating back to the Middle Eocene of North Africa. Hystricognath rodents (now represented only by caviomorph rodents) are so different from all other rodents that they clearly represent a separate early radiation of rodents. Both morphological and molecular evidence indicates that hystricognath rodents originate in South Asia (Jaeger

---

[1] This early presence of grasslands in South America is controversial. A 15-my gap that precedes grasslands elsewhere seems inexplicable. Grasslands are inferred from the high-crowned cheek teeth of herbivores, but this hypsodonty can be alternatively explained by airborne grit, dust, or volcanic ash.

et al., 2010b). Yet, where did the ancestors of the caviomorph rodents originate? In South America? In North America? In South Asia? In North Africa? The answer impacts on questions of platyrrhine primate origins, because both rodents and primates are traditionally thought to have immigrated into South America at roughly the same time. Although rodents and primates need not have immigrated from the same source area, any rough coincidence of their arrival may hold clues about conditions that favored rafting transport into South America. The oldest caviomorph rodents in South America are about 32 my old, and the oldest well-preserved platyrrhine primate cranium dates to the Miocene at 20.1 mya (Flynn & Wyss, 1998). This Early Miocene primate is *Chilecebus carrascoensis*. The species is less than 1 kg in size, and may have resembled living marmosets (Flynn *et al.*, 1995). Enigmatic teeth of the species *Branisella boliviana* indicate the earliest presence of primates in South America. The cheek teeth are very high-crowned, and suffer heavy, rapid wear. The M1 is considerably worn before the M2 erupts. Takai *et al.* (2000) suggest that *Branisella* may have been partly terrestrial and consuming grassland resources, based on the hypsodonty of the teeth. This is dubious. Rodents and ungulates in South America also show a very early shift to hypsodonty, so this trait is not unique to *Branisella*. Takai *et al.* further suggest that *Branisella* resembles the Fayum genus *Proteopithecus*, based on its reduced M_2 and large P_4. However, the authors are deliberately searching for African ties. They fail to address the fact that size differences along the postcanine tooth row are also found in North American omomyids. An orbital fragment of *Branisella* indicates that the species had small eyes. It is therefore interpreted as being diurnal (Kay *et al.*, 2001).

Based on the presence of *Branisella*, primates are thought to have appeared in South America by about 26 mya. Caviomorph rodents therefore antedate primates in South America by a considerable amount of time. Simpson's inclusion of rodents and primates into a second event horizon was meant to summarize the fossil history of South America as it was known in the last quarter of the twentieth century.

Two new caviomorph rodent taxa—one of which is the oldest chinchilla-like species—have been retrieved from the Tinguiririca Fauna from the Central Andes of Chile (Bertrand *et al.*, 2012). These fossils date to 32 mya, but are already diverse. This indicates that caviomorph rodents probably originated in South America during the Middle to Late Eocene, but were restricted to the equatorial tropics until later. The Tinguiririca Fauna occurs in middle to high latitudes, and the environment was arid, mirroring the global onslaught of colder, arid climates typical of the Oligocene. The well-dated rodent fossils indicate a much earlier presence of rodents in South America than primates. Thus, although both rodents and primates are part of Simpson's second horizon of waif immigrants to South America, they did not arrive simultaneously. The earliest primate (*Branisella*) is dated to 6 my after the earliest rodents. No one would be naïve enough to think that both rodents and primates waft over on the same mat of floating vegetation, but dates of rodent fossils are thought to constrain the arrival of primates. Yet, is this necessarily true? The fossil evidence indicates a considerable gap in time between the first rodents and the first primates. And there is a gap in the South American primate fossil record until the Miocene—even then, the record is impoverished. Another factor affecting the earlier dates of rodent fossils in South America is their ability to occupy terrestrial grasslands. This results in a greater probability of fossilization. Caviomorph rodents in the Tinguiririca Fauna have very hypsodont teeth, which is thought to be the earliest signal of the spread of open-country grassland ecosystems in earth history (Bertrand et al., 2012). Note that the presence of grassland is inferred from hypsodont teeth in rodents and ungulates (and *Branisella*!). An alternative interpretation is that airborne dust and grit during periods of arid climate trigger intense dental wear as grit is ingested with food items. Volcanic activity in the growing Andes produced ash that floated in air or coated vegetation. Ingestion of grit or ash quickly wears teeth, and thus selects for higher-crowned or hypsodont teeth.

Land mammals, such as primates and rodents, are notoriously poor at crossing water barriers (Lawlor, 1986). Islands—even island continents—are unlikely to acquire land mammals. There is no doubt that both primates and rodents are waif immigrants, colonizing South America by what G. G. Simpson first called "sweepstakes dispersal" (Simpson, 1940a). That is, these land animals cross substantial water gaps on floating mats or islands of vegetation either discharged from the deltas of great rivers or dislodged from a coast by violent storms. Winds and ocean currents propel the floating mass. Colonizing animals using these floating mats may cross water once from mainland to mainland, or cross through a series of islands (island-hopping). It is clear that the odds against land animals successfully colonizing an area in this way are astronomical—hence the "sweepstakes" designation. Crossing through a series of islands may either increase or decrease these odds. If the islands are widely separated, odds are increased. If the islands form part of a volcanic island arc system, odds are decreased. Volcanic activity within this system forms intermittent land surfaces and new islands. Over geological time, these promote wider penetration into an area. Plate tectonic movements within the Caribbean Basin have been particularly complex, and many volcanic island arc systems arose and disappeared as a series of small lithospheric plates jostled against each other while the Atlantic Basin opened and the Pacific Basin closed. Figure 13.2 illustrates the development of volcanic island arc systems in the Caribbean Basin during the Middle Eocene at approximately 49 mya. Small, island-hopping land mammals may have colonized South America from North America via sweepstakes dispersal through such island arc systems. In addition, islands of the Greater Antilles (Cuba, Jamaica, and Hispaniola) are ancient continental blocks that could also offer temporary haven to dispersing land mammals.

If a land mammal is a powerful swimmer (e.g. the elephant), this enhances the probability of crossing a water gap. However, most non-human primates are unable to swim, which guarantees that they must cross water on floating mats of vegetation. If this seems unlikely, at least one macaque species (genus *Macaca*)

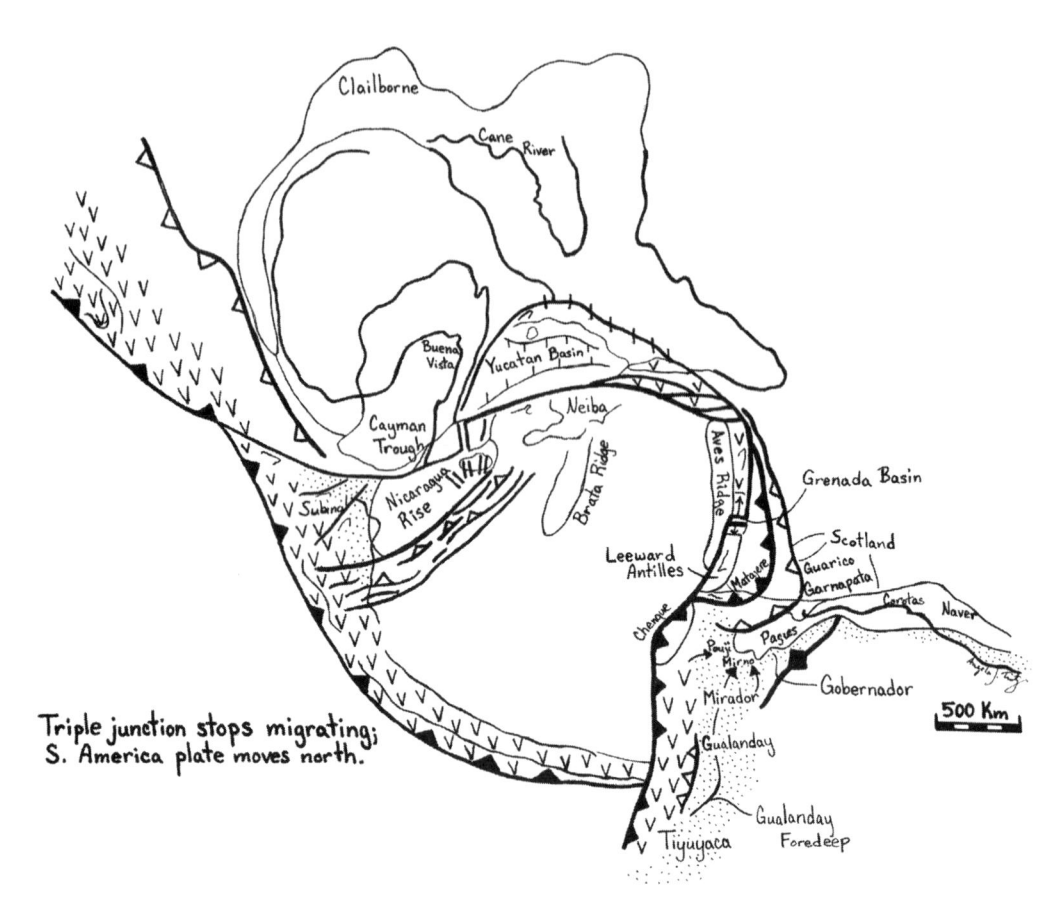

Figure 13.2. Plate tectonics of the Caribbean during the Middle Eocene, about 49 mya. This reconstruction illustrates the complexity of plate movements and the development of volcanic island arc systems. After a reconstruction by Dr. James Pindell. (http://www2.flu.edu/~draper/caribbean/Caribreconstr.html/). Illustration by Angela J. Tritz.

rafted over deep water when colonizing the islands of Southeast Asia (Abegg & Thierry, 2002)—and macaques are large-bodied primates, far from rodent-sized. Certain biological factors increase the likelihood of a successful sweepstakes colonization. These factors include small body size; the ability to subsist on the floating vegetation; the ability to get necessary water from food; the ability to enter a state of torpor or hibernation when temperature drops and food disappears; and the ability rapidly to build up population numbers through a high rate of reproduction once landfall is made. Theoretically, only a single pregnant female who survives the transit and bears a litter is needed for a species successfully to colonize an area in this way. Rodents and prosimian primates are small-bodied, herbivorous, and bear litters. And true hibernation is found in some living primates. Living fat-tailed lemurs (species in the genus *Cheirogaleus*) enter hibernation when climatic conditions are poor, even though they are living in the tropics. Extreme heat and drought trigger this state. During the period of torpor, they live off the fat stored at the base of their tails. But how long could an animal endure such a voyage? Recent work on the early Tertiary colonization of Madagascar by lemur-like prosimians and tenrecid insectivores indicates that the Mozambique Channel could have been crossed in 30 days or less, given very strong ocean surface currents from the west at 60 mya, and the possibility of major tropical cyclones (Ali & Huber, 2010; Krause, 2010). The narrowest width of the Mozambique Channel is 460 km, so this might establish the largest water gap that could be crossed by a primate via rafting.

Given that primates and rodents are waif immigrants, crossing a water gap to reach South America, where did they come from? Did they cross the Caribbean from North America (Simpson, 1978)? Did they cross the South Atlantic from Africa (Hoffstetter, 1971, 1972)? Was there a reverse vector to their movement? Did they cross the South Atlantic from South America to settle in Africa? Did ancestral platyrrhines invade South America twice over? This last idea, which is not widely held, is based on an immunological study of living platyrrhine primates that demonstrates a wide divergence between *Cebus* and *Lagothrix* (Bauer & Schreiber, 1997). This wide divergence is thought to indicate the presence of two invading ancestral stocks. Sweepstakes dispersal is so unlikely that some researchers have even argued that higher primates arise in a conjoined South America/Africa, and are then geographically separated into platyrrhines and catarrhines by plate tectonic movements that create the South Atlantic (Hershkovitz, 1977; Heads, 2010). However, the plate tectonic history of the South Atlantic is well known. If the common ancestors of platyrrhines and catarrhines originate in a conjoined South America/Africa, they would need to originate during the Cretaceous—a fact that is not supported by the fossil evidence.

In my opinion, the ancestors of the platyrrhines cross into South America from North America, crossing the ancient Caribbean (Cachel, 1981). The geological history of the Caribbean is fantastically complex (Iturralde-Vinent & MacPhee, 1999). However, one crucial part of this history is that both a series of Tertiary volcanic island arc systems form here, and isolated portions of continental lithosphere jostle along plate margins and accrete through time. And fossil primates are found on Caribbean islands. As far back as the Early Miocene, platyrrhine primate fossils are found on islands of the Greater Antilles (MacPhee & Iturralde-Vinent, 1995). How they reached these islands is unknown. They may have colonized these islands after crossing a water gap from South America. Alternatively, they may have formed part of a terrestrial fauna isolated on blocks of ancient lithosphere. Between 50 and 30 mya, islands of the Greater Antilles (Cuba, Jamaica, Hispaniola) were connected to the Aves Ridge, a now sunken land mass west of the Lesser Antilles. The Greater Antilles and the Aves Ridge formed a huge peninsula linked to mainland South America (Iturralde-Vinent & MacPhee, 1999). Given this geological history, it is not surprising to find fossil platyrrhines in the islands of the Greater Antilles—e.g. Cuba (*Paralouatta*), Jamaica (*Xenothrix*), and Hispaniola (*Antillothrix*).

I believe that the ancestral platyrrhine that utilized the Caribbean dispersal route was a prosimian, possibly an advanced omomyid similar to *Rooneyia viejaensis*. However, this implies that anthropoid primates evolve separately in the Old and New Worlds (Cachel, 1979a, 1981). Since the recognition of plate tectonics, most researchers fasten on the crossing of a narrow South Atlantic, with the founders of the platyrrhine stock originating in Africa. Sometimes the route is reversed, with the founding catarrhines originating in South America. In any case, most researchers are reluctant to admit the possibility of an independent origin for higher primates in the Old and New Worlds. It is possible that this reflects an unconscious bias towards perceiving higher primates as so advanced and specialized that they could only evolve once. I have presented a paleoecological model for the evolution of higher primates (see Figure 12.1 and Cachel, 1979b).

Do fossils or comparative anatomy provide any clues to the origins of platyrrhine primates? No. They appear in South America *de novo*. Hoffstetter (1977) argued that an unknown species of parapithecid rafted from Africa to South America during the Late Eocene. Yet, dental evidence does not link platyrrhines with catarrhines. Specifically, dental traits show no affinity between platyrrhines and fossil parapithecids from the Oligocene Fayum (Kay, 1980). The most recent investigation of the basicranial anatomy of the Fayum anthropoids, living catarrhines, and living platyrrhines sheds no light on platyrrhine origins (Kay *et al.*, 2008). Micro-CT scans were used to supplement traditional methods of anatomical study. Platyrrhine origins were not revealed. Specifically, no shared derived traits can be found in the bony ear region (temporal and tympanic) that link platyrrhines with fossil primates (*Aegyptopithecus, Apidium*) from the Egyptian Fayum.

Detailed investigation of the age of material that formed the ancient Caribbean island arc systems can hold clues to the vector of island formation. The age of islands in a volcanic island arc system reflects movement of the underlying mantle hotspot producing the islands. The Hawaiian Islands are a familiar example, but a more apropos example comes from the volcanic islands in the Gulf of Guinea that were produced as a result of the opening of the central Atlantic Ocean. In descending order of age, these are Principe (30 mya), São Tomé (13 mya), Annobon (5 mya), and Bioko, which was connected to the mainland by a land bridge until 10,000 yrs ago (Jones, 1994). If islands in an ancient Caribbean volcanic arc system could be dated, the dates would offer clues to the sequence of available land surfaces that island-hopping land mammals could colonize.

Platyrrhine diversity

Although research on New World monkeys tends to be slighted in favor of Old World monkeys, the remarkable diversity and variability of platyrrhines was noted by early researchers (Schultz, 1926). This diversity was recently made the subject of a special issue of *The Anatomical Record* on the evolution and functional morphology of platyrrhines (Rosenberger & Laitman, 2011). Primatologists (especially Marc van Roosmalen) continue to discover a startling number of hitherto unknown living platyrrhine species, especially among the small-bodied marmosets and tamarins (Shuker, 2012:100–104, 285). A logjam of 20 apparently new monkey species are currently being formally described (http://marcvanroosmalen.org/.). Yet, research on known species has been marred by serious taxonomic inflation (Chapter 2), principally by workers who reject the biological species concept. This has led to such taxonomic instability that a Tower-of-Babel-like situation has developed. Because of new genus and species names, assigned for the most trivial of reasons, even specialists are not sure which taxon is being written about, and it is impossible to follow the taxonomic literature (Rosenberger, 2012b).

There is a fundamental dichotomy in interpreting the fossil platyrrhine record. Traditionally, fossil material is interpreted as the first representatives of existing platyrrhine groups. This is a logical, "connect the dots" approach often seen in paleontology. Another mode of interpretation is to identify platyrrhine fossil material as enigmatic or bizarre—clearly very different from any living species. Both modes of interpretation are needed to explain the platyrrhine fossil record. Because this fossil record is very poor, DNA and molecular phylogenies have long been used to infer the evolutionary history of the platyrrhines. Comparative genomic analyses of 186 living primates distributed across about 90 percent of primate genera have been used to differentiate living platyrrhine groups and their time of origin (Perelman *et al.*, 2011). The Family Pitheciidae (*Callicebus, Pithecia, Chiropotes, Cacajao*) separates first at just over 20 mya, followed by the Family Atelidae (*Alouatta, Ateles, Brachyteles, Lagothrix*). At 20 mya, the Family Cebidae (*Cebus, Saimiri*) differentiates. In the succeeding 700,000 yrs, the *Aotus* species complex and the tamarins and marmosets (*Saguinus, Leontopithecus, Callimico, Callithrix, Cebuella, Mico*) arise.

In contrast to other primates, a number of living platyrrhine primates possess prehensile tails (Figures 13.3–13.4). These are members of the Family Atelidae, and include howler monkeys (*Alouatta*), the woolly monkey (*Lagothrix*), the woolly spider monkey (*Brachyteles*), and the spider monkeys (*Ateles*). Many Neotropical vertebrates develop prehensile tails, which can support the entire weight of the body. Besides the atelid primates, six other vertebrate families also evolve prehensile tails (Emmons & Gentry, 1983). Neotropical forests have an abundance of palms, and lianas are relatively thin and not woody. Forest canopies in the New World therefore offer particularly unsteady substrates and discontinuous routes for arboreal animals.

The marmosets and tamarins (Family Callitrichidae) were traditionally considered to be the most primitive of the living anthropoids (Hershkovitz, 1977). They are now considered to be derived anthropoids that have secondarily reduced their body size, and secondarily re-developed claws on every digit except the hallux from a progenitor that was histologically nail-like. Because body size is small, callitrichids can support their body weight using claws alone. They can therefore cling to vertical tree trunks and eat exudates like gum, sap, and resin that are oozing from damaged bark (Garber, 1992, 2007). Despite their squirrel-like size, the callitrichids are definitely not occupying squirrel-like niches in the Neotropics. The marmosets and tamarins are also specialized for a derived,

Figure 13.3. A howler monkey (*Alouatta palliata*), typical of the living platyrrhine primates. Note the prehensile tail. Illustration by Angela J. Tritz.

increased rate of reproduction, because they normally give birth to dizygotic twins (Figure 13.5). As a result, they demonstrate a complex of behaviors unique to primates: polyandry (the mating of one alpha female with multiple males), cooperative rearing of the young birthed by the alpha female, and reproductive suppression in subordinate females. When animals are provisioned under captive conditions, callitrichids can give birth to triplets or quadruplets. Females release additional eggs as more food is provided. These young would not survive under wild conditions, because they are dependent upon their mother's milk for survivorship. Because of their high intrinsic rate of increase, marmosets and tamarins can colonize disrupted environments, and quickly build up population numbers when open or disturbed patches in the tropical forest are re-forested. These specializations may have evolved relatively recently. The genus *Callithrix* is so young that reproductive boundaries between species are very permeable. As demonstrated by phenotypic characters and mtDNA, *Callithrix jacchus* and *Callithrix penicillata* re-introduced into the forests of eastern Brazil hybridize so readily that a hybrid swarm exists (Malukiewicz *et al.*, 2012).

Figure 13.4. The relative limb proportions in living platyrrhine primates: *Alouatta* (the howler monkey), *Lagothrix* (the woolly monkey), *Brachyteles* (the woolly spider monkey [muriqui]), and *Ateles* (the spider monkey). All genera are members of the Family Atelidae, and all possess prehensile tails. The ventral surface is hairless at the terminal one-third of these tails, and the skin surface is covered with dermatoglyphics. A prehensile tail can support the entire weight of the suspended body, and the tail is both highly innervated and muscular. The dermatoglyphics serve to increase friction and prevent slipping when the body is hanging from the tail. Note that none of the genera have specialized elongation of either the forelimb or the hindlimb. The forelimbs and hindlimbs are equal in length. A zygodactylous hand structure (major axis of the hand running between digit II and III) can be seen in *Alouatta* and *Lagothrix*. After Erikson (1963). Illustration by Angela J. Tritz.

Inferring locomotion without postcrania in fossil platyrrhines

Because most of the fossil platyrrhine taxa lack postcranial skeletons altogether, the semicircular canals of these taxa sometimes yield the only available evidence for locomotion. In contrast to the early catarrhines, the semicircular canal evidence of fossil platyrrhines indicates a very early adaptation to agile locomotor behavior, although none were as quick as modern primates that are specialized for leaping (Ryan *et al.*, 2012). The five platyrrhine taxa examined are *Dolichocebus*

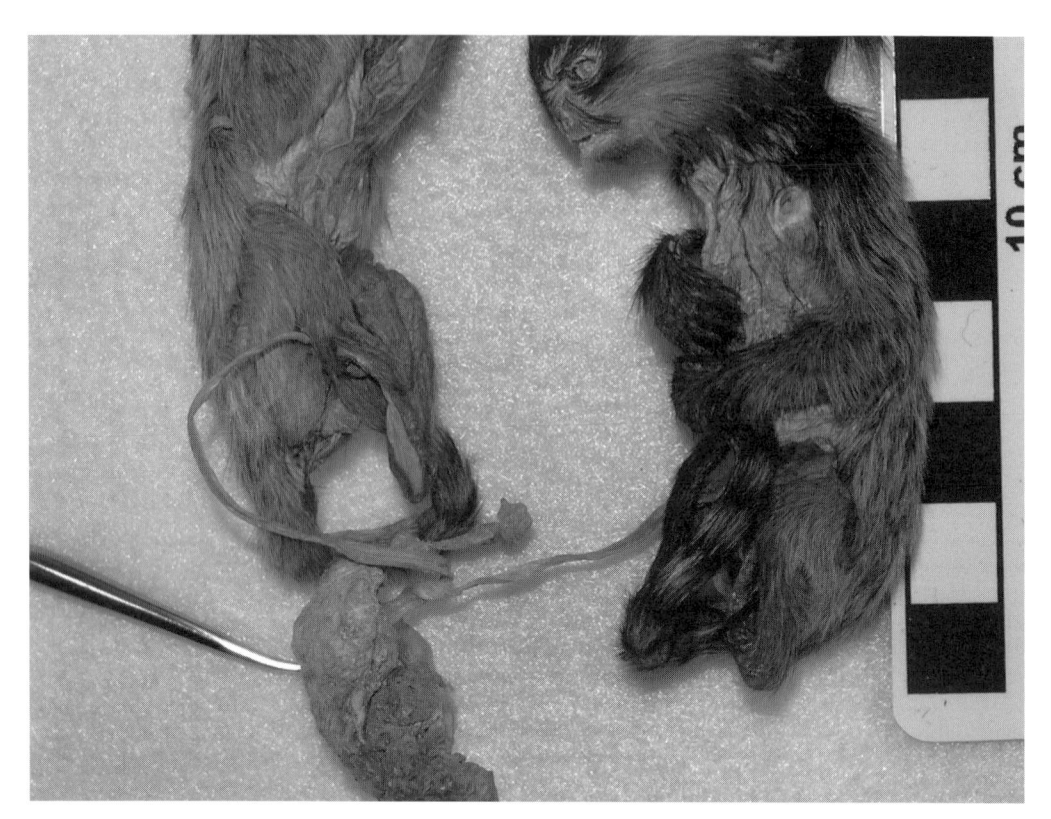

Figure 13.5. Fetuses of the common marmoset (*Callithrix jacchus*). These are dizygotic twins, produced by different sperm fertilizing two different eggs. Dizygotic twinning is the norm in callitrichids, and largely accounts for their high rate of increase and good colonizing potential. In theory, these twins should be only 50 percent genetically related—i.e., no more related than siblings of the same parents born from different pregnancies. However, note that the separate umbilical cords attach to a common placenta. Blood can be shared between the twins across the placenta. The blood stream of a twin may therefore contain "foreign" genetic material from the other twin, making these individuals genetic chimeras. The consequently greater degree of genetic relatedness between callitrichid twins has been used to explain the complex sociality observed in callitrichids.

gaimanensis, *Homunculus patagonicus*, *Lagonimico conclucatus*, *Tremacebus harringtoni*, and *Chilecebus carrascoensis*. These taxa appear to have had a mode of locomotion similar to that of small living platyrrhines like the tamarins *Leontopithecus rosalia* or *Saguinus oedipus* and the common marmoset *Callithrix jacchus*. That is, the fossil platyrrhines were generalized arboreal quadrupeds that probably engaged in some vertical clinging and leaping. Why do the fossil platyrrhines appear to be more acrobatic than any of the basal catarrhines are? It is possible that gum and resin eating, which is typical of living marmosets and tamarins, was established at an early stage in platyrrhine evolution. Collection of gum and resin is facilitated when an animal can cling vertically to a tree trunk. Vertical clinging demands a greater degree of arboreal agility than more staid, horizontal travel. The marmosets can even actively gouge tree bark and elicit the flow of resin, because their mandible is V-shaped, and their lower incisors are as long as their canines. This is the so-called "short-tusked" condition. Vigorous gouging movements on a vertical surface are even more dependent on balance and quick reflexes. In any case, exudate collection in the trees is a behavior that demands an instant dynamic response as the animal shifts its center of gravity. The clawed digits of living marmosets and tamarins (except for the hallux) allow these animals to cling to and maneuver on vertical surfaces. However, this ability is affected by body size. If marmosets and tamarins were large, claws could not support their body weight. Thus, gum and resin eating, vertical clinging, claws, and small body size form an integrated functional system.

Centers of platyrrhine diversity

In addition to extensive work on living New World primates, fossil prospecting continues in South America. As mentioned above, John Flynn and his colleagues have found abundant mammal fossils, including the 20.1 mya primate *Chilecebus*, in the Central Andes of Chile. Richard Kay and his colleagues have been searching for basal platyrrhines in the Rio Gallegos region of Argentina and other Argentine sites. The Atlantic Coast of Patagonia contains abundant late Early Miocene mammal fossils from the Santa Cruz Formation, dating to 18–16 mya (Kay *et al.*, 2012; Vizcaíno *et al.*, 2012). These include primate fossils. Because this area was in a high latitude, questions about climate, seasonal sunlight, ancient vegetation, and energy flow in paleocommunities become important. These concerns mirror, at the opposite end of the western hemisphere, concerns about reconstructing Paleocene mammal communities above the Arctic Circle (Chapter 9).

Amazonia could be presumed to be a center of fossil primate diversity, if lowland rainforest habitats were present. Did tropical rainforests exist in Miocene Amazonia? Yes, they did. Thanks to a serendipitous recent discovery, we know something about the degree of floral diversity in these forests. Amber found in the Middle Miocene Pebas Formation of northeastern Peru contains pollen and spores, as well as insect and spider inclusions (Antoine *et al.*, 2006). Species diversity in this Middle Miocene sample appears comparable to that found in modern rainforests.

Major fluctuations have occurred in the rainforests of Amazonia since the end of the Miocene. It is notable that much of the evidence for these fluctuations relies on modern animal and plant distributions, and not directly on paleovegetation data. Primates contribute to this reconstruction. For example, the living capuchin monkeys (genus *Cebus*) are taxonomically divided into a robust (tufted) group and a gracile group. Some taxonomists would like to separate these groups into two genera (*Cebus* for the gracile group and *Sapajus* for the tufted group). These are sympatric across much of their ranges. Genetic data reveal that these two capuchin groups diverged in the Late Miocene (Alfaro *et al.*, 2012). Capuchins initially radiated out of the Amazon Basin, and the ancestors of the gracile group become geographically isolated in the Brazilian Atlantic coastal forest when the rainforest contracts. During the Plio-Pleistocene, both of these groups diversify. When the rainforest spreads again, contact is re-established between the Amazon Basin and the Brazilian Atlantic coast.

The collection of fossils in the Santa Cruz Formation was an epic task. Because the fossils were embedded in sandstone beach rock in the intertidal zone, specimens needed to be hastily retrieved using both geological hammers and jack hammers. Blocks of sediment containing fossils had to be removed before the tide turned, and cold Atlantic sea water swept over the collecting area. These high-latitude Patagonian localities were warm and seasonally humid during the Early Miocene. Yet, the warm and seasonal climate soon began to cool, and they became increasingly more arid during the Middle Miocene. A mosaic existed of open, temperate and semiarid forests and warm, temperate humid forests. Seasonally restricted rainfall and sunlight limited exuberant plant growth.

Fossil mammals from Patagonia

Platyrrhine primates have been collected from coastal Patagonia for over 120 years. All of the fossils come from the genus *Homunculus*. The famous Argentine paleontologist Florentino Ameghino argued in 1891 that this genus was the first hominin, as indicated by the genus name ("little human"). This genus remains the most well known of the Miocene fossil platyrrhines, because three of the Miocene taxa are known only from a single distorted or fragmentary cranial specimen. These are *Chilicebus carrascoensis*, *Tremacebus harringtoni*, and *Dolichocebus gaimanensis*. Yet, *Homunculus* is known from fossil crania, mandibles, dentition, and postcranial elements, and three species occur in the Santa Cruz Formation (Figure 13.6).

The Santa Cruz Formation yields a myriad of microbiotherians, relatives of the endemic "monito del monte" ("little mountain monkey") and paucituberculates, relatives of the living caenolestids or shrew opossums. These arboreal marsupials were an important component of the ancient ecosystem, and were potential competitors with primates. Fossil taxa in these two marsupial orders are often much larger in body mass than their living representatives. The largest of the extinct taxa were frugivorous, and inhabited warm, forested habitats with pronounced seasonal rainfall (Abello *et al.*, 2012). These arboreal marsupials occupied niches and exploited resources that would otherwise be exploited by small South American monkeys. Note that the name "monito del monte" translates to "little mountain monkey," indicating that local people have remarked on a similarity with small platyrrhine primates. Competition with primates would be exacerbated in the Miocene, when larger arboreal marsupials overtopped the size range of marmosets and tamarins (120–600 g), and entered the size range of owl, titi, and squirrel monkeys (700–1,000 g). The living "monito del monte" engages in deep torpor or hibernation during the temperate zone winter—the only living South American marsupial to show this behavior. This might explain how small-bodied mammals in high-latitude Patagonia were able to survive pronounced seasonality: they hibernated just as living fat-tailed lemurs hibernate during the Malagasy

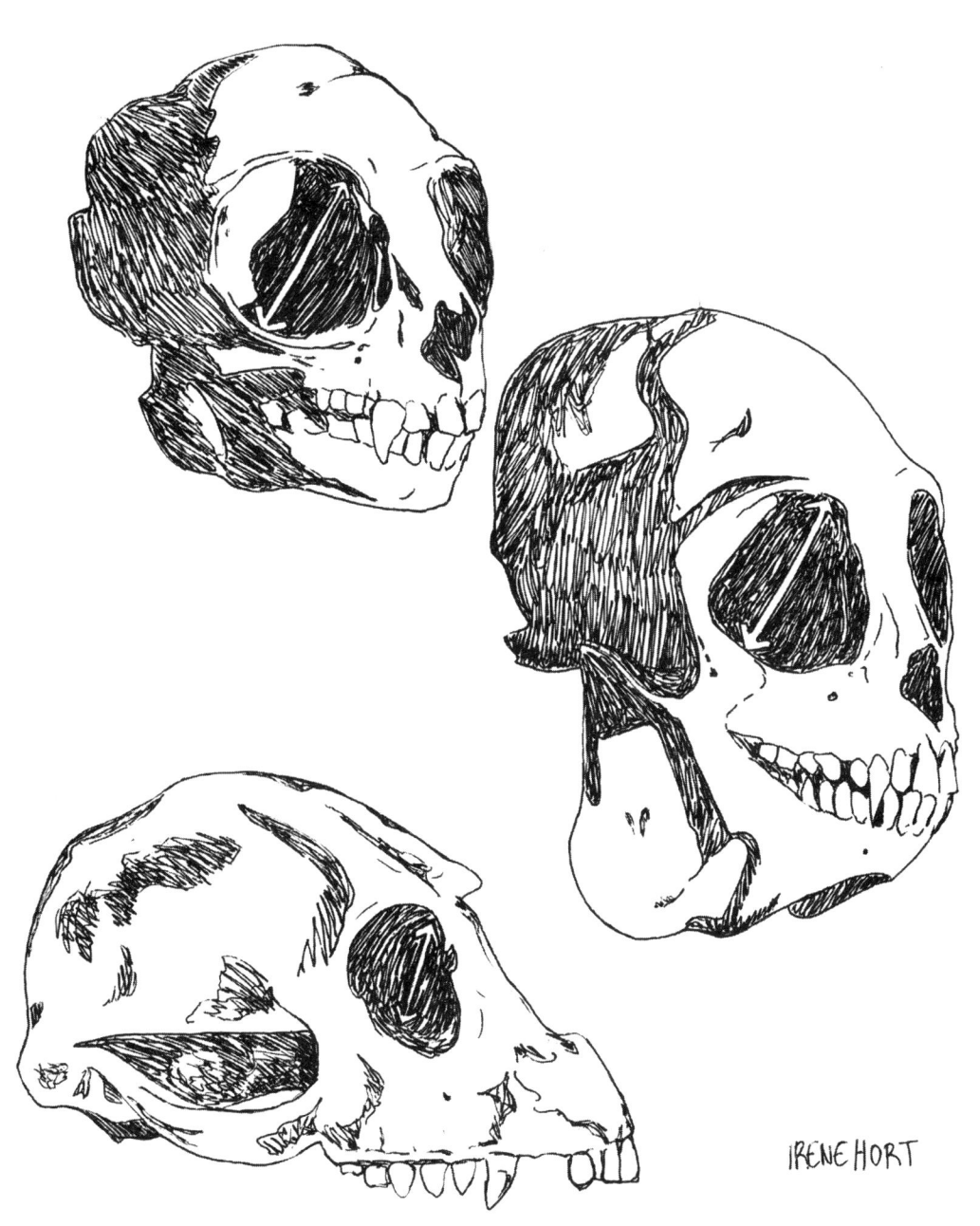

Figure 13.6. Orbital size in *Homunculus patagonicus* (bottom) contrasted with the nocturnal owl monkey *Aotus trivirgatus* (top) and the diurnal dusky titi monkey *Callicebus moloch* (middle). White arrows indicate orbital diameter. After Kay *et al.* (2012:Fig. 16.3). Illustration by Irene V. Hort.

dry season. The climate of late Early Miocene Patagonia became progressively drier through the Santa Cruz Formation, so that both seasonal darkness and drought would have been increasingly problematic for contemporary mammals (Figure 13.7). This undoubtedly triggered a retreat of ancient primates to forested

Figure 13.7. Reconstruction of a late Early Miocene community in Patagonia. *Homunculus patagonicus* and an opossum-like marsupial are arboreal. A giant ground sloth and two large, capybara-like caviomorph rodents are on the ground. Illustration by Irene V. Hort.

habitats near the equator. Such was the beginning of primate biodiversity in Amazonian rainforests.

Homunculus is now known to be a basal platyrrhine. It has no close evolutionary relatedness to any living platyrrhine group. Kay *et al.* (2012) reconstruct the paleobiology of this genus in fine detail. Using three fossil crania, they describe features of the cranial anatomy of *Homunculus patagonicus* from high-resolution CT scans and composite scans, and also present a mid-sagittal section through the cranium, along with a reconstructed endocast. The brain to body size ratio is remarkably low—far below that of any living higher primate. This confirms that relative brain size increase occurred independently in New and Old World anthropoids. Visual and olfactory capacities in *Homunculus* are assessed from endocast volumes, and are comparable to those of living diurnal platyrrhines. Relative orbital size also indicates that *Homunculus* was diurnal. Tactile sensitivity of the snout (assessed from blood vessels feeding the facial vibrissae) and auditory acuity are also similar to living diurnal platyrrhines. The semicircular canals of the inner ear indicate that *Homunculus* was an agile animal, and could execute quick movements like those of the living sifaka, which is a highly specialized leaper. Traits of the postcranium also indicate that leaping was important in this arboreal primate. Based on dental traits, *Homunculus* was a folivore/frugivore, and did not engage in insectivory or bark-gouging to elicit sap or resin flow. Body size in *Homunculus* averaged between 2 and 3.5 kg. This is comparable to that found in the living New World saki or uakari monkeys. Only a moderate degree of sexual dimorphism existed, which is typical for living platyrrhine monkeys.

Tremacebus harringtoni was once identified as a species of *Homunculus*. *Tremacebus* is slightly older in age, dating to about 20 mya. The type specimen of *T. harringtoni* is shown in Figure 13.8. Hershkovitz (1974) argued that this specimen illustrated the origin of the postorbital septum, because a large fissure is present in the inferior orbit. However, the CT scan in Figure 13.8 cannot differentiate between the original bone, rocky matrix, and reconstructive plaster that are present in this region. Hence, the presence of this fissure cannot be confirmed in the fossil. The crania of living owl monkeys (*Aotus* spp.) exhibit a large fissure in this region, but the existence of this fissure is explained by the enlargement of the eyes and orbits in this genus, which is secondarily nocturnal. CT scans of the olfactory fossa in *Tremacebus* indicate that this area was not large (Kay et al., 2004). The olfactory bulbs were therefore equivalent in size to those of living platyrrhines of the same body size. The orbits were not markedly enlarged. *Tremacebus* is therefore inferred to have been diurnal.

The islands of the Greater Antilles (Cuba, Jamaica, Hispaniola) yield enigmatic primate fossils unlike any species found in the New World today. It is important to note that these islands are tectonically ancient. They are not part of the recent volcanic island arc system that completed the formation of the Isthmus of Panama. Instead, these islands are blocks of continental lithosphere, which have jostled through a complex series of movements through the Tertiary. Nearly all of the Antillean fossil primates are Quaternary in age, and some may have survived until human settlement of these islands. Based on a fossil primate talus found in Cuba, MacPhee and Iturralde-Vinent (1995) argue that primates are present in the Greater Antilles as far back as the Miocene. If so, this is accounted for by the continental nature of these islands. It is possible that primates reached these

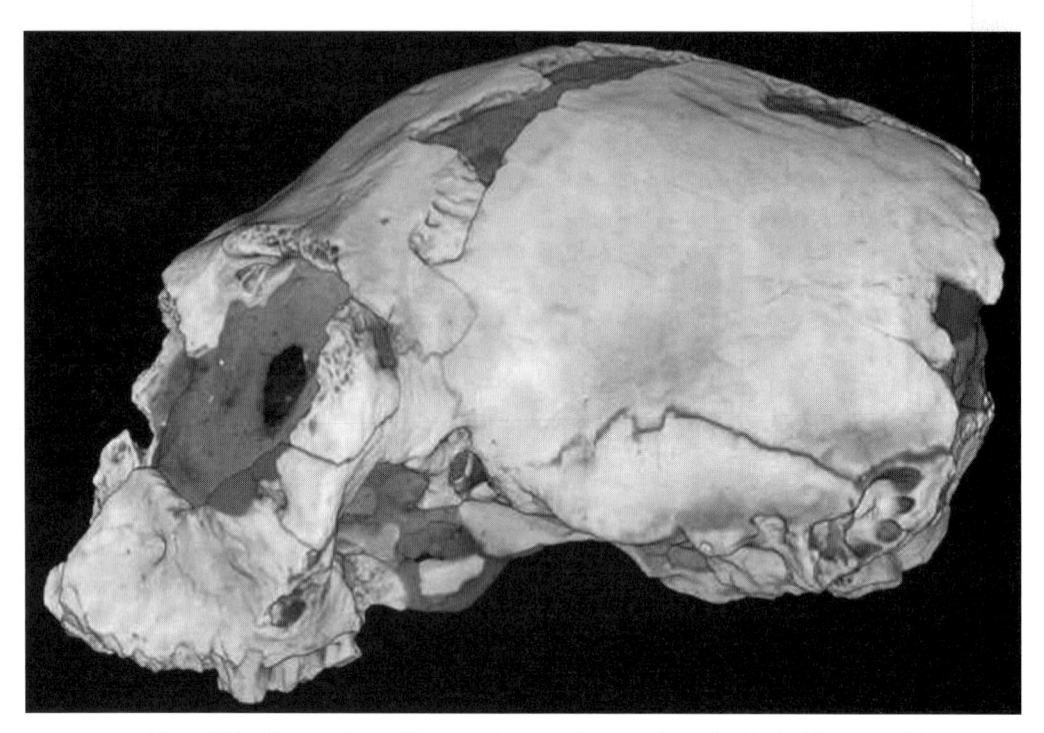

Figure 13.8. The cranium of *Tremacebus harringtoni*, from the Early Miocene of Argentina. This is the type specimen. Funding for the CT scan provided by a National Science Foundation grant to the University of Texas at Austin. Copyright © Digital Morphology Group, in conjunction with The University of Texas High-Resolution X-ray Computed Tomography Facility. http://digimorph.org/.

islands via a "Noah's Ark" dispersal—plate tectonic movements of the lithosphere passively disperse them through space (McKenna, 1973). Thus, primates need not cross a water gap.

Xenothrix mcgregori is found in Jamaica. It has a callitrichid dental formula of 2:1:3:2, but is far larger than any living marmoset or tamarin. It was between 2 and 4 kg in weight. The hip bone and other postcranial remains indicate that *Xenothrix* was a slow climber, and may have spent a good deal of time suspended in an inverted posture under tree branches, like living kinkajous (MacPhee & Fleagle, 1991). This type of locomotion is not seen in any living platyrrhine.

Other fossil platyrrhines occur in the Greater Antilles. *Antillothrix bernensis* is found in the Dominican Republic of Hispaniola, and *Paralouatta varonai* is found in Cuba. Scuba divers exploring underwater caverns have recently brought additional fossils of *Antillothrix* to the surface (Rosenberger, 2012a). In spite of the genus name, *Paralouatta* is now thought to have no affinities with the living howler monkeys. Its closest relative may have been *Antillothrix*—in fact, these three genera may have been closely related. MacPhee and Horovitz (2004) suggest that all of the fossil monkeys of the Greater Antilles were monophyletic, and were most closely linked to the living titi monkeys (*Callicebus* spp.). The disappearance of these fossil taxa is easily explained. Island species are particularly vulnerable to human hunting or human environmental disruption. They probably became extinct soon after Amerindians settled on these islands.

Although living platyrrhines are substantially smaller than catarrhines, fossil platyrrhines once existed that were far larger than living species. Pleistocene deposits within the enormous Toca de Boa Vista cave in the Brazilian state of Bahia yielded the nearly complete skeleton of *Protopithecus brasiliensis*. With a weight of 25–30 kg, it was twice the size of any living platyrrhine. Placed in the Subfamily Atelinae, it closely resembled a very large spider monkey, although it possessed the enlarged hyoid bone found in howler monkeys (Hartwig, 1995; Hartwig & Cartelle, 1996). The nearly complete skeleton of another unknown, slightly smaller species (*Caipora bambuiorum*) was discovered in the same cave deposits (Cartelle & Hartwig, 1996).

14 The Miocene hominoid radiation

Introduction

The earliest known hominoid is *Rukwapithecus fleaglei*, which comes from an Oligocene site in Tanzania dated to 25.2 mya (Stevens et al., 2013). It is known only from the type specimen. The genus *Kamoyapithecus* is far older (27.8–23.9 mya), but its attribution is uncertain—it is known only from isolated teeth. The first appearance of abundant hominoids is heralded by the genus *Proconsul*, which appears at 22.5–20 mya. Hominoids thus have a respectably ancient evolutionary history. Midway through this history, they experience an evolutionary radiation. As a set, they are currently teetering on the edge of extinction. Thus, the rise and fall of hominoids has been compressed within the last 15 million years.

The DNA of humans and their closest living relatives, all large-bodied hominoids, has obviously received intense scrutiny. Changes or mutations in mitochondrial and nuclear DNA have been used to construct hominoid phylogenetic trees. However, when hominoids are compared to other primate lineages, the branching length for hominoid lineages is much shorter. This is caused by a decrease in mutation rate per year in the hominoids. Humans, for example, have amassed 30 percent fewer mutations than baboons since their divergence from a last common ancestor (Scally & Durbin, 2012). The decrease in hominoid mutation rate has been termed the "hominoid slowdown," and it is known to have been caused by an increase in body size itself. Increasing body size increases generation length, and this alone creates a fall in mutation rate, because it lowers the population size of hominoid species. Hence the appearance of relatively large body size in hominoids, beginning in the Early Miocene about 20 mya, has fundamentally reduced the abundance of hominoid species.

Genome sequences have been completed for the large-bodied living hominoids—humans, common chimpanzees, bonobos, orangutans, and gorillas. Sequencing of the orangutan and gorilla genomes reveals what some taxonomists had long suspected: Bornean and Sumatran orangutans are two species (*Pongo pygmaeus* and *Pongo abelii*, respectively); lowland gorillas and mountain gorillas are also two species (*Gorilla gorilla* and *Gorilla beringei*, respectively). Orangutans appear to have been evolving at a significantly slower rate than other catarrhines, including rhesus macaques (Locke et al., 2011). About 500 protein-coding genes demonstrate heightened rates of parallel evolution in humans, gorillas, and chimpanzees (Scally et al., 2012). Genes involved in hearing are especially enhanced in humans and gorillas, indicating that selection for vocal communication is accelerated in these species. One entirely unexpected result of these genome comparisons is that chimpanzees and humans do not universally appear to be more similar to each other than either is to the gorilla. In about 30 percent of the genome, gorillas are more similar to humans than humans and chimpanzees are to each other (Scally et al., 2011). This is also true for the bonobo, where 1.6 percent of the genome is more similar to humans than to common chimpanzees (Prüfer et al., 2012). Common chimpanzees have an approximately equal degree of relatedness to humans, because 1.7 percent of the human genome shows a closer relationship to common chimpanzees than to bonobos. There are several ways to explain these results. One obvious answer is that gene flow or hybridization between ancestral species with incomplete species boundaries might have taken place. Because intense scrutiny of ancient and modern human DNA has revealed gene flow between modern humans, Neanderthals, Denisovans, and possibly the Red Deer Cave people in China, a complicated population history in closely related Late Miocene African hominoids is certainly possible.[1] In any case, these genetic data demonstrate how cladistic methodology and the phylogenetic species concept (Chapter 2) may not be congruent with the evidence of DNA or field biology.

After a decade of comparing the genomes of humans and other hominoids in order to discover the genetic bases for unique human traits, genetic duplications are now known to underpin human brain size. A gene called *SRGAP2* is active in brain development in all mammals. Like other mammals, non-human primates have one copy of this gene, but humans have three additional copies that are slightly different from each other. The ancestral gene stimulates the maturation of dendrites on the surface of neurons; one of the later copies of the gene increases the number of immature dendrites in humans. Delayed maturation of the dendrites increases their density. Comparing the human gene sequences with those of common chimpanzees and orangutans, it appears that the ancestral gene was duplicated at 3.4 mya, 2.4 mya, and 1 mya in the human lineage (Geschwind & Konopka, 2012). The copy of the gene that increases the number of immature dendrites, *SRGAP2C*, has a high level of protein expression and is essentially fixed in all human populations. It appears at 2.4 mya—a date that is associated with the origin of the human genus *Homo* from an australopithecine ancestor (Charrier et al., 2012; Dennis et al., 2012). This gene duplication may therefore be responsible for the increase in relative brain size that is characteristic of *Homo*, when compared to earlier hominid species.

It is ironic that great ape species are increasing through genomic research precisely when wild non-human hominoids are plummeting in numbers. We are living in a world in which hominoids are dying off. It is easy to overlook this fact, because the abundance and success of humans blinds us to the diminishment of

---

[1] An alternative explanation is that "incomplete lineage sorting" is occurring. That is, the use of DNA evidence from genetically mixed groups creates gene trees that do not reflect phylogenetic events. Inferring phylogeny is difficult under these conditions.

our fellow hominoids. Yet, they are dwindling down in numbers. Sometimes the die-off is so severe that it is impossible to pretend any more that these species will ever be able to exist outside zoos or colonies. For example, the Zaire strain of the Ebola virus was the confirmed cause of death for both gorillas and common chimpanzees in the Lossi Sanctuary in the northwestern Republic of the Congo. Gorilla mortality rates in 2002 and 2003 were 90–95 percent, and 5,000 gorillas were estimated to have died in the Lossi Sanctuary (Bermejo et al., 2006). There is no recovery from this rate of loss through disease—and other factors contribute to the loss. Besides pathogens, leopard predation, human poaching, and habitat loss and fragmentation contribute to the grim prospects of ape survivorship.

During the Miocene, however, hominoids experienced a great evolutionary radiation. Hominoids were widespread throughout the Old World, and occurred in many geographic areas (e.g. Europe, western Asia) where one would not expect to find apes today. Hominoids were speciose and widely dispersed. The major range expansion of hominoids occurred during the Middle Miocene, between 17 and 15 mya. The paleogeography of the earth during the Middle Miocene at 14 mya is shown in Figure 14.1. At this time, hominoid ranges expanded through much of Eurasia, and this dramatic spread led to increased speciation as newly encountered geographic barriers caused reproductive isolation. Hominoid range expansion is intimately linked to the spread of subtropical forests during the Middle Miocene. Macrobotanical and pollen data from Europe show the presence of subtropical and warm temperate tree species that flourished under warmer and wetter conditions than exist in Europe today. Furthermore, conditions were more equable, with far less seasonality than at present (Utescher et al., 2011). However, in order for subtropical forests to exist in Western and Central Europe, atmospheric CO_2 levels need to be higher than they are now. A computerized climate model that combines atmospheric composition and ocean circulation reveals details of paleoclimate in the Middle Miocene of Europe. The East Antarctic ice sheet was forming in the Middle Miocene, but was only 25 percent of its modern size. This reduced Antarctic ice sheet modified oceanic circulation, causing wetter conditions between 30° N and 50° N in Europe. The simultaneous interplay of higher atmospheric CO_2 and increased rainfall led to the spread of subtropical forests in Europe (Hamon et al., 2012). Declining atmospheric CO_2 levels and increasing Antarctic ice later in the Miocene cause the collapse of European subtropical forests and greater seasonality. Thus, at the end of the Miocene, hominoids, which were dependent on forested habitats and equable conditions, experienced a great extinction. They were not the only land mammals affected by major extinctions during the Late Miocene. In fact, one of the greatest extinction events in land mammal evolution occurs during the Late Miocene, beginning about 7 mya, with the worldwide expansion of grasslands. Nevertheless, Miocene hominoids—whose living representatives are generally celebrated for flexible and innovative behavior—were unable to shift habitats or diets in response to the elimination of subtropical forests and increasing seasonality. The principal cause

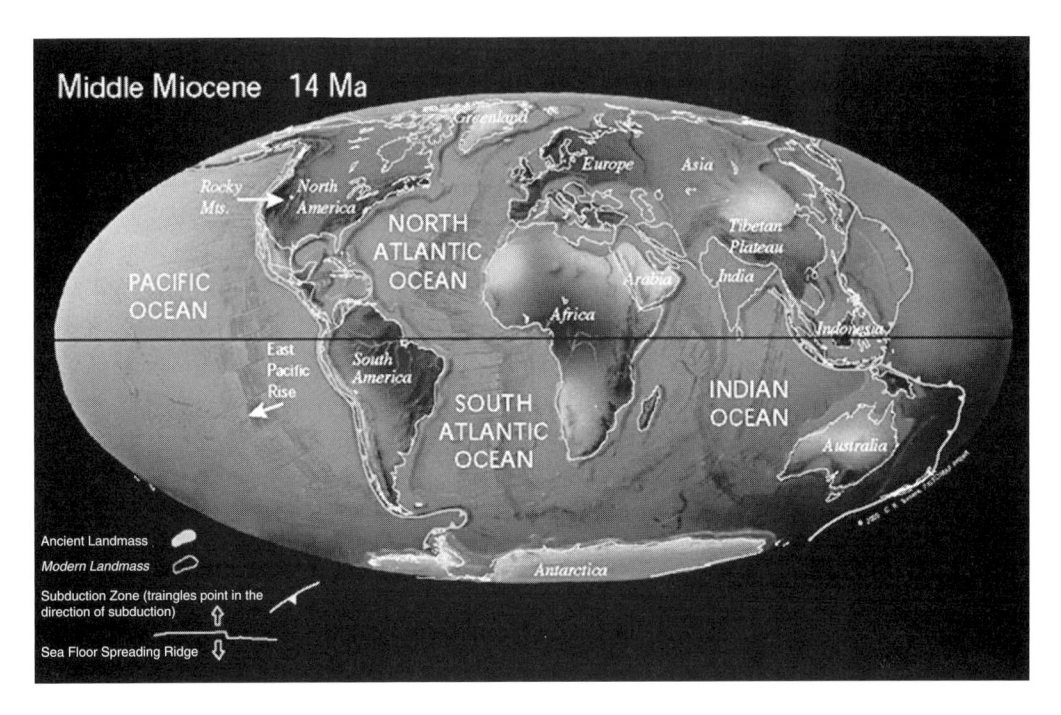

Figure 14.1. Paleogeography of the earth during the Middle Miocene, 14 mya. Courtesy of Dr. Christopher Scotese.

of hominoid decline during the Late Miocene was their inescapable dependence on warm, wet forested habitats and equable climate.

Postcranial anatomy of the crown hominoids

Crown hominoids have a suite of postcranial features that differentiate them from other primates, even the closely related cercopithecoids (Figures 14.2 and 14.3). The torso is broad, and is flattened anteroposteriorly. The ribs are strongly curved, which causes the vertebral column to protrude into the chest cavity. The scapula is located on the dorsum of the body, and the clavicle is long. There is virtually no bony support of the shoulder joint, so that hominoids are capable of raising and rotating the shoulder joint in a nearly 360° arc. The sternum is broad, and the sternebrae are fused in adults. The elbow joint allows for rotation of the forearm during pronation and supination. The styloid process of the ulna has retreated from the wrist joint, allowing for more lateral movement of the wrist. The iliac blades of the hip bone are shorter and broader. There is no tail. Ward (2013) suspects that many of these features are homoplasies that evolve independently in a number of lineages. She notes that even derived Late Miocene hominoids, such as *Pierolapithecus*, *Hispanopithecus*, and *Oreopithecus*, which were undoubtedly using suspensory posture and locomotion, have vertebrae that resemble those of hylobatids, rather than the great apes.

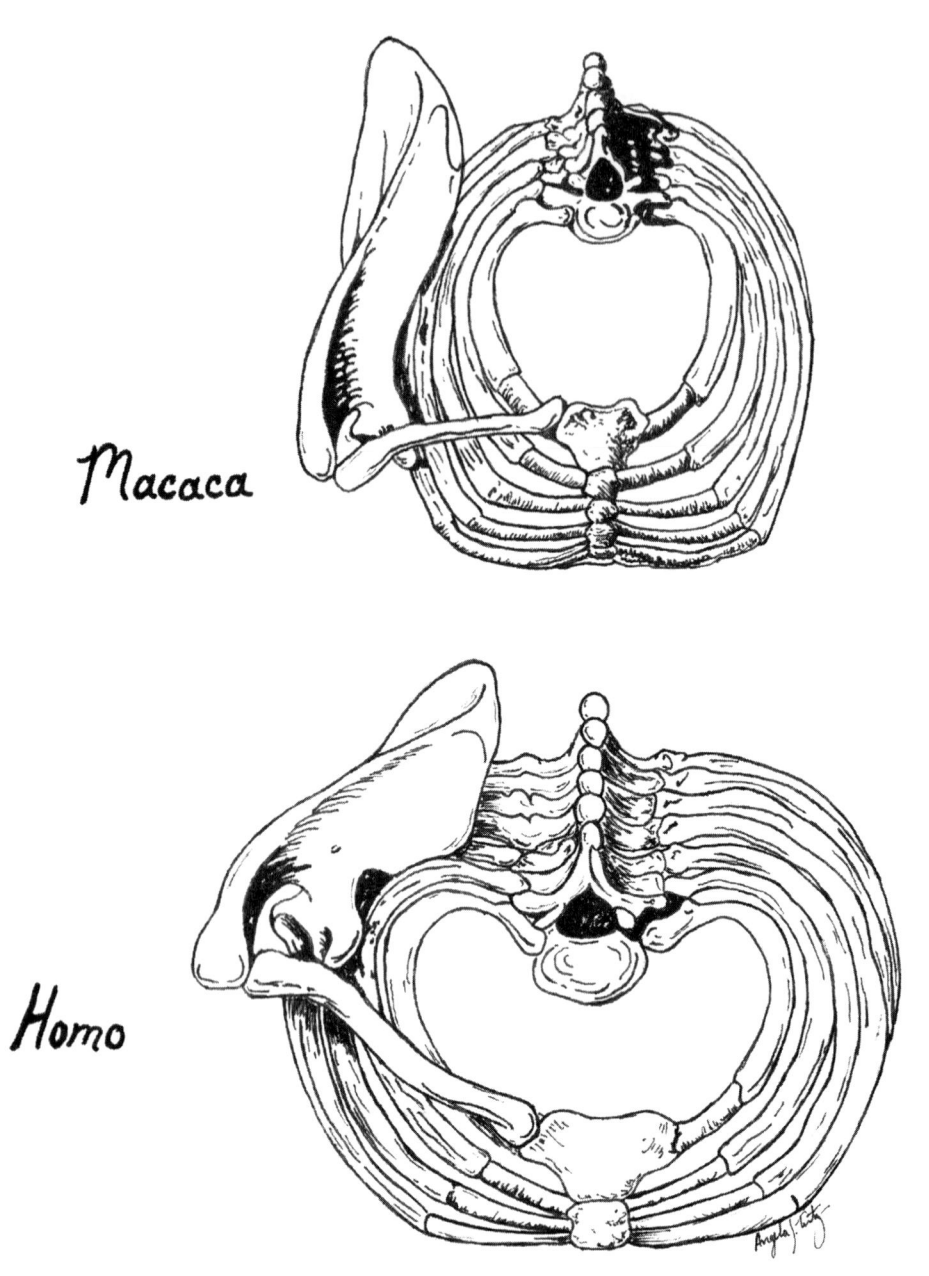

Figure 14.2. Superior view of the torso skeleton of a living hominoid (*Homo*) contrasted with that of a living cercopithecoid (*Macaca*). The cercopithecoid condition is primitive, and is found in most mammals. The hominoid torso is broad, and is anteroposteriorly flattened. The ribs are strongly curved, which results in the vertebral column protruding into the chest cavity. The scapula is located dorsally, rather than on the side of the body. There is virtually no bony support of the shoulder joint, which enables hominoids to rotate the shoulder freely in a nearly 360° arc. The clavicle is long. The sternum is broad, and the sternebrae fuse in adults. Most of these traits are not found in hominoids until the Late Miocene. Illustration by Angela J. Tritz.

Figure 14.3. The relative limb proportions in some living hominoid primates: *Hylobates* (the gibbon), *Pan* (the chimpanzee), *Pongo* (the orangutan), and *Gorilla* (the gorilla). Note the absence of a tail. All of the genera have elongated forelimbs, which is especially marked in *Hylobates*. In *Pongo*, the lower leg is shortened, which accentuates the length of the forelimbs. After Erikson (1963). Illustration by Angela J. Tritz.

Many of these hominoid postcranial features are thought to have evolved because of vertical climbing, or because of suspension of the body below a support and the use of the arms for propulsion. Body size relative to the diameter of an arboreal support affects posture and locomotion (Figure 14.4). Once body size has

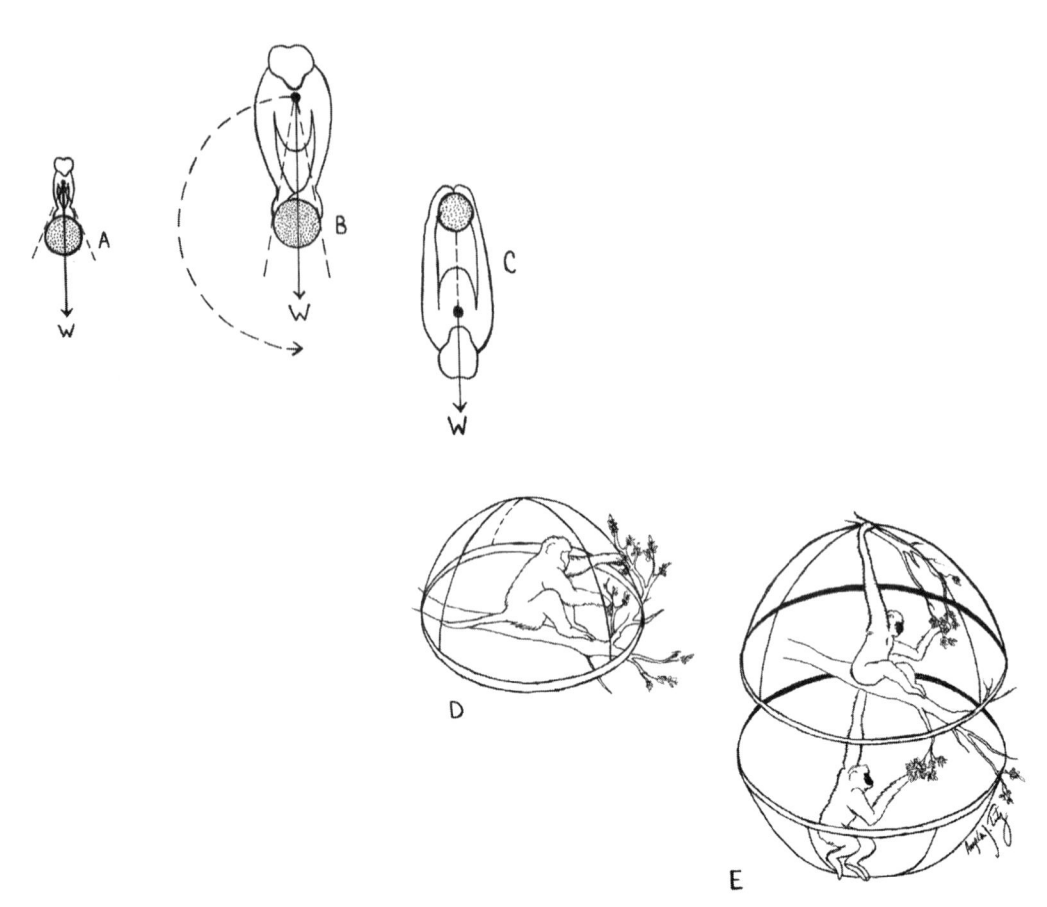

Figure 14.4. Foraging above and below branches is affected by body size. (A) An animal whose body is small relative to the diameter of an arboreal support can easily move quadrupedally over the support. (B) Instability occurs when body size increases relative to the diameter of the support. (C) Suspension of the body below the support is much more stable, but depends on grasping extremities. (D) A catarrhine primate that is a generalized arboreal quadruped (e.g. *Macaca*, *Papio*) can sit upright and forage with its hands. (E) As shown by a gibbon (*Hylobates*), elongated forelimbs increase foraging reach, regardless of whether an animal is sitting above or hanging below a branch. After Napier (1993: Figs 36 & 37). Illustration by Angela J. Tritz.

increased, suspension of the body below a support is a stable posture, but depends upon grasping extremities. Elongated forelimbs can increase the reach of a foraging animal, regardless of whether it is sitting above or hanging below a branch.

Discovery of the hominoid radiation

Knowledge of the existence of fossil apes extends deep into the history of paleontology. The first ape fossil, a mandible of *Dryopithecus fontani*, was unearthed in France, in 1856. This pre-dated the discovery of natural selection by Darwin and Wallace, and thus the occurrence of a fossil ape did not tempt paleontologists to muse about human origins. However, an ape in Europe was unusual, as was its habitat. Macrobotanical remains indicated the presence of a forest with temperate trees, such as oaks. This led to the ape's genus name: *Dryopithecus* translates to "tree ape" when rendered from the Greek. Further discoveries of this type of ape were made in Europe. In 1932, an expedition from Yale University led by G. Evelyn Hutchinson was sent to northern India to probe the mysteries of Gondwana geology and paleontology. The expedition succeeded in finding numerous ape jaws and teeth. These were given genus names based on those of the Hindu gods. Taxonomic inflation occurred, because sexual dimorphism and population-level variability was not taken into account when these fossils were described. Nevertheless, it was clear that apes had been much more speciose and widely dispersed than they are today.

In the early twentieth century, Gregory and Hellman (1926) conducted a major review of the dental and jaw fragments of these fossil apes, classified as members of the Family Pongidae, Subfamily Dryopithecinae. They argued that these dryopithecine apes had a fundamental role in hominoid evolution. The fossils had dental traits that were also found in living hominoids. Gregory and Hellman therefore concluded that the ancestors of all living hominoids were derived from fossil dryopithecines. In particular, the lower molar teeth of these fossil apes had a distinctive pattern of five cusps separated by a Y-shaped groove (Figure 14.5). This was the Y-5 or dryopithecine pattern, and it is a derived trait that links together all of the living hominoids. Later discoveries demonstrate that the dryopithecines also had the derived postcranial anatomy found in all living hominoids. These postcranial traits had been accreting in fossils since the Early Miocene, but dryopithecines exhibit the full constellation. The position of the dryopithecines is thus secure as a general ancestral group to living hominoids. However, there is no trail of fossils that lead clearly from living taxa to this Miocene group.

Nearly 40 years later, a major taxonomic revision of the dryopithecine apes took place. This revision drastically reduced the recognized number of dryopithecine genera and species (Simons & Pilbeam, 1965:142–143). The authors pared about 28 genera and over 50 species down to the genus *Dryopithecus* and several additional genera (*Gigantopithecus, Ramapithecus, Pliopithecus,* and *Prohylobates*). Fossils placed in the genus *Dryopithecus* were divided into subgenera (*Dryopithecus, Sivapithecus,* and *Proconsul*) that were linked to geography. *Dryopithecus (Dryopithecus) fontani* and *D. (D.) laietanus* occurred in Europe; *Dryopithecus (Sivapithecus) indicus* and *D. (S.) sivalensis* occurred in Asia; and *Dryopithecus (Proconsul) africanus, D. (P.) nyanzae,* and *D. (P.) major* occurred in Africa. Beyond reversing the trend toward taxonomic inflation, this revision accomplished two things. It promoted the idea that the dryopithecines had been a relatively uniform group of apes, in spite of their wide geographic dispersal. It also highlighted the distinctiveness of ape fossils assigned to different genera. This was particularly the case for *Gigantopithecus* and *Ramapithecus*. *Gigantopithecus* is still recognized as a peculiar outlier of the hominoid radiation

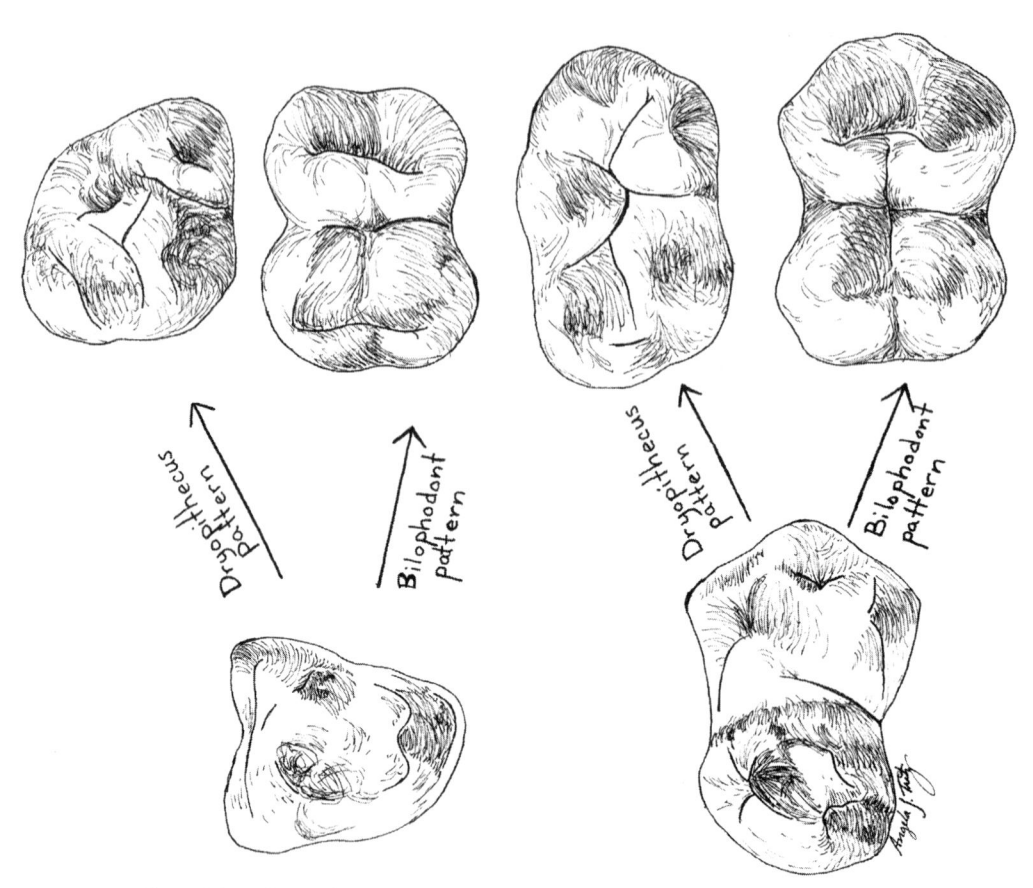

Figure 14.5. The evolution of molar patterns in hominoids and cercopithecoids. These patterns are derived from the ancestral morphology, shown at the bottom. Anterior (mesial) is at the bottom, external (buccal) is on the left. Upper molars are shown on the left of the top row. The bilophodont pattern of Old World monkeys exhibits two lophs or crests, one anterior (mesial) and one posterior (distal). Many species (e.g. species of *Macaca*, *Papio*, *Lophocebus*) preserve a hypoconulid in the lower M3. Cercopithecine molars are shown. Colobine molars have a more exaggerated bilophodont pattern, in which waisting occurs between the anterior and posterior lophs. The tooth then resembles a figure eight when seen from the occlusal surface. In addition, colobines have relatively higher (hypsodont) molar crowns, when measured against total height of the tooth. Relative depth of the mandibular ramus anchoring the roots of the lower teeth may also be greater in colobines. The lower molars of hominoids exhibit the Y-5 or "dryopithecine" pattern first recognized by Gregory and Hellman (1926). Illustration by Angela J. Tritz.

(see below). *Ramapithecus* was originally widely touted as the first hominid (Simons, 1961; Simons & Pilbeam, 1965), but has now lost that designation.

The history of the study of *Ramapithecus* illuminates how compulsive and faddish the search for human origins can be. In 1932, G. Edward Lewis, a Yale graduate student, discovered an upper jaw fragment of an ape from northern India that he named *Ramapithecus brevirostris*. In 1937, Lewis recognized this fragment as the first member of the Family Hominidae in his unpublished dissertation.

Small canine size was the derived trait that led to the hominid diagnosis. Simons (1961) resurrected this idea. *Ramapithecus* was also thought to have had a maxillary dental arcade that was parabolic in shape—another hominid-like trait. However, Vogel (1975) clearly showed that the maxillary dental arcade of *Ramapithecus* was not parabolic. In spite of the absence of postcranial fossils, *Ramapithecus* was thought to be a terrestrial biped by virtue of a circuitous argument. Because the canine teeth were small, *Ramapithecus* must have had its hands free to carry weapons for defense against predators. The taxon was therefore bipedal. African *Kenyapithecus* material was later subsumed into *Ramapithecus* (Simons, 1969), and the presence of some open-country grassland in Middle Miocene sites in Africa and northern India was also used as a criterion of hominid status (Tattersall, 1969a, 1969b, 1975). For nearly 20 years, *Ramapithecus* reigned supreme as the undoubted first hominid in the paleontological literature (Simons, 1977). Since the 1980s, it has been demoted to dryopithecine status, and is considered to have affinities to Asian dryopithecines. The demotion of *Ramapithecus* is often considered to be the result of the molecular clock (Lewin, 1987). Using the molecular clock (Chapter 6), separation of the ancestral hominids from the chimpanzee lineage is dated at 5–6 mya. This time range is far later than the 14–15 mya date for *Ramapithecus*. Because an undoubted hominid species (*Orrorin tugenensis*) has been found at 6 mya, hominid origins clearly precede the date given by the molecular clock, although there is no evidence for a divergence in the Middle Miocene.

Despite the taxonomic condensation of Simons and Pilbeam (1965), taxonomic inflation has again occurred among the Miocene apes. In particular, the dryopithecine subgenera that they created have been abandoned. Three major genera (*Dryopithecus, Sivapithecus, Proconsul*) take their place in Europe, Asia, and Africa, respectively. This is a reversion to the classification used before the Simons and Pilbeam revision. There are two reasons for this taxonomic inflation. Because of chauvinism, some paleoanthropologists champion the unique status of fossils discovered in their country by giving them new genus or species names. Hence, Late Miocene fossils originally identified as the widespread species *Dryopithecus laietanus* now become *Hispanopithecus laietanus* in Spain and *Rudapithecus hungaricus* in Hungary. And there have also been discoveries of genuinely novel fossil material that warrant new taxonomic designations. For example, a much larger species of *Sivapithecus* (*Sivapithecus parvada*) dated to 10 mya has been discovered in Asia.

Inferring lifeways in fossil hominoids

The posture and locomotion of fossil hominoids have been extensively analyzed and debated. A major difficulty is that the postcranial anatomy typical of all living or crown hominoids appears late in time. In fact, most of these traits appear only in the Late Miocene, and are first identified as a suite in the Spanish taxon *Pierolapithecus catalaunicus*, dating to 13–12.5 mya. Derived hominoid anatomy is present, although the hand is short, with short, relatively straight fingers (Moyà-Solà *et al.*, 2004). *Hispanopithecus laietanus*, a later Spanish taxon, is even more advanced. It has a hand with long metacarpals and phalanges, and it has a long forelimb with a straight humeral shaft (Moyà-Solà & Köhler, 1996). Thus, most of the Miocene hominoid postcranial material outside Europe is more aptly described as anatomically primitive, coming from basal catarrhines.

Problems in interpreting the Miocene hominoid postcranium are exemplified by changes in description of the postcranial skeleton of the genus *Proconsul* (Figure 14.6). When originally described, *Proconsul* was thought to be an

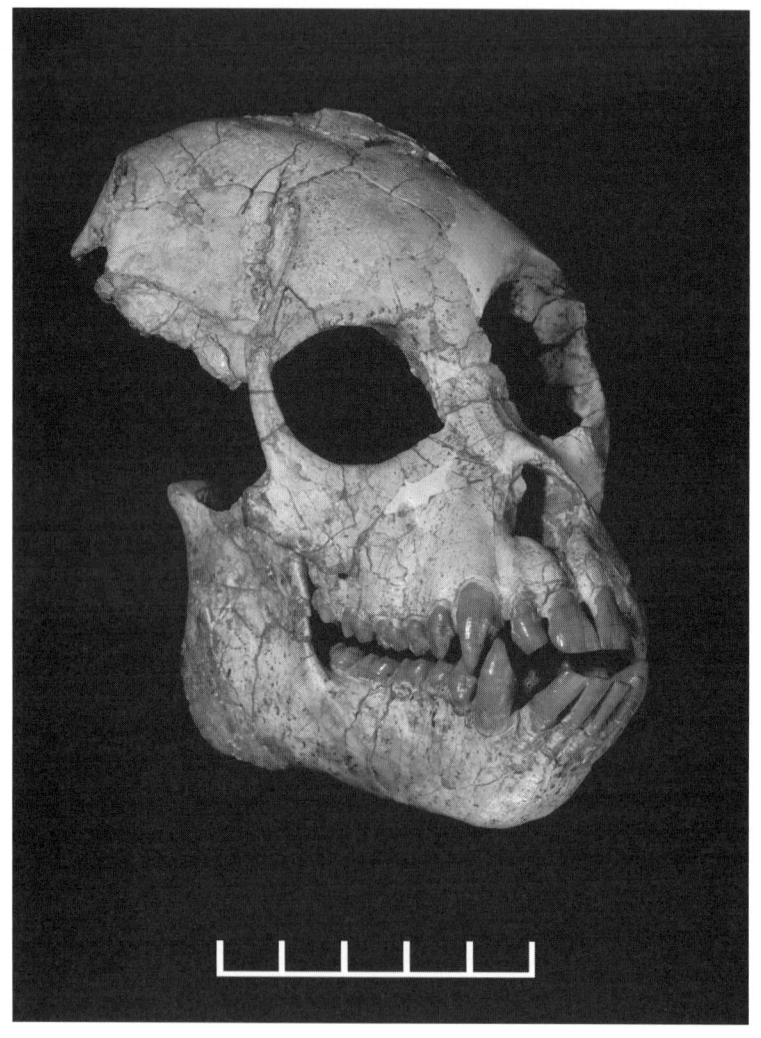

Figure 14.6. *Proconsul heseloni* (KNM-RU 7290), formerly classified as *Proconsul africanus*, from Rusinga Island, Western Kenya, about 18 mya. In 1981, additional fragments of the posterior neurocranium were recovered. Courtesy of the Mary Evans Picture Library/ Natural History Museum.

undoubted hominoid, and a possible ancestor of the living African great apes. The genus name is derived from the name of a chimpanzee named Consul, who was a famous denizen of the London Zoo at the time of the fossil's discovery. This idea persisted. During the 1960s, three taxa in the genus were thought to be directly linked to living hominoids, based largely on body size: *Proconsul major* was ancestral to *Gorilla*, *Proconsul nyanzae* was ancestral to *Pan*, and it was strongly hinted that *Proconsul africanus* (now *P. heseloni*) was ancestral to *Homo* (Pilbeam, 1969). However, when analyzed in detail, the postcranial anatomy of *Proconsul* revealed a troubling lack of traits typical of living hominoids. In fact, the postcranium of *Proconsul* was startlingly like that of cercopithecoid monkeys, leading some to the conclusion that these species were merely "dental apes" (Tuttle, 1974). The semicircular canals of *Proconsul heseloni* reveal that it engaged in moderately fast locomotion, similar to that of Old World monkeys like the rhesus macaque (*Macaca mulatta*) or the Japanese macaque (*Macaca fuscata*). These species are generalized arboreal quadrupeds that can sometimes erupt into agile leaping and climbing behaviors (Ryan *et al.*, 2012). This reconstruction for *Proconsul heseloni* is at odds with most studies of the postcranial material, which imply that it was a generalized arboreal quadruped, and moved in a slow, purposeful way. However, Ruff (2002) used the internal bony structure of the shafts of the humerus and femur to compare the relative strength of these long bones in *P. heseloni*. Because the femur was stronger than the humerus, he inferred that this species engaged in some leaping behavior. This confirms the semicircular canal evidence.

Inferring diet in fossil hominoids has long occupied the minds of paleontologists. Before the advent of stable isotope analysis, alternative methods of investigating diet were applied to fossil hominoids. The relative thickness of the enamel on the molar crowns was first used to argue whether tough, hard-object feeding was an important dietary component (Martin, 1985).

Niche structure in sympatric Early Miocene East African hominoids

In addition to species of *Proconsul*, many other hominoid taxa are present in the Early Miocene of East Africa (Figures 14.7–14.9). Because a number of these taxa occur together at the same site, it is possible to examine other aspects of the lifeways of these sympatric species. If these species are sympatric, how are they dividing up the resource space? This is a fundamental question that field biologists address when they study closely sympatric species today. This question involves niche structure and competition. This question might appear to be unsolvable to paleontologists, but one way that competition and niche structure has been studied in fossils is through Hutchinsonian ratios (Hutchinson, 1959; Cachel, 2006). These ratios, empirically discovered through field biology evidence, may specify the minimal amount of separation needed to decrease competition and create niche separation between closely related species. For linear measurements, the ratio of minimal separation is 1.3; for areal measurements, the ratio of

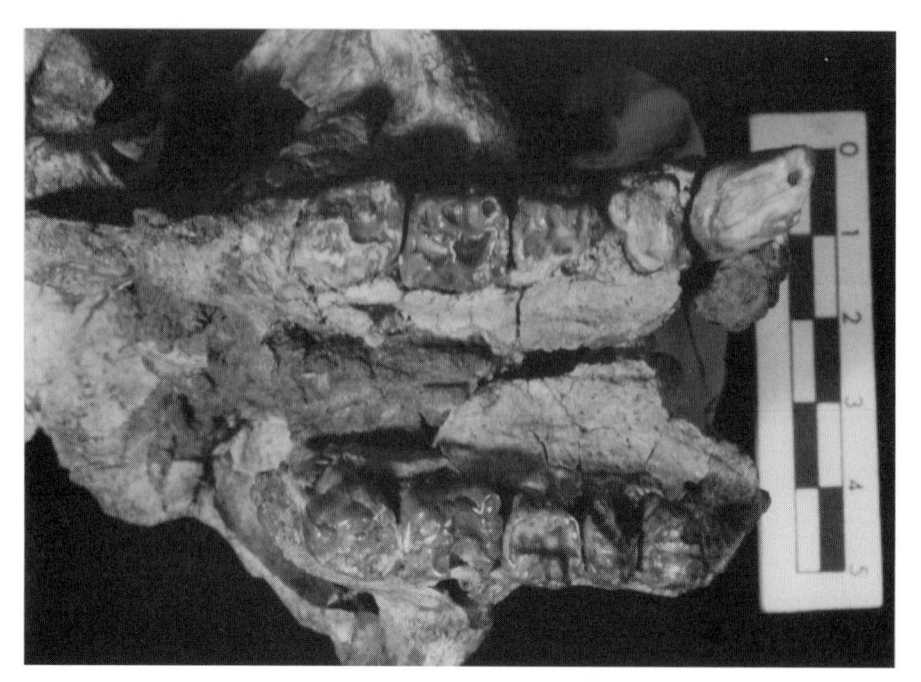

Figure 14.7. Maxillary dentition of *Turkanapithecus kalakolensis* (KNM-WK 16050A, type specimen). Centimeter scale.

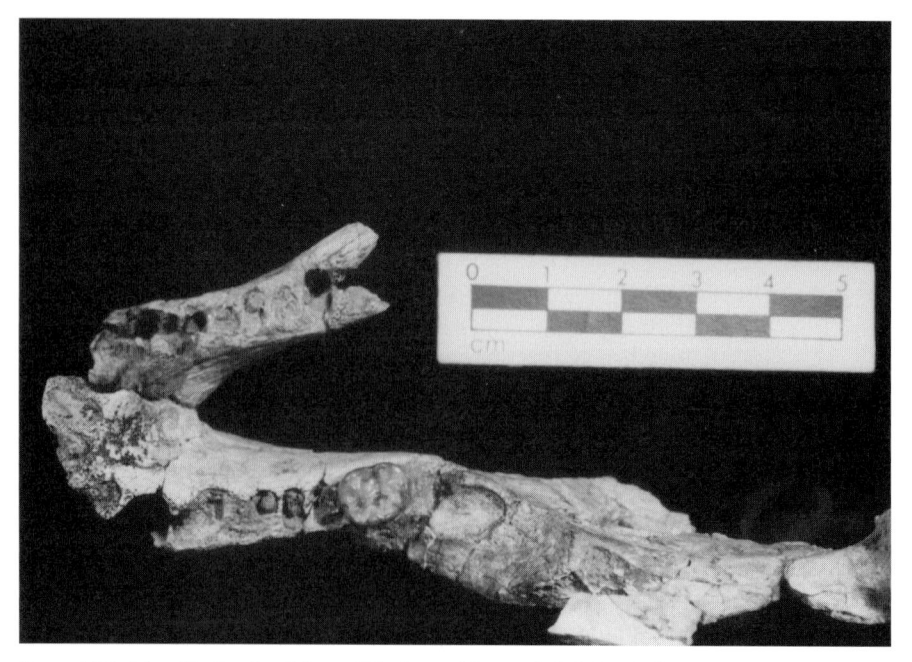

Figure 14.8. Mandibular dentition of *Turkanapithecus kalakolensis* (KNM-WK 16050B, type specimen). Centimeter scale.

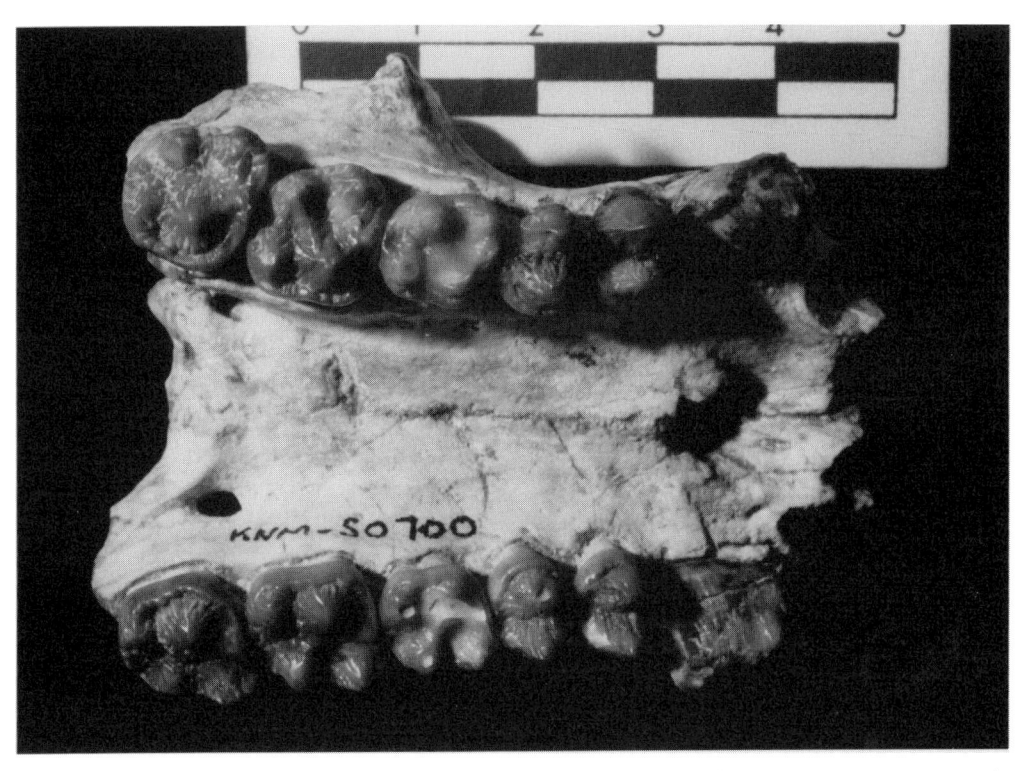

Figure 14.9. Maxillary dentition of *Rangwapithecus gordoni* (KNM-SO 700, type specimen). Centimeter scale.

minimal separation is 1.69; and for volume measurements, the ratio of minimal separation is 2.2. Note that 1.69 is the square of 1.3, and 2.2 is approximately the cube of 1.3. A statistical test determines whether sympatric taxa are at the minimal distance. If they exceed this distance, their niches are divergent enough to avoid competition.

I examined sympatric Early Miocene hominoid taxa at the East African Miocene sites of Legetet, Songhor, Rusinga Island, and Kalodirr (Cachel, 1996). Using linear and areal measurements of dentition, I determined that minimal distance separated *Afropithecus turkanensis* and *Turkanapithecus kalakolensis* at Kalodirr. At Songhor, *Rangwapithecus gordoni* and *Proconsul major* show minimal distance, as do *Rangwapithecus gordoni* and *Dendropithecus macinnesi*. Because *Rangwapithecus* shows minimal niche separation with two sympatric taxa, it appears to be a linchpin species among the hominoids at Songhor. In this light, it is interesting that the earliest known hominoid, *Rukwapithecus fleaglei*, from an Oligocene site in Tanzania dated to 25.2 mya, is thought to be ancestral to *Rangwapithecus* (Stevens *et al.*, 2013). The antiquity of the *Rangwapithecus* lineage could signal its generalized niche structure, and hence account for the minimal niche separation with two other taxa. Later hominoids in East Africa develop more advanced dietary traits.

Ancestors for the living apes?

As is true for many living primates, there is a long history of publications trumpeting the discovery of the first chimpanzee, gorilla, orangutan, gibbon, and human. In part this is caused by the general tendency for paleontologists to play "connect the dots" when confronted by fossil evidence. We know that living species exist, and must have had ancestors. If a fossil can be located in a suitable time frame, there is a bias towards connecting fossil species to living species. This tendency is exacerbated when dealing with primate fossils, because humans are primates, and we wish to discern the outlines of our own evolutionary history.

Evidence of the first chimpanzee appears late in time, at 500,000 yrs B.P. This unequivocal evidence is based on isolated teeth found in west Kenya (McBrearty & Jablonski, 2005). *Chororapithecus abyssinicus* (10.5–10.0 mya) from Ethiopia has been suggested as the first gorilla. There are nine isolated teeth, which are comparable in size to those of the living gorilla. The molar teeth are not well preserved, but enough remains to show that they have higher crowns and thicker enamel than in modern gorillas. These may be primitive hominoid traits, which would not preclude *Chororapithecus* from gorilla ancestry. The major point about these two genera is that the fossils occur in areas where their presumed descendants do not. This further highlights the relict nature of the living hominoids.

Orangutan ancestry is now straightforward. *Sivapithecus indicus* was long thought to be the ancestral orangutan for two reasons: it was an Asian dryopithecine, and became a putative first orangutan because it occurred in the same continent, and because its facial architecture resembled that of the living orangutan (Pilbeam, 1982). However, the shape of the orbits, the supraorbital tori, and the shape of the lower face are convergent between *Sivapithecus indicus* and the orangutan. *Ouranopithecus* from Macedonia also has a lower face and incisor morphology that resembles that of the orangutan. This morphology may have been widespread among Eurasian Miocene hominoids. The most telling point is that *Sivapithecus* has terrestrial adaptations in its postcranial anatomy—not what one would expect in an ancestral orangutan, which is the most arboreal large, modern hominoid. *Sivapithecus* has low humeral torsion, which implies a narrow thorax, and a scapula that is ventrally located; the phalanges are also short and straight, and the hallux is relatively small (Ward, 2013). The best current candidate for first orangutan is the genus *Khoratpithecus*. The 9-7-my-old ape *Khoratpithecus piriyai* dates from the Late Miocene of Thailand, which also has a Pleistocene record for the orangutan. *Khoratpithecus* shares both dental and mandibular traits with the orangutan (Chaimanee et al., 2006). The teeth of *Khoratpithecus* have marked similarities to those of the orangutan, in terms of enamel thickness, degree of sexual dimorphism, and enamel wrinkling. However, the most important derived trait that it shares with living orangutans is the lack of an anterior digastric muscle (Chaimanee et al., 2004). This trait is unique to orangutans among living primates (Cachel, 1984). An earlier Miocene species, *Khoratpithecus chiangmuanensis*, dates to 13.5–10 mya in Thailand (Chaimanee *et al.*, 2003). The fossil evidence thus indicates that the orangutan lineage diverged from other great apes during the Middle Miocene. This is confirmed by an estimated divergence date of 16–12 mya based on nucleotide differences from comparative primate genomics (Locke *et al.*, 2011:Fig. 1). The molecular clock yields a different result. Divergence dates based on the molecular clock (Chapter 3) nearly always precede the dates given by the fossils themselves. This is also true for primates, including hominoids (Chapter 7). Yet recent work on the genomes of individual humans clearly demonstrates that the mutation rate in modern humans is only about half the mutation rate used for nearly 50 years to construct molecular clocks. If the new, slower human mutation rate and known generation time in wild African apes are combined to measure hominoid evolution, the rate of evolutionary change then sluggishly creeps along (Langergraber *et al.*, 2012). The divergence of the orangutan lineage occurs between 46 and 34 mya (Gibbons, 2012:190), creating a gap of 32–20 my between the time of origin given by the molecular clock and the *Khoratpithecus* fossils. Because the molecular clock does not measure time with a metronome-like regularity, well-dated fossils currently trump the evidence of the molecular clock, even if the fossils are not those of the original founding population. *Khoratpithecus* pins orangutan origins at 13.5 mya.

Where are the ancestral hylobatids? Various taxa have been advanced and then dropped as contender for the earliest hylobatid. "*Aeolopithecus*" *chirobates* (now *Propliopithecus chirobates*) was once offered as the earliest gibbon (Simons, 1967), but is now recognized as a stem catarrhine. At present, the best contender for earliest hylobatid is the genus *Yuanmoupithecus*, dating between 8.2 and 7.1 mya from Leilao, in the Yuanmou Basin, Yunnan Province, China (Harrison *et al.*, 2008). No detailed descriptions of this material have been published. The 13.7 mya genus *Kenyapithecus wickeri* from Fort Ternan, Kenya, has also been identified as a possible stem hylobatid (Harrison, 2010a). An earlier species of this genus (16.5–16 mya) is found at the site of Paşalar, Turkey. However, a distal humerus from Fort Ternan, identified as possibly belonging to *Kenyapithecus*, shows indications of coming from a semi-terrestrial quadruped, even though the elbow joint has the rotational ability of living hominoids. It is difficult to imagine a hylobatid ancestor with such terrestrial propensities. Of course the distal humerus may not belong to *Kenyapithecus*. However, other features militate against a hylobatid classification for *Kenyapithecus*. McCrossin and Benefit (1993, 1997) argue that adaptations of the anterior teeth indicate that *Kenyapithecus* was a specialized seed predator like modern sakis.

Late Miocene extinctions

There are a number of fossil hominoid sites that allow for a detailed reconstruction of paleoclimate and paleoenvironment. An example of such a site is Rudabánya, in northern Hungary (Figures 14.10–14.13). At 10 mya, the site was located on a peninsula along the shoreline of a gigantic lake, the Pannonian Lake. Fossils of

Figure 14.10. Students at the University of Toronto Paleoanthropology Field School conduct excavations at the Miocene site of Rudabánya, Hungary, under the direction of Dr. David Begun. Courtesy of Ms. Darshana Shapiro.

the derived hominoid *Rudapithecus hungaricus* and a more primitive pliopithecid hominoid, *Anapithecus*, occur in deposits indicating the existence of a subtropical swamp forest with hackberry (*Celtis*) and swamp cypress (*Taxodium*) trees (Kordos & Begun, 2002).

However, humid, subtropical forests soon disappear in Central and Western Europe. The record of vegetation change is best preserved in Spain. A major shift in forest composition occurs here, beginning at 9.6 mya. Nearly 45 percent of the taxa after this shift are deciduous trees that are now familiar components of temperate forests, such as alders, oaks, maples, elms, and walnuts (Agustí & Antón, 2002). Macrobotanical remains at the site of Can Llobateres 1 in northern Spain preserves the best evidence of this shift in vegetation (Marmi *et al.*, 2012). *Hispanopithecus laietanus* occurs here, and is found in other Western European sites dating from 12–9 mya. Initial conditions at Can Llobateres 1 are very humid, with local development of marshes; forests are dense, with palms, laurels, and figs, as well as deciduous trees. However, deciduous trees become dominant starting in the Late Vallesian. This signals the end of the largely frugivorous *Hispanopithecus*.

Figure 14.11. Rutgers graduate student Ms. Darshana Shapiro triumphantly uncovers the scapula of a fossil rhinoceros during excavations at Rudabánya, Hungary. Courtesy of Ms. Darshana Shapiro.

In southern Pakistan, *Sivapithecus* also becomes extinct in the Late Miocene, and is supplanted by colobine monkeys. In this area, another source of habitat and climatic disruption was the rise of the Tibetan Plateau, caused by the sliding of the Indian tectonic plate under the southern borders of the Asian plate (Figures 14.14 and 14.15). The dramatic elevation of Tibet altered climate worldwide through the decline of global temperatures, but the elevation of this plateau created local mountainous topography, and may have triggered the onset of modern monsoonal rain cycles. Precipitation would thus become very seasonal.

In China, three species of the genus *Lufengpithecus* survive to the end of the Miocene. These hominoid species have been found in localities in Yunnan Province, in southwestern China. The earliest record of the genus is at 11–10 mya. The taxon *Lufengpithecus lufengensis* from the site of Shihuiba dates to 6.9–6.2 mya, and represents the latest surviving species in the genus. The available fossil evidence, including juvenile crania, reveals that *Lufengpithecus* is not a member of any extant hominoid lineage—in particular, it has no relationship to the living orangutan. Juvenile cranial materials from two species of *Lufengpithecus* are markedly different. Because this variability exists within a narrow geographic range and time interval, it appears that endemic species were evolving within a

Figure 14.12. Reconstruction of *Rudapithecus hungaricus*, shown in a Miocene swamp forest at Rudabánya, Hungary. Note the orthograde position of the trunk when the animal is resting. This is typical for all living (crown) catarrhines. Illustration by Angela J. Tritz.

landscape undergoing mountainous uplift at the eastern edges of the Tibetan Plateau (Ji et al., 2013). Complex mountain topography caused local, *in situ* evolution. Suvivorship of hominoids to this late date was possible because the forests of southwestern China were not as strongly affected by climatic oscillations as the forests of Europe and western Asia were. However, *Mesopithecus*, the first cercopithecoid monkey genus in Asia, is found at 6.9–6.2 mya in Shuitangba, Yunnan Province. Contemporary with *Lufengpithecus lufengensis*, *Mesopithecus* signals the looming rise of the cercopithecoid primates and the decline of the hominoids in the Old World.

The Late Miocene of Africa has only rare and fragmentary hominoid fossils, such as *Nakalipithecus nakayamai*, dating to about 10 mya (Figure 14.16). At this late date, hominoid fossils in Africa are principally scrutinized as potential ancestors for the African great apes or as potential ancestors for hominids. However, no convincing evidence for such ancestor–descendant relationships exists.

At the end of the Miocene, 74 percent of land mammal genera and 18 percent of land mammal families go extinct—truly, the scythe of the Grim Reaper was active at this time. Further extinctions of large mammals (the "megafaunal" extinctions) take place in the Pleistocene. Hence, the world of mammals that we observe

Figure 14.13. A swamp cypress (*Taxodium distichum*), typical of swamp forests in the southern USA. If these trees are permanently inundated, their trunks develop pronounced buttresses. Swamp cypress remains are found at Rudabánya, Hungary. Swamp cypress trunks with buttresses are also found buried in later, 8-my-old Miocene sediments in Hungary. Botanical Gardens, University of Bonn.

today is essentially a relict world. Our understanding of ecosystem structure and niche division in living mammals needs to recognize that much has been winnowed out.

Oreopithecus bambolii

The European species *Oreopithecus bambolii* is one of the last survivors of the Miocene hominoid radiation. From 9–6.5 mya, *Oreopithecus bambolii* inhabited a large Mediterranean island composed of northern Italy (Tuscany), Corsica, and Sardinia. This was known as the Tusco-Sardinian Island. Along with *Sivapithecus parvada*, *Lufengpithecus*, and *Gigantopithecus*, *Oreopithecus* became one of the few European and Asian hominoids to survive those paleoenvironmental changes at the end of the Miocene known as the "Vallesian Crisis" (Chapter 16). Crushed and flattened bones of *Oreopithecus* were first unearthed in the 1870s by Tuscan coal miners working in lignite deposits at a number of sites in northern Italy. Many of the gruesomely crushed bones were discarded by the miners, who

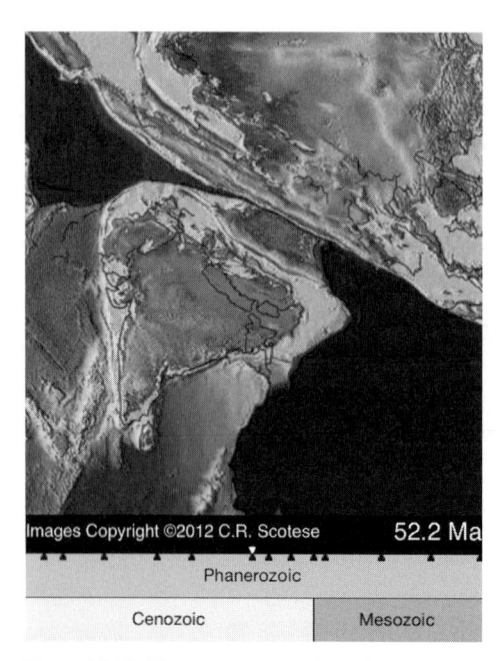

Figure 14.14. The near approach and eventual collision of the Indian lithospheric plate with the Asian lithospheric plate. This map is dated to 52.2 mya. The movement of the northern portion of India sliding underneath the southern portion of the Asian plate has resulted in the creation of the Tibetan Plateau. The rise of this plateau has contributed to the decline of global temperatures and increasing seasonality of the Late Cenozoic. Courtesy of Dr. Christopher Scotese.

apparently discovered a mother lode of these ancient apes. They are the most abundant mammal species in the deposits. The taxonomic status of this species has been highly controversial (Delson, 1986). Since its discovery, it has been identified as a cercopithecoid monkey, a hominoid, a hominid, and a late descendant of the Fayum genus *Apidium*. The *Apidium–Oreopithecus* connection was established by virtue of the fact that both taxa possess a centroconid, a centrally positioned extra cusp on the lower molars (Simons, 1960). Many researchers classified *Oreopithecus* as an aberrant cercopithecoid monkey, principally because of the transverse pairing of the molar cusps, cresting on the upper and lower molars, and absence of the Y-5 pattern (Delson, 1979; Szalay & Delson, 1979). It was sorted into a separate Family Oreopithecidae within the Superfamily Cercopithecoidea. Gregory (1951) argued that it was a relic of a basal catarrhine radiation that occurred before the separation of cercopithecoids and hominoids. Detailed description of a new slab and counter-slab containing a crushed skeleton of *Oreopithecus* emphasized a possible connection to hominids, because the face was short, the canines were small, and the hip bones were short and broad (Hürzeler, 1958, 1960). Other researchers assessed the new material and denied that the taxon was a hominid (Straus, 1963). Current assessments of the

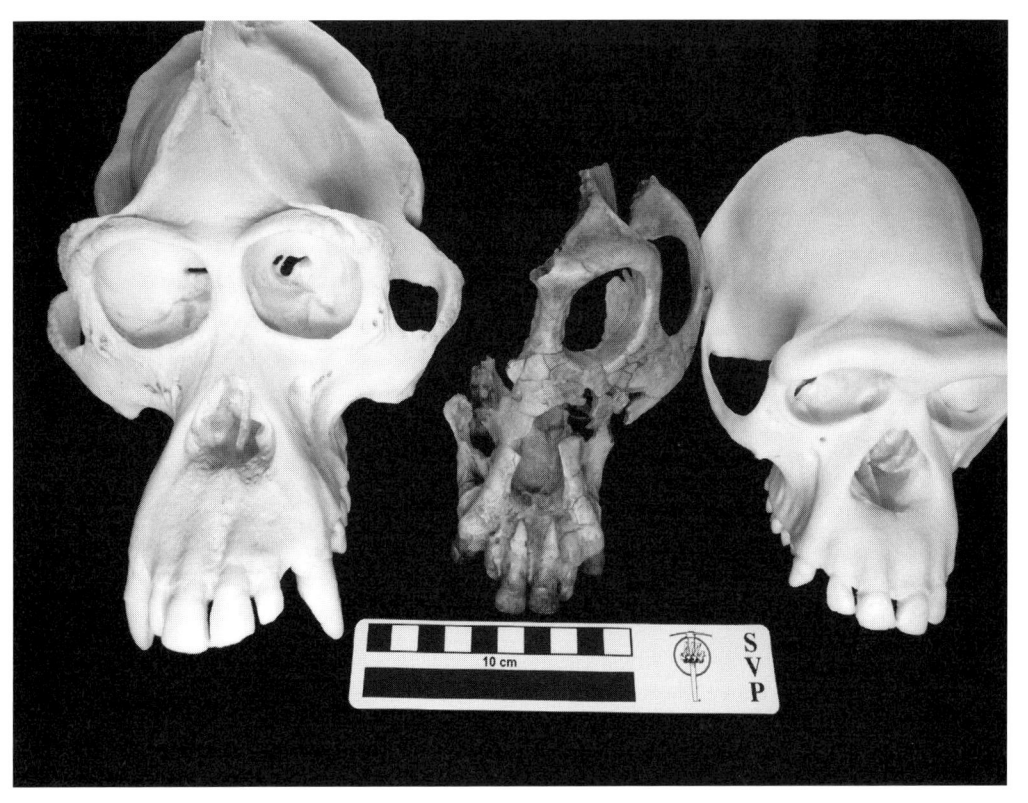

Figure 14.15. Cast of the cranium of *Sivapithecus indicus* (GSP 15000), Potwar Plateau, Pakistan, 8 mya. Note how the conformation of the orbit, lower face, and procumbent incisors resembles that of the adult male orangutan (left), rather than that of the adult female common chimpanzee (right).

taxonomic status of *Oreopithecus* recognize its unique status by classifying it in a separate Family Oreopithecidae within the Superfamily Hominoidea.

Like the modern orangutan, *Oreopithecus* had short hindlimbs and long forelimbs. However, its pelvic anatomy and foot anatomy indicate that it had evolved specializations for a unique type of terrestrial bipedalism (Köhler & Moyà-Solà, 1997). That *Oreopithecus* engaged in bipedal behavior is revealed by the structure of the cortical bone of the ilium. The internal trabecular structure of the ilium strikingly approximates that of living humans and fossil humans known as the australopithecines (Rook *et al.*, 1999). This trabecular orientation indicates that body weight was transferred through the ilium in a manner appropriate for a biped. However, there was no lordosis in the lumbar region, and the overall lumbosacral region does not appear to be adapted for habitual bipedalism (Russo & Shapiro, 2013). The foot structure of *Oreopithecus* was unique among primates – it had very short lateral rays and a widely abducted hallux. The foot consequently functioned as a tripod. The three points of the tripod were formed by the short lateral rays, the hallux, and the calcaneus. *Oreopithecus* thus slowly stumped about on the ground, using feet that functioned as tripods. The predator-free

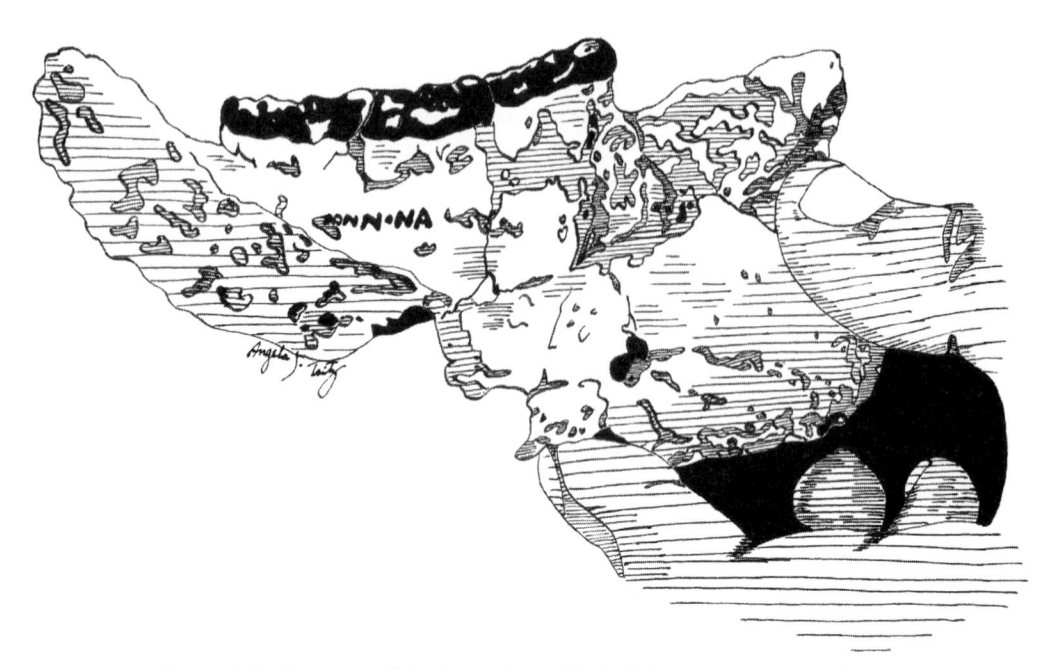

Figure 14.16. Fragment of the lower jaw of *Nakalipithecus nakayamai*, from the Late Miocene of Kenya, about 10 mya. Illustration by Angela J. Tritz.

Tusco-Sardinian Island was especially conducive for experimentation with facultative terrestrial bipedal locomotion. Because the unique anatomy of *Oreopithecus* is not ancestral to that of the australopithecines, this taxon unequivocally indicates that a Late Miocene hominoid was independently experimenting with terrestrial bipedalism. Moyà-Solà *et al.* (1999) argue that the hand of *Oreopithecus* was capable of a precision grip, but this is denied by Susman (2004), who argued that the hand morphology indicated a power grip and arboreal locomotion and suspension. The hand phalanges are curved, indicating powerful flexor muscles in the hand, but the distal phalanges are wide. This indicates that some manipulation was occurring. Presumably food items of some sort were being handled.

Oreopithecus weighed approximately 30–35 kg. There are locomotor specializations indicating arboreality, but, as discussed above, there are also indications of terrestrial bipedality and extensive use of the hands in manipulation. Because the remains of *Oreopithecus* are found in lignite deposits, members of this species apparently lived in swamp forests. After dying, these animals plummeted from the trees into the underlying swamp. The Tusco-Sardinian Island was devoid of large predators, except for otters and bears. However, when the island was re-connected to the European mainland at 6.5 mya, *Oreopithecus* quickly went extinct. The vulnerability of this bipedal, slow-moving terrestrial primate to predation illuminates circumstances surrounding the origins of hominids.

The earliest undoubted hominids, the australopithecines, also had a mosaic of arboreal and bipedal specializations. Their bipedality, especially as revealed by the anatomy of the lower limb and foot, was clearly more advanced than that of *Oreopithecus*, and was obviously antecedent to that seen in genus *Homo*. Australopithecine anatomy was stable for about 3 my—this group was therefore not a short-lived and aberrant stage of human evolution. During this time, australopithecines were engaging in terrestrial bipedal locomotion in an environment rife with large terrestrial carnivores. Unlike *Oreopithecus*, they did not go extinct. This clearly indicates that the australopithecines possessed antipredator strategies that most other primates do not have. From this, I have inferred the existence of complex sociality and behaviors in the australopithecines (e.g. vigilance and sentinel behavior) that are found in later hominids (Cachel, 2006).

Last survivor: *Gigantopithecus*

The largest primate that ever lived was the extinct giant ape *Gigantopithecus blacki*, found in the extreme Far East. Besides its large body size, the species is interesting for a number of reasons. It is the descendant of a much smaller species (*Gigantopithecus giganteus* [*G. "bilaspurensis"*]). Based on a single molar tooth, *G. giganteus* was originally described as a new species (*Dryopithecus giganteus*) in 1915. A mandible of *G. "bilaspurensis"* was described in 1969 (Simons & Chopra, 1969). Following the International Code of Zoological Nomenclature (Chapter 2), the official species name reverted to *G. giganteus* when researchers understood that the mandible belonged to the same species as the original type specimen.

Gigantopithecus giganteus is found in the Late Miocene of northern India, and dates to 8.5 mya. Thus, the lineage demonstrates size increase through time, a trend common in mammalian evolution. *Gigantopithecus blacki* itself lived from the Early Pleistocene, 1 mya, to the Late Pleistocene. Its last representatives became extinct about 100,000 years ago. It thus has the dubious distinction of being the latest known survivor of the great Miocene hominoid radiation (Figure 14.17). The species was sympatric with the hominid species *Homo erectus* and *Homo sapiens* in China, and some paleoanthropologists speculate that hominids hunted *Gigantopithecus* to extinction. There is no evidence for this, however—one would need to find bones with cut-marks or percussion marks signaling hominid butchery, or bones charred by fire. Yet, postcranial bones of *Gigantopithecus*, and, indeed, craniofacial bones, are absent—strangely absent. One would expect that bones of such size and density would survive taphonomic winnowing. However, the occurrence of *Gigantopithecus* fossils in caves and fissure-fills indicates that smaller specimens (teeth, jaw fragments) were being washed into subterranean traps. Larger specimens may have been left to weather on the ancient land surfaces. In any case, the body shape and form of *Gigantopithecus blacki* remain a mystery. Furthermore, the species has dental traits and dental wear patterns that are unique among primates. Their premolars and even their canines appear to have become molarized. And, since their postcanine teeth

Figure 14.17. Reconstruction of *Gigantopithecus blacki*, shown confronting a band of early modern humans (*Homo sapiens sapiens*) in a dipterocarp forest in China, about 100,000 years ago. This is the Last Appearance Datum (LAD) for the fossil ape. Illustration by Angela J. Tritz.

do not wear flat, there is an extraordinary difference on each tooth between the worn and unworn sides of the tooth. In the mandible, the buccal side of the tooth wears more; in the maxilla, the lingual side of the tooth wears more. These dental traits must be the result of a unique diet. Again, some paleoanthropologists argue that *Gigantopithecus* may have had a niche similar to that of the living giant panda. This aberrant bear has become specialized for feeding on bamboo. There is some evidence to support this interpretation for *Gigantopithecus*, as I will detail later. Lastly, two renowned paleoanthropologists (Franz Weidenreich and John Robinson) argued that members of the genus *Gigantopithecus* were the first true hominids. This genus has thus influenced ideas about human evolution (Weidenreich, 1946; Robinson, 1972).

Fossil teeth of *Gigantopithecus* in China were first discovered in a Chinese apothecary's shop by the Dutch paleontologist G. H. R. (Ralph) von Koenigswald. Why was Ralph von Koenigswald searching for fossils in a drugstore? Traditional Chinese medicine valued (and still values) powdered dragon bones, which could be added to food or drink, and acted as a general cure-all. These so-called dragon bones were actually the remains of fossil mammals. Thus, Western paleontologists soon realized that a visit to a local Chinese drugstore was a quick way of assessing the fossil-bearing potential of a certain area, because dealers, mining and selling dragon bones, had already identified nearby fossil localities. The teeth of *Gigantopithecus blacki* were very large, so it quickly became clear that an enormous fossil ape had once roamed the Pleistocene landscape of China. Yet, how large was this species? It is generally thought that *Gigantopithecus* was about 400 kg in weight. This qualifies as a large animal by any definition, and *Gigantopithecus* reigns as the largest primate that ever lived.

Teeth of *Gigantopithecus blacki* have also recently been discovered in Vietnam, which extends the known geographic range of this species (Ciochon *et al.*, 1990b). The hypodigm of *Gigantopithecus* consists of over a thousand fossil teeth and several fragmentary jaws, but nothing else remains. Extensive survey in China and Vietnam has yielded no postcranial fossils. This has not prevented scientists from reconstructing the complete body shape and form of this species. It is usually reconstructed as looking gorilla-like, doubtless because the living gorilla is currently the largest living primate, and this unconsciously influences our thinking about the species. King Kong, for example, was fashioned as a giant gorilla, and not an orangutan, or some completely different type of unknown primate.

One striking artist's reconstruction of *Gigantopithecus* shows a social group of these animals feeding quietly on bamboo, and then suddenly encountering a band of *Homo erectus* humans across the banks of a small stream (Ciochon, 1988:32–33). An orangutan slowly climbs away in the distant forest, reminding us that orangutans lived on the Asian mainland in historic times. A *Homo erectus* male ominously brandishes a hand-ax at the group of feeding apes. Things do not bode well for the apes. Human fantasizing about these giant apes is irrepressible. For reasons unknown, a life-size bronze sculpture of *Gigantopithecus* has been erected on the grounds of the State University of New York at Oneonta, NY, and many an American believes that *Gigantopithecus* haunts the forests of the Pacific Northwest in the guise of Bigfoot or Sasquatch. *Gigantopithecus* and *Homo erectus* appear to have coexisted for several hundred thousand years. We certainly have no evidence that hominids hunted and butchered this extinct species of ape. However, Ciochon (2009) has since argued that *Homo erectus* occupied only open-country ecosystems in the Far East, and did not penetrate the dipterocarp lowland tropical rainforest occupied by *Gigantopithecus*, the giant panda, and *Stegodon*, an extinct genus of elephant. Dipterocarp lowland tropical rainforest still persists in Southeast Asia today, and is the typical vegetation of undisturbed habitats (Morley, 2000). Global sea levels were lowered during the Pleistocene, and shallow continental shelves emerged to connect Borneo and the Indonesian islands with Asia proper. Dipterocarp tropical rainforest was the dominant vegetation of this land mass.[2]

Phytoliths have been used to infer diet in *Gigantopithecus*. After scanning electron microscopy analysis, about 30 phytoliths were discovered adhering to fossil tooth surfaces (Ciochon *et al.*, 1990a). They are not there because of accidental incorporation of plant material into the matrix surrounding the teeth, because the phytoliths are bonded to the tooth enamel. The phytoliths are therefore thought to come from ingested food items. In addition, there are striations on the enamel itself that are created by the phytoliths as they are drawn over the tooth surface. The phytoliths are subdivided into two plant types: those from grass (i.e. bamboo) and those from dicotyledons. The dicotyledon phytoliths may come from fruits of a tree belonging to the Family Moraceae. In fact, they may come from durian fruits, which are a member of this family. These fruits are highly favored by living orangutans.

The origin of hominids

The importance of Ethiopia as a font for early hominid discoveries cannot be overestimated. Sites in the Afar region have yielded the vast majority of Pliocene hominid fossils. Furthermore, the geology and paleoecology of the sites are well known, and chronometric dating techniques yield absolute dates. The importance of this region was underscored by the discovery at Dikika, Ethiopia, of a remarkably complete juvenile specimen of *Australopithecus afarensis* (DIK-1-1), dated to 3.3 mya (Alemseged *et al.*, 2006). The specimen is a female of about 3 years of age whose rate of growth resembles that of living great apes. It preserves skeletal elements hitherto unknown, and allows body proportions to be examined in a single individual. Bones are articulated, indicating that an intact skin surrounded the bones prior to burial—burial was therefore swift. The geology and paleoecology of the Dikika site indicates that the environment was slightly less forested than contemporary sites at Hadar (Wynn *et al.*, 2006).

In fact, most of the sites in East Africa dating back to 7 mya that contain fossil hominids have less than 40 percent canopy cover (Cerling *et al.*, 2011). They are consequently identified as grasslands. Hominids therefore occur in grasslands, which is very rare for primates, and the isotopic chemistry of their tooth enamel indicates that they early on begin to incorporate C_4 resources into their dietary regime. This is also rare for primates, even for contemporary savanna baboons and common chimpanzees living in relatively open environments. Only the living gelada baboon has such a dietary signature. Living apes (including savanna chimpanzees) do not eat C_4 resources like grasses or sedges. Nor do they eat animals that eat these resources. The diet of early hominids is clearly different

---

[2] I cannot resist pointing out that the forests of the Pacific Northwest of North America are not dipterocarp tropical rainforests. If humans sighting Sasquatch in these forests are sighting the lurking descendants of *Gigantopithecus*, then that ancient ape must have made an immense habitat shift as it migrated into North America.

from that of apes. The hominid C_4 dietary shift occurred at about 3.5 mya (Lee-Thorp & Sponheimer, 2013; Sponheimer *et al.*, 2013). Although hominids unequivocally emerge 6 mya with the species *Ororrin tugenensis*, and have a reasonable fossil record before 4 mya—particularly for the species *Australopithecus afarensis*—carbon isotopes retrieved from tooth enamel indicate a preponderance of C_3 resources in hominid diet prior to 3.5 mya.

One aspect of South African hominid diets is their variability. Lee-Thorp and Sponheimer (2013:295–296) note that hominid carbon isotope values are more variable than nearly all modern or fossil taxa that have been studied in South Africa. This is not affected by environmental change or time, because hominid carbon isotope values remain variable irrespective of locality. Furthermore, they are variable within any given stratum or time interval. Yearly or seasonal differences in diet, detected within a single tooth, are also striking. The theoretical implication of this variability is enormous. Without variability, there is no evolution. Because natural selection can only work on phenotypic variability, there is a potential for dramatic shifts in selection intensity and selection direction on early hominid diet. After 3.5 mya, dietary evolution could have rapidly shifted and increased.

In summary, hominids early on demonstrate a habitat shift towards open country. When they emerge, they are creatures that are exploring and exploiting open country. They are not lurking in the forests or woodlands. Hominids also eventually demonstrate a dietary shift towards C_4 resources. These C_4 resources can either be the plants themselves—probably the case for *Australopithecus* (*Paranthropus*) *boisei* in East Africa—or animals that eat C_4 plants. The incorporation of vertebrate meat into early hominid diet signifies a radical departure from the diet observed in other primates. Although primates as an order are often characterized as omnivores, they are more accurately identified as generalized herbivores that sometimes eat small prey items like insects. The big question is this: what precipitated the entry of hominids into the C_4 world? What events occurring at 3.5 mya initiated this shift? It was not simply a matter of encountering open country, because this happened as early as 7 mya. Nor was it a matter of adopting tool behavior, because the archaeological evidence for stone tools only appears 2.6–2.5 mya—a million years after the C_4 shift. Entry into the C_4 world was probably triggered by a phase change involving cognitive and social differences not seen in other primates. The fundamental dietary mode of hominids is wildly different from that of other primates, and it therefore necessitates novel adaptations in cognition and sociality (Cachel, 2006). Thus, the unthinking use of living non-human primates as direct, referential models for the ecology and behavior of early hominids is a theoretical pitfall that plagues evolutionary anthropology.

Hominoid locomotor adaptations

Living hominoids are highly diverse in their locomotion. The living hylobatids are the most acrobatic of all primates. However, the semicircular canal evidence of

Table 14.1. The Pattern of Hominoid Locomotor Evolution.

STAGE I Stem catarrhine Body size ~5 kg Africa, ~30 mya e.g. *Aegyptopithecus* Generalized arboreal quadruped Establishment of typical catarrhine sexual dimorphism		
STAGE II Body size increase Body size ~10 kg	→	Greater arboreal climbing and grasping; tail loss
Orthograde resting posture Africa, ~25 mya	→	Ischial callosities → *Proconsul, Morotopithecus*
STAGE III Body size ~ 15–30 kg	→	Arboreal suspension, suspensory arm locomotion
Arboreal suspension, suspensory arm locomotion Facultative arboreal bipedalism Flexible shoulder, elbow, wrist Broad thorax, dorsal scapula Curved ribs. shortened ilium	→	*Pierolapithecus* (Spain) 13–12.5 mya
Larger hands	→	*Hispanopithecus laietanus* (Spain) *Rudapithecus hungaricus* (Hungary) 9.5 mya

STAGE IIIA (alternate route 1)
Body size decrease

Body size ~ 5 kg
Social monogamy
Very reduced sexual dimorphism

→ Ricochetal brachiation → Stem hylobatid → *Yuanmoupithecus* (China) 8.2–7.1 mya

STAGE IIIB (alternate route 2)

Body size ~ 25 kg

Facultative terrestrial bipedalism
Bipedal traits in pelvis, tripodal feet
Late Miocene global cooling

→ Facultative terrestrial foraging

(Island environment, no large terrestrial predators)

→ Selection for increased body size

→ *Oreopithecus* (the Tusco-Sardinian Island) 9–6.5 mya

→ Diet more diverse → *Ouranopithecus turkae* (Turkey) 8 mya
→ *Gigantopithecus* (S & E Asia) 8.5 mya-100,000 yrs

STAGE IIIC (alternate route 3)
Body size increase

Body size ~ 50 kg
Increased extremity size, long lateral rays, reduced pollex and hallux

→ Obligate arboreality

(Swamp forest habitat)

→ Slow, quadrumanous climbing → *Pongo*

STAGE IIID (alternate route 4)
Body size increase

→ Facultative terrestrial foraging → Independent acquisition of knuckle-walking → *Gorilla*, *Pan*

Table 14.1. (cont.)

Body size ~ 50 kg
Hand, wrist, & forearm
Develop knuckle-walking traits

STAGE IIIE (alternate route 5) → Ardipithecus
Body size ~ 50 kg   5.8–4.4 mya
Reduction in dominance hierarchies

Facultative terrestrial foraging
Facultative terrestrial bipedalism
(Natural history intelligence, sentinel behavior reduce likelihood of predation by large terrestrial carnivores)

Very reduced sexual dimorphism

Bipedal traits in pelvis,
lower limb

STAGE IIIF (alternate route 6) → Orrorin → Australopithecus
Body size ~ 50 kg   6 mya
Reduction in climbing, clambering

Facultative terrestrial foraging
Facultative terrestrial bipedalism
(Natural history intelligence, sentinel behavior reduce likelihood of predation by large terrestrial carnivores)

Reduction in dominance hierarchies
Very reduced sexual dimorphism
More manipulation   Vertebrate meat and marrow
Hand size reduced   a regular dietary component
Large apical tufts on manual
phalanges
Bipedal traits in pelvis,   Increased terrestrial ranging
lower limb, foot   because of bipedal efficiency

fossil hominoid taxa indicates that, with the exception of *Proconsul heseloni*, which engaged in some leaping, locomotion was slow and deliberate (Ryan et al., 2012). Taxa examined include *Oreopithecus bambolii*, *Hispanopithecus laietanus*, and *Rudapithecus hungaricus*. Agility scores for *Hispanopithecus* and *Rudapithecus* measured on a 6-point scale imply that they had the slowest locomotion of all the 16 examined fossil taxa (Ryan et al., 2012:Table 2). From this result, one might infer very purposeful and calculated movements. The large hands of these taxa probably functioned in climbing, or in suspending the animal below supports, and did not merely grasp horizontal or near horizontal supports. There is no sign of the explosive richochetal brachiation seen in the living hylobatids. Yet, the semicircular canals of *Proconsul heseloni* and *Victoriapithecus*, the earliest cercopithecoid, demonstrate that these taxa were more agile than the Late Miocene hominoids. This implies that the Late Miocene hominoids were specializing in slow, deliberate movements. This was not the generalized arboreal quadrupedalism found in the early catarrhines (e.g. *Aegyptopithecus*, *Saadanius*). The postcranial evidence indicates that climbing, suspensory postures, and slow arm-swinging locomotion were adopted by the Late Miocene hominoids.

Diverse locomotion in hominoids has been linked to low levels of morphological integration in the pelvis, which allows for greater ease of evolutionary change. However, when the entire primate order is examined, all primates show a lack of morphological integration in the pelvis (Lewton, 2012). This is probably the cause of the varieties of locomotion observed in living primates (Chapter 7). Members of other mammalian orders are much more limited in their range of locomotion.

One can summarize the general outline of hominoid evolution by examining the pattern of locomotor evolution in this group (Table 14.1).

The anatomical traits associated with knuckle-walking in the African great apes do not show a high degree of morphological integration—they are fundamentally different in *Gorilla* and *Pan* (Williams, 2010). This indicates that knuckle-walking developed independently in each of these genera, and that hominids did not evolve from a knuckle-walking ancestor (Kivell & Schmitt, 2009). This has implications for the evolution of Late Miocene apes in Africa, regardless of whether reasonable fossil evidence is ever discovered from this time range. Not only does the independent evolution of knuckle-walking in gorillas and chimpanzees demonstrate a more diverse fossil history than traditionally reconstructed, but the independent origin of a terrestrial form of locomotion in gorillas and chimpanzees also indicates that powerful selection forces were operating to select for terrestriality in both these lineages. These powerful selection forces were also operating at a time when another terrestrial form of locomotion—bipedalism—was evolving in the proto-hominid lineage. In one case, terrestrial quadrupeds emerged. In the other case, terrestrial bipeds emerged. If body size were equivalent in the two cases, the quadrupeds must have had a higher center of gravity or longer forelimbs in order for knuckle-walking to evolve. The bipeds must have had a lower center of gravity or shorter forelimbs in order for bipedalism to evolve. Alternatively, the ancestral bipeds may have had a smaller body size. Anatomical differences between *Ardipithecus* and *Australopithecus* indicate that several lineages may have independently been experimenting with bipedality.

15 The cercopithecoid radiation

Old World monkeys are called cercopithecoids, after the Superfamily Cercopithecoidea. The earliest cercopithecoid dates to the Oligocene at 25.2 mya. This is *Nsungwepithecus gunnelli* from the Rukwa Rift in Tanzania (Stevens *et al.*, 2013). The species is represented by a partial mandible with a single tooth. However, this M3 is a bilophodont tooth (Figure 14.5). The bilophodont molar crown pattern of this species heralds both its cercopithecoid status, and its presumed ability to masticate tough, mature leaves. However, microwear on the teeth of the widespread Eurasian Late Miocene colobine *Mesopithecus pentelicus* indicates that it was a hard-object feeder, like the modern platyrrhine tufted capuchin *Cebus apella* (Merceron *et al.*, 2009). Hard seeds or fruit may have constituted the preferred diet of *Mesopithecus*, rather than mature leaves. Folivory could therefore initially have been a fallback dietary strategy for cercopithecoid monkeys. Analysis of the dentition of *Victoriapithecus*, the most abundant and well known of early cercopithecoids, indicates that it was probably more frugivorous than folivorous (Benefit, 1999).

Detailed analysis of stomach contents of cercopithecoids and hominoids demonstrates that the bilophodont teeth of cercopithecoids allow them to shred leaves into fine strips (Walker & Murray, 1975). Leaves appear to have been pulled across a grater. The bunodont teeth of hominoids merely rumple the leaves—they enter the stomach relatively intact, with only a few puncture marks. The cercopithecoid ability to shred leaves during mastication hastens digestion once leaves enter the stomach. The leaf-eating cercopithecoids or colobines (Subfamily Colobinae) have higher-crowned molar teeth and more pronounced bilophodonty than the cercopithecines. This allows for heavier mastication. Colobines also possess a sacculated stomach in which endogenous bacterial colonies can break down cellulose. This bacterial activity is known from many experiments on economically valuable domesticated cattle. In cattle, anaerobic bacteria in the rumen chamber of the stomach are responsible for breaking down otherwise indigestible plant cellulose while releasing calories and vitamins. The rumen is a center for biofermentation, and the anaerobic bacteria that it contains are therefore indispensable for digesting plant foods. Similar bacterial colonies in the sacculated stomachs of colobines could not only break down indigestible plant fibers, but also de-toxify natural plant alkaloids that would be poisonous to other primates.[1] For example, colobines are known to ingest high levels of deadly strychnine from plant foods without visible consequences. Ingesting high levels of plant toxins sometimes leads colobines to forage terrestrially for and eat soils high in kaolin. Black and white colobus monkeys that I observed in the Eastern Congo near Lake Edward returned so often to a particular site high in kaolin that they excavated a pit over 2 meters in depth as they dug out and ate this soil. Red colobus monkeys in Zanzibar routinely forage for and eat charcoal left in abandoned village hearths. Charcoal, like kaolin, has the ability to bind to and de-toxify poison.

The first well-known cercopithecoid taxon is the Miocene species *Victoriapithecus macinnesi*, dating to 19 mya. Note that an approximately 7 my gap occurs between the appearance of *Nsungwepithecus* and the appearance of *Victoriapithecus*. This confirms the rarity of early cercopithecoids. Cercopithecoid monkeys do not show a sudden burst of abundance and species richness immediately after their appearance. In fact, except for *Victoriapithecus*, known from about 2,500 specimens, cercopithecoid monkeys remain rare until the Late Miocene. This is strange, because the living cercopithecoid monkeys account for much of the species diversity of living primates. Yet, they do not begin to flourish until the Late Miocene, and they do not appear to have left Africa until the very Late Miocene. Their dental and skeletal morphology remains conservative, so that variation in fossils reflects mainly species-level distinctions (Jablonski, 2002; Jablonski & Frost, 2010). The relatively unvarying dental and skeletal morphology reflects a successful adaptation to a generalist niche. Species-level variation occurs upon this generalized morphology, and this variation explodes in the Late Miocene. The South African record is especially well preserved in fissure-fill deposits, which record an explosion of baboon fossils beginning in the Pleistocene (Freedman, 1957). Cercopithecoids increase in abundance and diversity and become widely distributed throughout the Old World precisely when the hominoid radiation suffers a major extinction event. Only the pliopithecid hominoids survive for a time, and it is probably not an accident that cresting on the molar teeth of pliopithecids allows for some mastication of mature leaves.

What is known about *Victoriapithecus*, the most successful of the early cercopithecoid monkeys? *Victoriapithecus* differed significantly from living cercopithecoids in dental and craniofacial morphology, and it is therefore often sorted into a different family (Family Victoriapithecidae). Bilophodonty was not exaggerated in *Victoriapithecus*, as it is in modern colobine or Old World leaf-eating monkeys. Hence, reliance on mature leaves was similar to that of modern baboons or macaques. The facial skeleton of *Victoriapithecus* indicates that a relatively narrow separation occurred between the orbits. In addition, the face was

---

[1] Every leaf on a tree is an organ for photosynthesis. Animals that eat leaves are preying on these valuable organs. Plants respond to this predation by possessing mature leaves that are high in indigestible cellulose and low in protein, or by having these leaves contain poisonous compounds to deter predation.

long, the cranial vault was low, the nasal bones were long and narrow, and the zygomatic arches were broad. This confounds expectations about the face of primitive catarrhines, based on comparative morphology, which inferred a colobine-like facial structure in the earliest catarrhines (Vogel, 1966). Hence, the cranium of *Victoriapithecus* resembled those of basal catarrhines, such as *Aegyptopithecus* and Early Miocene hominoids of Africa (Benefit, 1999). Among living cercopithecoids, the craniofacial morphology of *Victoriapithecus* most resembled genus *Macaca*, although scaled down considerably in size. *Victoriapithecus* also appears to be a miniaturized version of a macaque in terms of generalized adaptations. It was about 3–5 kg, approximating that of living vervet monkeys (*Cercopithecus aethiops*). Postcranial anatomy also suggests that *Victoriapithecus* engaged in terrestrial locomotion as often as living vervet monkeys do, and these animals are often referred to as savannah guenons, because of their widespread presence in open-country environments. Terrestrial running, accentuating anteroposterior movements of the thigh, might have been emphasized (Benefit, 1999). In fact, the semicircular canals of *Victoriapithecus* indicate that it was much quicker and more agile in locomotion than earlier catarrhines, such as *Aegyptopithecus* or *Saadanius* (Ryan et al., 2012).

In summary, one of the most important things to understand about cercopithecoid monkeys is that, although they appear at an early date, their fossil record remains meager until very late in time. *Mesopithecus*, the first cercopithecoid monkey genus in Asia, is found at 6.9–6.2 mya in Shuitangba, Yunnan Province, China (Ji et al., 2013). Contemporary with the Chinese fossil ape *Lufengpithecus lufengensis*, *Mesopithecus* signals the looming rise of the cercopithecoid primates and the decline of the hominoids in the Old World. Cercopithecoids experience a notable evolutionary radiation during the Plio-Pleistocene. Climate change during the Plio-Pleistocene has been most intensively investigated in Africa, primarily because of the effort to discern the influence of climate on human evolution. Fortunately, these efforts also impact on the study of other African catarrhine primates. There are three narrow intervals in the climate records when more variable, drier climates are accompanied by shifts to more open habitats and some change in faunal assemblages (de Menocal, 2004). These intervals are 2.9–2.4 mya, 1.8–1.6 mya, and 1.2–0.8 mya. Plio-Pleistocene climatic fluctuations certainly affected speciation in African cercopithecoids. Thus, climatic fluctuations that accompanied diversity and abundance in hominids also witness the rise of the cercopithecoids. Habitats and climates that favored hominids also favored cercopithecoids. The increasing spread of grasslands was concomitant with the evolution of obligate bipedalism in hominids and colonization of open country by terrestrial cercopithecoids.

However, unlike most other primates, the speciation rate of cercopithecoid monkeys has not declined: the genera *Cercopithecus*, *Macaca*, and *Presbytis* contain far more species than other primate genera. Global climatic changes of the Plio-Pleistocene that abruptly expand and contract rainforest habitat are partly responsible for this, because remnant rainforest patches apparently serve

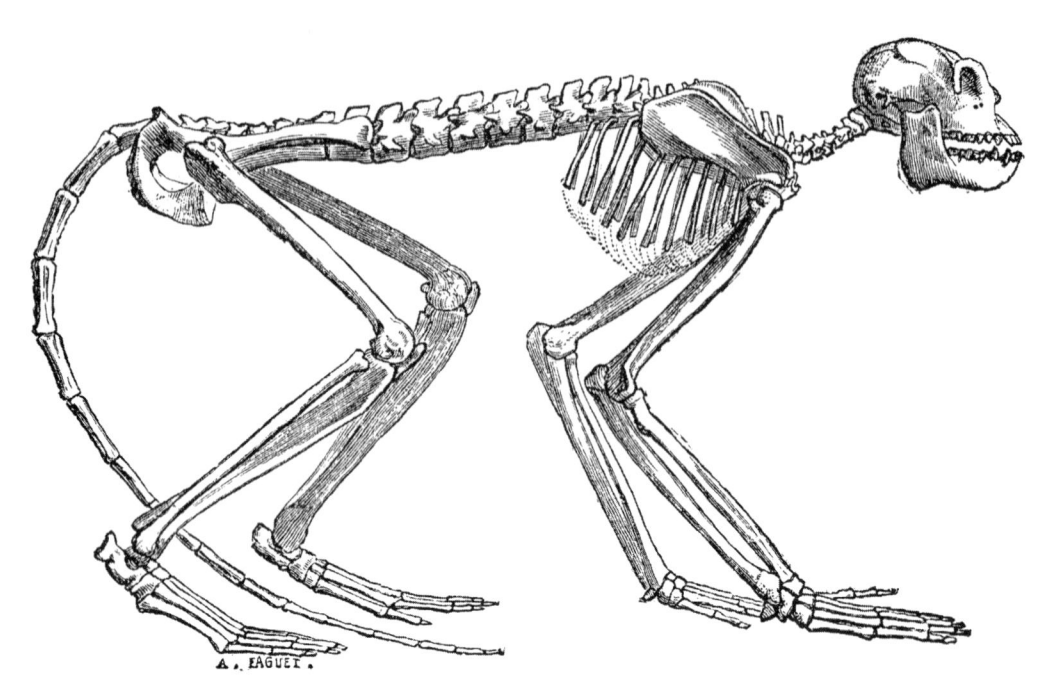

Figure 15.1. The Old World monkey *Mesopithecus pentelicus*. This Eurasian colobine was first described in 1839. It shows adaptations for terrestrial locomotion (e.g. its digitigrade hands), and probably was as terrestrial as the living Hanuman langur of Asia (*Presbytis entellus*). Courtesy of the Mary Evans Picture Library.

as nuclei for generating new species. When rainforest expands again, geographic barriers disappear, and closely related species become sympatric. Hybridization may occur. The retreat of rainforest also promoted experimentation with terrestrial life. Baboons (*Papio*, *Theropithecus*, *Mandrillus*) and macaques (*Macaca*) are among the handful of primates that have successfully experimented with terrestrial adaptations.

The Eurasian species *Mesopithecus pentelicus* appears to be a typical colobine in dentition and cranial anatomy. However, microwear on its teeth indicates hard-object feeding (Merceron et al., 2009). Its postcranial anatomy is well known (Figure 15.1), and demonstrates terrestrial adaptations as advanced as that of any living colobine, such as the Hanuman langur (*Presbytis entellus*). European localities in which *Mesopithecus* is found are open-country sites (Delson, 1973).

Of all living cercopithecoids, the teeth of macaques (genus *Macaca*) appear to be the most primitive. In fact, because the genus demonstrates very little morphological change through time, macaques might be considered a legitimate higher-primate living-fossil taxon (Delson & Rosenberger, 1984). The Old World monkey genus *Macaca* is now among the most species-rich of all living primate genera. The genus is also noteworthy for two other features. It qualifies as a primate living fossil, since it appears in the Late Miocene, about 5.5 million years ago (Delson & Rosenberger, 1984). And it is distributed across three continents (Africa, Europe, and Asia), from northern Africa to eastern Indonesia. At least one macaque species crossed an ocean gap on a floating mat of vegetation during their colonization of Southeast Asia (Abegg & Thierry, 2002). Some of the living macaque species are highly tolerant of human presence, and do not seem to suffer from human alteration of the environment. In fact, some of these species thrive in environments that have been disrupted by humans. These species have been labeled "weed macaques," because of their ability to invade novel or altered habitats, and their ability to steal and feed on domesticated plants (Richard et al., 1989). In fact, many local farmers consider these macaque species to be vermin.

Body size is variable in living macaque species. Yet, there does not appear to be a general evolutionary trend for body size changes in the genus. Although many mammal lineages undergo increasing body size changes through time, this does not appear to be the case for the genus *Macaca* (Rook et al., 2001). However, a large-bodied species (*Macaca robusta*) existed during the Pleistocene of Europe. This is true for many Ice Age mammals, from muskrats, beavers, and cougars, to the prosimian aye-aye. The living Japanese macaque (*Macaca fuscata*) is quite tolerant of cold temperatures. In the depths of winter it can eat bark, mushrooms, and other low-quality food items. Its ability to survive winter weather is not determined by ambient temperature, but by the depth of snow cover. Macaques must be able to dig below the snow for terrestrial food items.

Gelada baboons (genus *Theropithecus*) were both speciose and remarkably abundant during the Pleistocene throughout Africa, as well as colonizing Spain and India (Jablonski, 1993). The genus first appears in the fossil record between 4 and 3.5 mya (Delson, 2000). Many fossil geladas were far larger than the single surviving living species (Figure 5.4). At some sites (e.g. Olorgesailie, in central Kenya), nearly 50 percent of the vertebrate fossils are gelada baboons. I can testify to this, since I was able to identify gelada teeth and jaw fragments in approximately half of the vertebrate fossils that I encountered as surface finds while traversing the Olorgesailie Basin in Central Kenya.[2] This fact both epitomizes gelada abundance and illustrates the ubiquity of geladas in the fossil ecosystems of Africa. Their biomass must have rivaled that of modern small to medium-sized bovid species, such as impalas. One can imagine the spectacle of a typical open-country grassland habitat about 2 mya, dotted with herds of zebras, gazelles, antelopes, and geladas.

Fossil geladas almost certainly would have existed in relatively small social groups. The large aggregations that occur in living geladas today take place when small social groups that have been foraging separately during the day accumulate en masse at scarce sleeping cliffs at sundown to shelter from predators. But modern geladas live in the treeless highlands of Ethiopia, where only cliffs and

---

[2] The Olorgesailie Basin contains many preserved open-air archaeological sites, with associated vertebrate fossils. A multitude of both stone tools and fossil bones continue to erode from the badlands sediments. They can be easily spotted and examined by visitors. Olorgesailie is classed as a Kenyan National Monument, and is maintained by the National Museums of Kenya.

rock faces offer shelter at night. The multilevel social organization of living geladas, where harems congregate into bands, and bands congregate into herds of 600 or more individuals, was the basis of a novel idea about social complexity and language origins. Robin Dunbar (1996), who studied wild geladas in Ethiopia, advanced the idea that complex vocal communication contributes to survivorship and reproductive success within intricate multilevel societies.

Modern geladas are severely depleted in number. Grass constitutes over 95 percent of their diet, and they occur today only where soft, high-protein fescue grasses thrive in the mountains of Ethiopia. It is important to note that stable isotope analysis of ancient enamel shows that the extinct gelada species *Theropithecus darti* and *Theropithecus oswaldi* had diets resembling those of living geladas (Fourie et al., 2008). Enamel samples from 4–1-my-old fossil geladas in Kenya demonstrate an increasing reliance on grasses. The C_4 signal is dominant at 4 mya in *T. brumpti*, but is almost 100 percent in 1-my-old *T. oswaldi* (Cerling et al., 2013). One hitherto unrecognized problem is that the strength of the C_4 signal in fossil geladas rivals that of the C_4 signal in the sympatric fossil hominid *Australopithecus* [*Paranthropus*] *boisei* (Sponheimer et al., 2013). Figure 15.2 shows the triumvirate of *Theropithecus oswaldi*, *A.* [*P.*] *boisei*, and *Homo erectus*—taxa that are sympatric in many East African sites. *T. oswaldi* also disperses out of Africa with *Homo erectus* in the Plio-Pleistocene. Just as *Theropithecus* is the most abundant cercopithecoid, *A.* [*P.*] *boisei* is the most abundant fossil human in East African sites. Although traditionally paleoanthropologists have striven to differentiate the niches of sympatric fossil humans in terms of competitive interactions between them, or between humans and carnivores, *A.* [*P.*] *boisei* may well have been experiencing more resource competition with fossil geladas than with other fossil humans or with sympatric carnivores. This completely alters traditional reconstructions of fossil human behavior. A fossil human competing with geladas? Of course, *A.* [*P.*] *boisei* might have engaged in tool behavior, which would modify its ability to acquire and modify food resources. There is anatomical and archaeological evidence that a close relative in South Africa, *Australopithecus robustus*, exhibited tool behavior. And *A.* [*P.*] *boisei* is the fossil human species that is most often found in association with stone tools in East African sites. In the history of paleoanthropology, the grass seed-eating of gelada baboons was used as the basis for an influential model of human origins— the seed-eating hypothesis (Jolly, 1970). Gelada comparisons were certainly prescient with regard to *A.* [*P.*] *boisei*.

Geladas began staging a retreat into higher altitudes when a warming African climate reduced fescue grasses in lowland areas. Thus, the spread of arid tropical grasses after the Late Pleistocene not only drove fossil gelada species to extinction, but reduced the living species to a relict distribution. Might this also have been responsible for the extinction of *A.* [*P.*] *boisei*? Modern geladas seek shelter from predators at night on rocky slopes and vertical rock faces high in the Ethiopian mountains. The ubiquitous lowland geladas of the Pleistocene must also have sheltered from predators at night, but where did this occur? Trees that lined

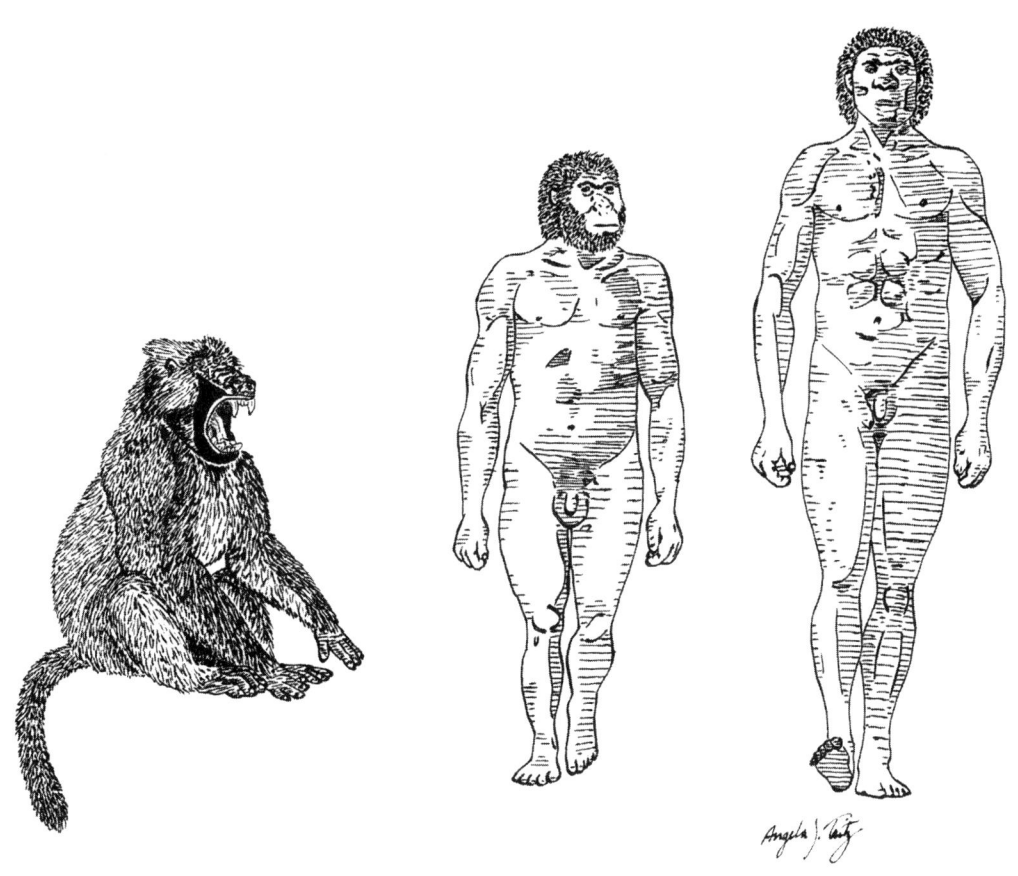

Figure 15.2. The giant fossil baboon *Theropithecus oswaldi*, shown with two sympatric fossil humans, *Australopithecus* [*Paranthropus*] *boisei* (middle) and *Homo erectus* (right). After Turner and Antón (2004:Fig. 5.14). Illustration by Angela J. Tritz.

ancient waterways would have groaned under the mass of fossil geladas, especially because these extinct species were up to twice the size of living geladas. Rocks and rock faces would have been limited in distribution, and therefore difficult to rendezvous at after a day of protracted foraging.

Plio-Pleistocene diversity and sympatry

Because their evolutionary radiation was so recent, fossil cercopithecoids, like their modern counterparts, are a speciose group. For example, the Plio-Pleistocene site of Makapansgat Limeworks in South Africa contains 12 sympatric species of cercopithecoid monkeys, separated into three genera (Fourie et al., 2008). This diversity is real, because the recent date of the site and excellent preservation mean that paleontologists are not inflating species numbers due to fragmentary and cryptic fossils. The presence of these sympatric species allows one to study how resources like food and space were divided—that is, what is determining niche differentiation, and how species interactions affect the evolution of any given species.

These questions about niche structure and species interactions have been studied in detail for some animal groups, such as birds (Brown & Mauer, 1987). This has not been done for primates. Cercopithecoid monkeys offer a magnificent test group for such study. Platyrrhine primates also occur in species-rich assemblages, but living platyrrhines are evolutionarily disparate. They are diverse in phylogeny. But all cercopithecoid monkeys (indeed, all catarrhines) are morphologically and genetically more coherent, and their lineages are less divergent than platyrrhine lineages. Furthermore, because of their species richness, fossil cercopithecoid monkeys can be studied to examine how primate communities evolve. That is, given constraints of food and space on sympatric species, how do diet and locomotion change in individual lineages? Do these changes coincide with changes in sympatric lineages? Are these changes shifting in the same direction, or are these changes diverging? How do abundance and distribution change in these lineages through time? Is the rarity and extinction of one lineage associated with a pulse in abundance of another or several other lineages? Thus, sympatric species of fossil cercopithecoids can allow one to study fundamental problems of macroecology and evolutionary processes.

16 Late Cenozoic climate changes

Major climatic oscillations that are characteristic of the Pleistocene actually begin far earlier, in the Late Cenozoic. Seasonality becomes more pronounced. Temperature and precipitation are no longer equably distributed throughout the year. Aridity increases, and tree cover is lost in many places. Plants that use the C_4 photosynthetic pathway (largely tropical grasses that can tolerate prolonged drought) spread very widely, heralding the shift to a "C_4 world" (Cerling & Ehleringer, 2000). This cooler and drier global climate foreshadows the outright appearance of continental glaciation during the Pleistocene.

Geochemical signatures of atmospheric carbon dioxide have been followed over the last 20 million years using boron/calcium ratios from fossil foraminifera. During the Middle Miocene (14–10 mya), atmospheric carbon dioxide was roughly similar to that of the present, even though global temperatures were about 3–6°C warmer, and sea levels were about 25–40 m higher (Tripati *et al.*, 2009). A major Late Miocene land mammal extinction event termed the "Vallesian Crisis" occurred in Western and Central Europe at the end of the Vallesian Land Mammal Stage beginning 9.6 mya. Many rhinoceroses and tapirs disappeared, and pig diversity declined. A turnover occurred among the rodents. After the Vallesian Crisis, murid rodents, which include modern mice and rats, become the dominant rodents in Late Miocene communities. Atmospheric carbon dioxide fell during the Late Miocene. Ice sheets in West Antarctica and Greenland grew when atmospheric carbon dioxide fell significantly below modern levels. Glacial conditions thus intensify during the Late Miocene (after ~10 mya) and Late Pliocene (3.3–2.4 mya).

One major geographical distinction of the Late Miocene is that the Mediterranean Sea completely dried up at about 5.6 mya. This desiccation was caused by the closure of the Gibraltar Strait, which cut the Mediterranean off from the Atlantic Ocean. Evaporation rates exceeded both local rainfall and the amount of river water debouching into the Mediterranean from the Nile, Rhone, and other major rivers. The Mediterranean Basin thus became a gigantic salt pan, inimical to life. Geologists name this event "The Messinian Salinity Crisis."[1] At 5.33 mya, Atlantic waters finally incised their way through the closed Gibraltar sill, resulting in an abrupt, catastrophic flooding of the Mediterranean Basin. This event is known as

---

[1] A website is devoted to publications and discussions of this event: http://www.messinianonline.it/.

the Zanclean Flood, after the first stage of the Pliocene epoch. It occurred virtually instantaneously in terms of geological time, taking place between a few months to 2 years. During this geologically fleeting period, 90 percent of the water in the current Mediterranean was transferred into the dry sea basin. In the most catastrophic flood ever recorded in earth history, sea levels may have risen as quickly as 10 m a day (Garcia-Castellanos et al., 2009).

The desiccated Mediterranean Basin affected not only the geography but also the local climate of the circum-Mediterranean region. Although animal and plant dispersal could have occurred through the basin at the beginning of desiccation, and before the full onslaught of the Zanclean Flood, conditions within the basin during the height of the desiccation were absolutely inimical to life. Conditions would have superficially resembled the alien, flat, salt desert landscape west of the Great Salt Lake in northwestern Utah, USA (Figure 16.1). However, evaporitic sediments in the Mediterranean Basin would have been kilometers thick, and the basin itself would have sat 3–5 km below sea level. This formed the largest salt basin known to geologists. Its white and radiant, shining surface would have been visible from Mars. The massively deep evaporites were mostly composed of halite (rock salt), gypsum, and anhydrite. Small brine lakes would occasionally appear, but the absence of fresh water and the unearthly heat on the abyssal plain were incompatible with multicellular life. Anhydrite evaporates out from water that is warmer than 35°C. These temperatures, coupled with the blast-furnace heat

Figure 16.1. Landscape west of the Great Salt Lake, northwestern Utah, USA. The salt flats of the desiccated Miocene Mediterranean Basin would have resembled this landscape, although the Mediterranean salt deposits were piled up to a depth of several kilometers. Courtesy of Dr. Craig S. Feibel.

radiating from the bright, reflective salt and the great depth below sea level—which accentuates ambient temperatures—would make a formidable barrier to animal and plant dispersion through the desiccated basin. Great rivers like the Nile and Rhone were rapidly cutting down their beds as they debouched into the empty basin, and some oases of life could have endured in such isolated waterfall/river delta areas. Apart from these islands of life, animal and plant life could not have endured the extreme conditions of the Mediterranean Basin.

The desiccated basin would have had an impact on the climate of the entire circum-Mediterranean region. In fact, signals of arid climate and atmospheric dust associated with the Messinian Crisis are found in deep-sea cores off the coast of West Africa, the Red Sea, and the Persian Gulf (de Menocal, 1995). In addition, one could infer the hostile effect of salt storms, or blowing clouds of airborne salt, on local animals and plants. Salt storms occur today around the Great Salt Lake of Utah and the rapidly shrinking Aral Sea of Central Asia, when high winds sweep across the surrounding salt flats. Salt-encrusted soils resulting from these storms can become too saline to support local plant species. The vast majority of land plants are intolerant of salinity. Yet, salt-resistant plant species may not be edible for local animal species. Community composition alters as animal and plant species change to accommodate saline soil composition.

During the Late Cenozoic, climatic deterioration occurred worldwide. Major extinctions of land mammals took place. This was especially pronounced in North America, where the fossil record is dense and well dated. The Messinian Crisis coincides with the North American Late Hemphillian Land Mammal Stage. This is an extinction event caused by a sudden burst in aridity and drastic increases in extremes of seasonal temperature. Many herbivorous mammals in North America exhibited catastrophic extinctions, especially browsing taxa. Some mammal groups (horses, camels, tapirs) survived only along relict woodland habitats of the Gulf Coastal Plain (Webb *et al.*, 1995). An astounding 74 percent of mammal genera and 18 percent of mammal families went extinct in the Late Hemphillian beginning about 6 mya. This major North American mammal extinction parallels extinction events in Eurasia. Eurasian hominoid primate extinctions occurred against the general background of Late Miocene mammal extinctions. The hominoid fragility of species, and their absolute ties to disappearing woodland and forest, explains the hominoid extinctions of the Late Miocene.

With the advent of the Pleistocene, climatic fluctuations that occurred since the Late Miocene become markedly more frequent and increase in amplitude. Details of the last 2 my of global climate change are very well known, and have also been correlated to pronounced fluctuations in global sea level. The link between global temperature and sea level is caused by the encapsulation of sea water in continental ice sheets when global temperatures are low. The sea level changes are referred to as oxygen isotope stages (OIS), because they are established by the ratio of isotopes O_{16} to O_{18} retrieved from ice cores or deep-sea sediment cores. These oxygen isotope stages are counted back from the present stage (OIS 1), which is occurring during an interval of warm global climate. Oxygen isotope

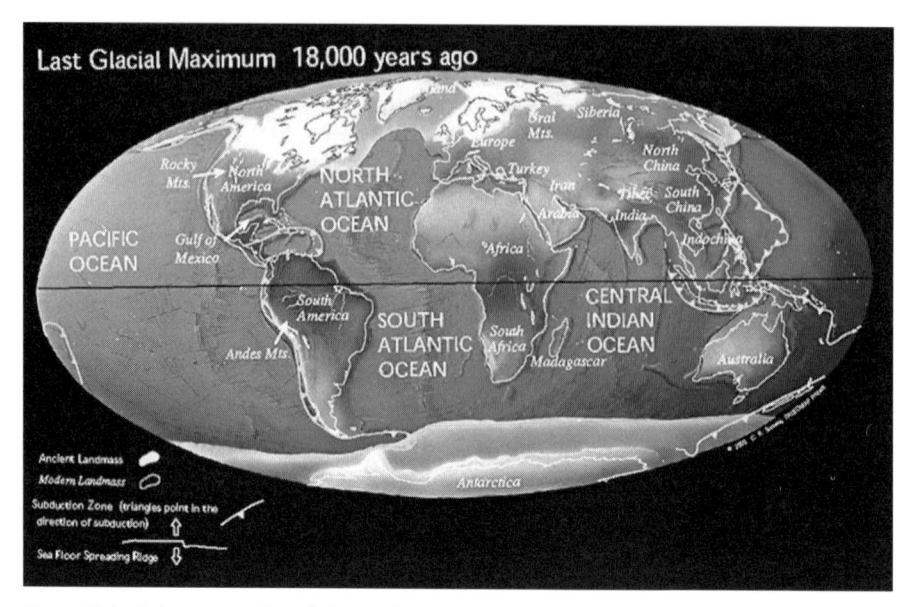

Figure 16.2. Paleogeography of the earth during the Last Glacial Maximum (LGM), 21,000–18,000 years ago. Courtesy of Dr. Christopher Scotese.

stages that have odd numbers are therefore associated with warm climate, and even numbers with cold climate.

Pleistocene land bridges sometimes connected areas now separated by water gaps, because the lowering of global sea level exposed large areas of continental shelf. The effects of these land bridges on land mammal dispersion are particularly strong in the region of the Bering Strait and Australasia. In Australasia, most of the islands of Indonesia have Pleistocene land connections, and are connected in turn to the Asian mainland. Eventual sundering of Pleistocene land connections has sometimes caused local island species of primates to evolve, e.g. the endemic macaque species of Sulawesi, Taiwan, or Japan. This vicariance phenomenon is directly attributable to allopatric speciation. In fact, one can argue that the great species diversity in the catarrhine genera *Macaca*, *Presbytis*, and *Hylobates* is caused by the sundering of Pleistocene land bridges.

The last great worldwide eruption of cold is referred to as the Last Glacial Maximum or LGM (Figure 16.2). Because of the recent time period, fine details of paleogeography and paleoclimate are known for the Pleistocene, particularly the LGM. Time-averaging has not had a chance to eradiacate or collapse the data. A special advantage of studying the Pleistocene is that many living species of animals and plants have Pleistocene representatives. This allows researchers to study topics such as rates of evolutionary change from ancestor to descendant, the emergence and dispersion of species from refuge areas (refugia) after the dissipation of continental ice sheets, and the morphological response of organisms to major climate change. The Pleistocene is therefore a time period producing rich investigation into evolutionary theory and processes.

17 Conclusions

Primate fossils record the history of a group that showed a precipitous rise in diversity and abundance at the beginning of the Cenozoic. Plesiadapoid primates also demonstrated disparity—as many as 12 zoological families are currently recognized. The plesiadapoid primates were an overwhelming presence during the Paleocene because they were occupying a series of open niches for small-bodied generalized herbivores. After the origin and dispersal of true rodents in the Late Paleocene, primates decline, and they never again achieve the abundance and diversity that they had during the Paleocene. After the extinction of the first major primate radiation (the plesiadapoids), the second radiation of euprimates also suffers an extinction event at the end of the Eocene.

The primate order never recovers from these two major extinction events. Its subsequent history has been a history of decline, even though primate groups that fascinate the public (monkeys and apes) had not yet evolved. Given the prominence of non-human primates on lists of endangered species, one could argue that most living primates are doomed to extinction. Both paleontology and historical records illustrate how rapid the decline in primate numbers can be.

Gelada baboons (*Theropithecus gelada*) exemplify the dire threat of extinction to non-human primates. At many Pleistocene African sites, they constitute half of all the vertebrate fossils discovered. Beginning 1 mya, they suffer a precipitous decline. Geladas now are found only in the mountains of Ethiopia, and they are fast diminishing in numbers even within this last stronghold. During the 1970s, researchers estimated their numbers as being between 100,000 and 200,000 animals; their current numbers in the wild are estimated to be 20,000—a decline of as much as 90 percent in about 40 years (Tucker, 2009)

The danger that disease poses to non-human primate survival cannot be overestimated. Two outbreaks of the Ebola virus among common chimpanzees of the Taï Forest in the Côte d'Ivoire accounted for 66 percent of the deaths recorded, and Ebola was the principal cause of chimpanzee mortality (Boesch & Boesch-Achermann, 2000). Mortality rates for lowland gorillas were 90–95 percent in the Lossi Sanctuary of the Republic of the Congo; 5,000 gorillas were estimated to have died during the outbreak (Bermejo *et al.*, 2006). Neither common chimpanzee nor lowland gorilla populations are sustainable with these rates of attrition. In addition to Ebola, both chimpanzees and gorillas are succumbing to anthrax in the Taï Forest (Ivory Coast) and Dja Biosphere Reserve (Cameroon). Anthrax spores infect these forest apes from either contaminated water, antelope carcasses, or cropland (Leendertz *et al.*, 2006). This is the first known report of gorilla susceptibility to anthrax, and both chimpanzees and gorillas are dying across a broad swathe of landscape. Because of the difficulty of monitoring animals in the tropical rainforest, the number of deaths is thought to be greatly underestimated. Thus, the forecast of African ape survival becomes increasingly dire.

Conservationists are applauding the fact that extreme conservation measures have led to a recovery in numbers of the mountain gorillas in the Virunga volcanoes of East Africa. However, these conservation efforts would generally be considered radical: human guards were increased to protect the animals, gorilla groups were monitored daily, and animals suffering from respiratory conditions and other ailments received veterinary care (Robbins *et al.*, 2011). Not only are these radical conservation measures very expensive, but one might also argue that these mountain gorillas are being managed nearly as much as a domesticated species. Do they therefore exist in the "wild?" Does their behavior and ecology faithfully represent unaltered mountain gorilla behavioral ecology? For example, feeding competition has not been observed in the Virunga gorillas (Robbins *et al.*, 2011). This is assumed to be caused by their much diminished numbers. Yet, this condition—reflecting the recent population bottleneck—is clearly a result of human poaching and disease introduced by humans.

In 2002, world political leaders meeting in Johannesburg, South Africa, held a World Summit on Sustainable Development. They resolved to do their utmost to slow the rate of species loss. They expected to see some results by 2012—a grossly over-optimistic expectation. Some factors that contribute to the loss of biodiversity were already known. For example, it had been recognized for a long time that the geographic area of threatened habitats and their degree of fragmentation were important. These factors are so well known that researchers interested in preserving biodiversity commonly use the acronym SLOSS (Single Large Or Several Small fragments) to remember them. But fragmentation overrides total area. Thus, if one must select between a single large or several small preserves, one should opt for a single large one, even if the several small preserves equal it in area. In general, the larger the area and the less fragmented the area, the better are the prospects for species survival.

However, the precipitous slide of biodiversity worldwide may be unstoppable. It is known, for example, that "charismatic" species like mammals, birds, and flowering plants receive an inordinate amount of public attention and research funding. Humble arthropods, fungi, and bacteria receive almost no notice at all. These species are perceived as being intrinsically "boring." Does anyone care about ants, algae, mushrooms, or protozoa? Actually, someone should. The position of such species is fundamental to the normal functioning of an ecosystem. These species are involved in primary production or ecosystem energy flow. As wonderful as roses and langur monkeys are, they do not function in this fashion. It is actually more important to consider what bacteria, nematode worms, and ants are doing. Alas for primatology!

If one should then make crucial decisions about the preservation of biological diversity, those species that contribute to primary production or nutrient cycling ought to be favored. Primates contribute to neither of these areas. They have no obvious economic value (e.g. food, fuel, prospecting for novel biochemicals), and do not contribute to soil formation or water quality. The Convention on Biological Diversity (CBD) was established as a result of the 2002 World Summit on Sustainable Development, and researchers were challenged both to create methods of monitoring biodiversity and to halt its loss. Dobson (2005:Table 1) lists fundamental categories of service that species provide, such as soil formation, nutrient cycling, climate regulation, etc. Although primates provide bushmeat to local people, the inclusion of primate meat in local diets is not crucial to survival. Among the categories listed by Dobson (2005:Table 1), primates would fall only in the more nebulous class: cultural services, ecotourism, and education. If funding decisions for species preservation are made on objective grounds, primates would fail the test.

However, non-human primates do serve a function generally unnoticed by ecosystem biologists and biodiversity researchers. Non-human primates serve as reservoirs of disease and infection for local humans (Cachel, 2006). No one needs to be reminded of the danger of AIDS and worldwide research efforts to combat it. Less well known is that simian immunodeficiency virus (SIV) is ancestral to human HIV. African monkey species, such as vervets and sooty mangabeys, are natural hosts of SIV. Although they are infected with the virus, they never become ill or develop any signs of an AIDS-like disease, and infected females rarely pass SIV on to their offspring (Trivedi, 2010). AIDS is thus a classic example of a zoonosis—a human disease with an ancestral counterpart in an animal species. SIV spread to humans about a hundred years ago, during the late nineteenth or early twentieth centuries, because humans were hunting and consuming chimpanzees. Yet, chimpanzee SIV is a hybrid virus formed from the SIV of red-capped mangabeys and the SIV of greater spot-nosed guenons. Chimpanzee hunting of these forest monkeys led to the novel chimpanzee virus. Long-term study of SIV-infected chimpanzees at the Gombe Reserve now reveals that they eventually sicken and die of a disease resembling human AIDS, and that the fertility of infected females is diminished (Wolfe, 2011). It is ironic that the interplay of chimpanzee and human hunting results in novel diseases in both species.

This account illustrates the terrifying role that non-human primates play as generators of new human diseases and global pandemics. In fact, almost 20 percent of major human infectious diseases have their origins in non-human primates (Wolfe *et al.*, 2007). Twenty million people worldwide are infected with T-lymphotropic viruses, which can cause illnesses like leukemia or paralysis. These viruses have the potential for becoming pandemics. They originate in non-human primates, and spread to humans through bushmeat hunting. Monitoring pathogens in humans who hunt forest primates thus becomes a practical way to prevent the spread of dangerous new human diseases. One major researcher of novel viruses even argues that human penetration of open-country grasslands was a selective advantage because it freed early hominids from hazardous contact with disease ridden non-human primates in the African rainforests (Wolfe, 2011).

Malaria is another human disease with a non-human primate ancestry. Common chimpanzees and gorillas can be infected with *Plasmodium falciparum*, the deadliest of the malarial parasites that infect humans. These species are therefore reservoirs of continuing human infection. Even if this pathogen were eliminated from human populations, humans could be re-infected with the disease through their contact with common chimpanzees and gorillas (Duval *et al.*, 2010). This is true for other diseases, such as Ebola and AIDS, as well as undiscovered plagues that will continue to emerge in tropical forests worldwide.

Primate species may still be preserved in captive conditions. One slight hope for the preservation of chimpanzees in captivity lies in a recent decision by the US Institute of Medicine that deems most biomedical research on chimpanzees to be unnecessary (Anonymous, 2011a). The few acceptable uses of chimpanzees in medicine now involve research such as comparative genomics, or the study of social and behavioral factors that prevent or contribute to disease. Zoos remain the most likely source for preserving all non-human primate species in the foreseeable future.

In the modern world, habitat restriction and habitat loss are the single principal cause of animal and plant extinctions. Data from many sites across Africa document a recent precipitous decline in habitat for the great apes. Between the years 1995 and 2010, suitable habitat has been reduced to the following extent for the African great apes: 59 percent Cross River gorillas, 52 percent mountain gorillas, 32 percent Western gorillas, 29 percent bonobos, 17 percent Central chimpanzees, and 11 percent Western chimpanzees (Junker *et al.*, 2012). The major cause of loss of suitable habitat is patch size reduction and fragmentation.

Food stress may be another problem for large primates, and is now well documented in one species of orangutan. The knife-edge survivorship of surviving hominoids is exemplified by the status of the Bornean orangutan (*Pongo pygmaeus*). Borneo retains the ancient dipterocarp forests that are characteristic of Southeast Asia, and the local soils are very weathered. Plants respond to the nutrient-poor soils by producing fruit on a seasonal, but irregular, basis. Mast fruiting occurs—something that is rare in the tropics. Fruit production is spectacular when it occurs, but is otherwise very limited and highly unpredictable. This makes frugivory a difficult dietary regime, especially for a large-bodied frugivore like the orangutan. The orangutan has responded by developing an extraordinarily low metabolic rate (Pontzer *et al.*, 2010). Despite this response, orangutans often catabolize their fat reserves, and a study of orangutan protein cycling shows that they recycle protein during periods when fruit is rare (Vogel *et al.*, 2012). These animals therefore appear to be at the extreme physiological limits of existence.

Natural habitat fragmentation and diminution occur, but human alteration of environments has become increasingly obvious. Certainly humans have had a major impact on the surface of the earth since their ability to control fire, but their impact has become still more pronounced since the advent of agriculture. Human cutting and burning of forests to create open country for farmland has been a key feature of both Old World and New World landscapes since about 5,000 years ago. Because of the alarming, documented loss of animal and plant species that occurs with the loss of rainforest habitats, humans now heroically attempt to curb habitat loss in tropical forests. This is laudable. Ironically, however, the efforts to go "green" and to utilize renewable energy alternatives has only resulted in devouring more land and destroying habitats at an intensified pace. Using landscapes to create new power sources devastates habitats and annihilates species beyond the rate that traditional slash and burn agriculture does. Renewable or "green" energy sources adversely affect habitats and biodiversity. An example of how a "green" biofuel impacts a primate species can be seen in Borneo. Here, peatland swamp forest is cleared to grow palm-oil for diesel fuel. As a result, hundreds of orangutans are killed every year during the creation of new palm-oil plantations (Ridley, 2010). Bornean orangutans will soon be extinct in the wild. Thus, a large-bodied primate, spared from predation and disease, can still be threatened by habitat loss. The irony in this case is that the habitat loss is driven by human efforts to spare the environment and the natural world of human exploitation by going "green." Going "green" merely intensifies habitat disruption. And habitat disruption involving both humans and non-human primates has the potential for creating new diseases and setting the stage for terrifying global pandemics.

REFERENCES

Abegg, C., & Thierry, B. (2002). Macaque evolution and dispersal in insular South-East Asia. *Biological Journal of the Linnean Society* 75:555–576.

Abello, M. A., Ortiz-Jaureguizar, E., & Candela, A. M. (2012). Paleobiology of the Paucituberculata and Microbiotheria (Mammalia, Marsupialia) from the late Early Miocene of Patagonia. In *Early Miocene Paleobiology in Patagonia. High-Latitude Paleocommunities of the Santa Cruz Formation*, Vizcaíno. S. F., Kay, R. F., & Bargo, M. S., eds. Cambridge: Cambridge University Press, pp. 156–167.

Ackermann. R. R. (2007). Hybrids, spandrels, and the evolution of development: insights from baboons. *American Journal of Physical Anthropology* 132 (supplement 44):60.

Ackermann, R. R. (2010). Phenotypic traits of primate hybrids: recognizing admixture in the fossil record. *Evolutionary Anthropology* 19:258–270.

Ackermann, R. R., Rogers, J., & Cheverud, J. M. (2006). Identifying the morphological signatures of hybridization in primate and human evolution. *Journal of Human Evolution* 51:632–645.

Agusti, J., & Antón, M. (2002). *Mammoths, Sabertooths, and Hominids: 65 Million Years of Mammalian Evolution in Europe*. New York: Columbia University Press.

Ajmone-Marsan, P. A., Garcia, J. F., Lenstra, J. A., & The Globaldiv Consortium. (2010). On the origin of cattle: how aurochs became cattle and colonized the world. *Evolutionary Anthropology* 19:148–157.

Alemseged, Z., Spoor, F., Kimbel, W. H., et al. (2006). A juvenile early hominin skeleton from Dikika, Ethiopia. *Nature* 443:296–301.

Alexander, R. McN. (1985). Body size and limb design in primates and other mammals. In *Size and Scaling in Primate Biology*, Jungers, W. L., ed. New York: Plenum Press, pp. 337–343.

Alexander, R. McN. (1989). *Dynamics of Dinosaurs and Other Extinct Giants*. New York: Columbia University Press.

Alexander, R. McN., Jayes, A. S., Maloiy, G. M. O., & Wathuta, E. M. (1979). Allometry of the limb bones of mammals from shrews (*Sorex*) to elephant (*Loxodonta*). *Journal of Zoology, London* 189:305–314.

Alexander, R. McN., Jayes, A. S., Maloiy, G. M. O., & Wathuta, E. M. (1981). Allometry of leg muscles of mammals. *Journal of Zoology, London* 194:539–552.

Alfaro, J. W., Boubli, J. P., Olsen, L. E., et al. (2012). Explosive Pleistocene range expansion leads to widespread Amazonian sympatry between robust and gracile capuchin monkeys. *Journal of Biogeography* 39:272–288.

Ali, J. A., & Huber, M. (2010). Mammalian biodiversity on Madagascar controlled by ocean currents. *Nature* 463:653–656.

Allman, J. (1982). Reconstructing the evolution of the brain in primates through the use of comparative neurophysiological and neuroanatomical data. In *Primate Brain Evolution*, Armstrong, E., & Falk, D., eds. New York: Plenum Press, pp. 13–28.

Alroy, J. (1998). Diachrony of mammalian appearance events: implications for biochronology. *Geology* 26:23–26.

Altenhoff, A. M., Studer, R. A., Robinson-Rechavi, M., & Dessimoz, C. (2012). Resolving the ortholog conjecture: orthologs tend to be weakly, but significantly, more similar in function than paralogs. *PLoS Computational Biology* **8**. doi:10.1371/journal.pcbi.1002514.

Anonymous (2011a). Chimpanzees in research: statement on Institute of Medicine report by NIH Director Francis Collins. http://www.sciencedaily.com/releases/2011/11121545719.htm.

Anonymous. (2011b). Origin of species. *Nature* **475**:424.

Antoine, P.-O., De Franceschi, D., Flynn, J. J., et al. (2006). Amber from Western Amazonia reveals Neotropical diversity during the middle Miocene. *Proceedings of the National Academy of Sciences USA*. doi:10.1073/pnas.0605801103.

Arnold, M. (1997). *Natural Hybridization and Evolution*. Oxford: Oxford University Press.

Asher, R. J., Novacek, M. J., & Geisler, J. H. (2003). Relationships of endemic African mammals and their fossil relatives based on morphological and molecular evidence. *Journal of Mammalian Evolution* **10**:131–194.

Austin, C., Smith, T. M., Brodman, A., et al. (2013). Barium distributions in teeth reveal early-life dietary transitions in primates. *Nature* **498**:216–219.

Avise, J. (2000). Cladists in wonderland. *Evolution* **54**:1828–1832.

Basmajian, J. V. (1963). Control and training of individual motor units. *Science* **141**:440–441.

Basmajian, J. V. (1972). Electromyography comes of age. *Science* **176**:603–609.

Basmajian, J. V. (1974). *Muscles Alive. Their Functions Revealed by Electromyography*, 3rd edn. Baltimore, MD: Wiliiams & Wilkins Co.

Bauer, K., & Schreiber, A. (1997). Double invasion of Tertiary island South America by ancestral New World monkeys? *Biological Journal of the Linnean Society* **60**:1–20.

Beard, K. C. (1990). Gliding behaviour and palaeoecology of the alleged primate family Paromomyidae (Mammalia, Dermoptera). *Nature* **345**:340–341.

Beard, K. C. (1998a). A new genus of Tarsiidae (Mammalia: Primates) from the middle Eocene of Shanxi Province, China, with notes on the historical biogeography of tarsiers. *Bulletin of the Carnegie Museum of Natural History* **34**:260–277.

Beard, K. C. (1998b). East of Eden: Asia as an important center of taxonomic origination in mammalian evolution. *Bulletin of the Carnegie Museum of Natural History* **34**:5–39.

Beard, K. C. (2004). *The Hunt for the Dawn Monkey. Unearthing the Origins of Monkeys, Apes, and Humans*. Berkeley, CA: University of California Press.

Beard, K. C., Krishtalka, L., & Stucky, R. K. (1991). First skulls of the Early Eocene primate *Shoshonius cooperi* and the anthropoid-tarsier dichotomy. *Nature* **349**:64–66.

Beard, K. C., Qi, T., Dawson, M. R., Wang, B., & Li, C. (1994). A diverse new primate fauna from middle Eocene fissure-fillings in southeastern China. *Nature* **368**:604–609.

Behrensmeyer, A. K., & LaPorte, L. F. (1981). Footprints of a Pleistocene hominid in northern Kenya. *Nature* **289**:167–169.

Bell, P. R., Snively, E., & Shychoski, L. (2009). A comparison of the jaw mechanics in hadrosaurid and ceratopsid dinosaurs using finite element analysis. *The Anatomical Record* **292**:1338–1351.

Benefit, B. R. (1999). *Victoriapithecus*: the key to Old World monkey and catarrhine origins. *Evolutionary Anthropology* **7**:155–174.

Benjamin, M., Toumi, H., Ralphs, J. R., Bydder, G., Best, T. M., & Milz, S. (2006). Where tendons and ligaments meet bone: attachment sites ('entheses') in relation to exercise and/or mechanical load. *Journal of Anatomy* **208**:471–490.

Bennett, M. R., Harris, J. W. K., Richmond, B. G., *et al.* (2009). Early hominin foot morphology based on 1.5-million-year-old footprints from Ileret, Kenya. *Science* 323:1197–1201.

Benton, M. J. (2010). Naming dinosaur species: the performance of prolific authors. *Journal of Vertebrate Paleontology* 30:1478–1485.

Bermejo, M., Rodríguez-Teijeiro, J. D., Illera, G., Barroso, A., Vilà, C., & Walsh, P. D. (2006). Ebola outbreak killed 5000 gorillas. *Science* 314:1564.

Berner, R. A., VandenBrooks, J. M., & Ward, P. D. (2007). Oxygen and evolution. *Science* 316:557–558.

Berry, A., & Browne, J. (2008). The other beetle-hunter. *Nature* 453:1188–1190.

Bertrand, O. C., Flynn, J. J., Croft, D. A., & Wyss, A. R. (2012). Two new taxa (Caviomorpha, Rodentia) from the early Oligocene Tinguiririca fauna (Chile). *American Museum Novitates*, no. 3750.

Blanco, M. B., Dausmann, K. H., Ranaivoarisoa, J. F., & Yoder, A. D. (2013). Underground hibernation in a primate. *Scientific Reports* 3:1768. doi:10.1038/srep01768.

Bloch, J. I., & Boyer, D. M. (2002). Grasping primate origins. *Science* 298: 1606–1610.

Bloch, J. I., & Boyer, D. M. (2003). Response to comment on "Grasping primate origins". *Science* 300:741. [DOI:10.1126/science.1082060].

Bloch, J. I., Silcox, M. T., Boyer, D. M., & Sargis, E. J. (2007). New Paleocene skeletons and the relationship of plesiadapiforms to crown-clade primates. *Proceedings of the National Academy of Sciences USA* 104:1159–1164.

Blois, L. J., & Hadly, A. E. (2009). Mammalian response to Cenozoic climate change. *Annual Review of Earth and Planetary Sciences* 37(8):1–28.

Boesch, C., & Boesch-Achermann, H. (2000). *The Chimpanzees of the Taï Forest. Behavioural Ecology and Evolution.* New York: Oxford University Press.

Bown, T. M., & Rose, K. D. (1987). Patterns of Dental Evolution in Early Eocene anaptomorphine primates (Omomyidae) from the Bighorn Basin, Wyoming. *Journal of Paleontology* 61(supplement), Memoir no. 23.

Boyer, D. M., & Bloch, J. I. (2008). Evaluating the mitten-gliding hypothesis for Paromomyidae and Micromomyidae (Mammalia, "Plesiadapiformes") using comparative functional morphology of new Paleogene skeletons. In *Mammalian Evolutionary Morphology. A Tribute to Frederick S. Szalay*, Sargis, E. J., & Dagosto, M., eds. Dordrecht: Springer, pp. 233–284.

Boyer, D. M., Yapuncich, G. S., Chester, S. G. B., Bloch, J. I., & Godinot, M. (2013). Hands of early primates. *Yearbook of Physical Anthropology* 57:33–78.

Brncic, T. M., Willis, K. J., Harris, D., & Washington, R. (2007). Culture or climate? The relative influences of past processes on the composition of the lowland Congo rainforest. *Philosophical Transactions of the Royal Society B* 362:229–242.

Brooks, T. M., & Helgen, K. M. (2010). A standard for species. *Nature* 467:540–541.

Brown, J. H., & Maurer, B. A. (1987). Evolution of species assemblages: effects of energetic constraints and species dynamics on the diversification of the North American avifauna. *The American Naturalist* 130:1–17.

Brown, J. H., & Maurer, B. A. (1989). Macroecology: the division of food and space among species on continents. *Science* 243:1145–1150.

Brown, J. H., Marquet, P. A., & Taper, M. L. (1993). Evolution of body size: consequences of an energetic definition of fitness. *The American Naturalist* 142:573–584.

Budd, G. E. (2008). The earliest fossil record of the animals and its significance. *Philosophical Transactions of the Royal Society B.* doi:10.1098/rstb.2007.2232.

Burney, D. A., Burney, L. P., Godfrey, L. R., *et al.* (2004). A chronology for late prehistoric Madagascar. *Journal of Human Evolution* 47:25-63.

Bush, E. C., Simons, E. L., & Allman, J. M. (2004). High-resolution computed tomography study of the cranium of a fossil anthropoid primate, *Parapithecus grangeri*: new insights into the evolutionary history of primate sensory systems. *Anatomical Record A* 281A:1083-1087.

Cachel, S. (n.d.) Evolutionary processes and interpretation of the archaeological record. In *Apocalypse Then and Now*, Fernandez, D., *et al.*, ed. Calgary: University of Calgary Press.

Cachel, S. (1979a). A functional analysis of the primate masticatory system and the origin of the anthropoid post-orbital septum. *American Journal of Physical Anthropology* 50:1-18.

Cachel, S. (1979b). A paleoecological model for the origin of higher primates. *Journal of Human Evolution* 8:351-359.

Cachel, S. (1981). Plate tectonics and the problem of anthropoid origins. *Yearbook of Physical Anthropology* 24:139-172.

Cachel, S. (1984). Growth and allometry in primate masticatory muscles. *Archives of Oral Biology* 29:287-293.

Cachel, S. (1992). The theory of punctuated equilibria and evolutionary anthropology. In *The Dynamics of Evolution. The Punctuated Equilibrium Debate in the Natural and Social Sciences*, Somit, A., & Peterson, S. A., eds. Ithaca, NY: Cornell University Press, pp. 187-220.

Cachel, S. (1996). Megadontia in the teeth of early hominids. *Kaupia* 6:119-128.

Cachel, S. (2006). *Primate and Human Evolution*. Cambridge: Cambridge University Press.

Cachel, S. (2009). Using sexual dimorphism and development to reconstruct mating systems in ancient primates. In *Primatology: Theories, Methods and Research*, Potocki, E., & Krasiński, J., eds. New York: Nova Science Publishers, pp. 75-93.

Cande, S. C., & Stegman, D. R. (2011). Indian and African plate motions driven by the push force of the Réunion plume head. *Nature* 475:47-52.

Carotenuto, E., Barbera, C., & Raia, P. (2010). Occupancy, range size and phylogeny in Eurasian Pliocene to recent large mammals. *Paleobiology* 36:399-414.

Cartelle, C., & Hartwig, W. C. (1996). A new extinct primate among the Pleistocene megafauna of Bahia, Brazil. *Proceedings of the National Academy of Sciences USA* 93:6405-6409.

Cartmill, M. (1972). Arboreal adaptations and the origin of the Order Primates. In *The Functional and Evolutionary Biology of Primates*, Tuttle, R. H., ed. Chicago, IL: Aldine, pp. 97-122.

Cartmill, M. (1974a). Daubentonia, Dactylopsila, woodpeckers and klinorhynchy. In *Prosimian Biology*, Martin, R. D., Doyle, G. A., & Walker, A. C., eds. London: Duckworth, pp. 655-670.

Cartmill, M. (1974b). Rethinking primate origins. *Science* 184:436-443.

Cartmill, M. (1975). *Primate Origins*. Minneapolis: Burgess.

Cartmill, M. (1992). New views on primate origins. *Evolutionary Anthropology* 1:105-111.

Cartmill, M., Lemelin, P., & Schmitt, D. (2002). Support polygons and symmetrical gaits in mammals. *Zoological Journal of the Linnean Society* 136:401-420.

Cerling, T. E., & Ehleringer, J. R. (2000). Welcome to the C4 world. In *Phanerozoic Terrestrial Ecosystems*, Gastaldo, R. A., & DiMichele, W. M., eds. Paleontology Society Papers, no. 6, pp. 273–286.

Cerling, T. E., Harris, J. M., Leakey, M. G., Passey, B. H., & Levin, N. E. (2010). Stable carbon and oxygen isotopes in East African mammals: modern and fossil. In *Cenozoic Mammals of Africa*, Werdelin, L., & Sanders, W. J., eds. Berkeley: University of California Press, pp. 941–952.

Cerling, T. E., Wynn, J. G., Andanje, S. A., et al. (2011). Woody cover and hominin environments in the past 6 million years. *Nature* **476**:51–56.

Cerling, T. E., Chritz, K. L., Jablonski, N. G., Leakey, M. G., & Manthi, F. K. (2013). Diet of *Theropithecus* from 4 to 1 Ma in Kenya. *Proceedings of the National Academy of Sciences USA* **110**:10507–10512.

Chaimanee, Y., Jolly, D., Beammi, M., et al. (2003). A Middle Miocene hominoid from Thailand and orangutan origins. *Nature* **422**:61–65.

Chaimanee, Y., Suteethorn, V., Jintasakul, P., Vidthayanon, C., Marandat, B., & Jaeger, J.-J. (2004). A new orang-utan relative from the Late Miocene of Thailand. *Nature* **427**:439–441.

Chaimanee, Y., Yamee C., Tian P., et al. (2006). *Khoratpithecus piriyai*, a Late Miocene hominoid of Thailand. *American Journal of Physical Anthropology* **131**:311–323.

Chaimanee, Y., Chavasseau, O., Beard, K. C., et al. (2012). Late Middle Eocene primate from Myanmar and the initial anthropoid colonization of Africa. *Proceedings of the National Academy of Sciences USA*. doi:10.1073/pnas.1200644109.

Chaitra, M. S., Vasudevan, K., & Shanker, K. (2004). The biodiversity bandwagon: the splitters have it. *Current Science* **86**:897–899.

Charles-Dominique, P. (1977). *Ecology and Behavior of Nocturnal Primates*. New York: Columbia University Press.

Charrier, C., Joshi, K., Coutinho-Budd, J., et al. (2012). Inhibition of *SRGAP2* function by its human-specific paralogs induces neoteny during spine maturation. *Cell* **149**:923–935.

Ciochon, R. (1988). *Gigantopithecus*: the king of all the apes. *Animal Kingdom* **91(2)**:32–39.

Ciochon, R. (2009). The mystery ape of Pleistocene Asia. *Nature* **459**:910–911.

Ciochon, R, Piperno, D. R., & Thompson, R. G. (1990a). Opal phytoliths found on the teeth of the extinct ape, *Gigantopithecus blacki*: implications for paleodietary studies. *Proceedings of the National Academy of Sciences USA* **87**:8120–8124.

Ciochon, R., Olsen, J., & James, J. (1990b). *Other Origins. The Search for the Giant Ape in Human Prehistory*. New York: Bantam Books.

Collard, M., & Wood, B. (2000). How reliable are human phylogenetic hypotheses? *Proceedings of the National Academy of Sciences USA* **97**:5003–5006.

Conniff, R. (2006). For the love of lemurs. *Smithsonian* **37(1)**:102–109.

Conniff, R. (2010). Unclassified. *Discover* **31(5)**:52–57.

Conroy, G. C. (1987). Problems of body-weight estimation in fossil primates. *International Journal of Primatology* **8**:115–137.

Conroy, G. C. (1990). *Primate Evolution*. New York: W. W. Norton.

Conroy, G. C., & Vannier, M. W. (1984). Noninvasive three-dimensional computer imaging of matrix-filled fossil skulls by high-resolution computed tomography. *Science* **226**:456–458.

Conway Morris, S. (1998). *The Crucible of Creation. The Burgess Shale and the Rise of Animals*. New York: Oxford University Press.

Cooper, K. L., & Tabin, C. J. (2008). Understanding of bat wing evolution takes flight. *Genes & Development* **22**:121-124.

Copeland, S. R., Sponheimer, M., de Ruiter, D. J., *et al.* (2011). Strontium isotope evidence for landscape use by ancient hominins. *Nature* **474**:76-79.

Coxall, H. K., Wilson, P. A., Pälike, H., Lear, C. H., & Backman, J. (2005). Rapid stepwise onset of Antarctic glaciation and deeper calcite compensation in the Pacific Ocean. *Nature* **433**:53-57.

Cracraft, J. (1989). Speciation and its ontology. In *Speciation and Its Consequences*, Otte, D., & Endler, J. A., eds. Sunderland, MA: Sinauer, pp. 28-59.

Crampton, J. S., Beu, A. G., Cooper, R. A., Jones, C. M., Marshall, B., & Maxwell, P. A. (2003). Estimating the rock volume bias in paleobiodiversity studies. *Science* **301**: 358-360.

Cretekos, C. J., Wang, Y., Green, E. D., *et al.* (2008). Regulatory divergence modifies limb length between mammals. *Genes & Development* **22**:141-151.

Dabney, J., Knapp, M., Glocke, I., *et al.* (2013). Complete mitochondrial genome sequence of a Middle Pleistocene cave bear reconstructed from ultrashort DNA fragments. *Proceedings of the National Academy of Sciences USA* **110**:15758-15763.

Dagosto, M., Gebo, D. L., & Beard, K. C. (1999). Revision of the Wind River faunas, early Eocene of central Wyoming. Part 14. Postcranium of *Shoshonius cooperi* (Mammalia, Primates). *Annals of the Carnegie Museum* **68**:175-211.

Dalén, L., Nyström, V., Valdiosera, C., *et al.* (2007). Ancient DNA reveals lack of postglacial habitat tracking in the Arctic fox. *Proceedings of the National Academy of Sciences USA* **104**:6726-6729.

Dashzeveg, D., & McKenna, M. C. (1977). Tarsioid primate from the Early Tertiary of the Mongolian People's Republic. *Acta Palaeontologica Polonica* **22**:119-137.

Dashzeveg, D., Novacek, M. J., Norell, M. A., *et al.* (1995). Extraordinary preservation in a new vertebrate assemblage from the Late Cretaceous of Mongolia. *Nature* **374**: 446-449.

de Chardin, P. T. (1922). Les mammifères de l'éocène inférieure francais et leurs gisements. *Annales de paléontologie.* **10**:169-176; **11**:1-108.

de Menocal, P. (1995). Plio-Pleistocene African climate. *Science* **270**:53-59.

de Menocal, P. (2004). African climate change and faunal evolution during the Pliocene-Pleistocene. *Earth and Planetary Science Letters* **220**:3-24.

de Wit, M., & Masters, J. C. (2004). The geological history of Africa, India and Madagascar: dispersal scenarios for vertebrates. *Folia primatologica* **75**:117.

Delisle, R. G. (2006). *Debating Humankind's Place in Nature, 1860-2000: The Nature of Paleoanthropology.* Upper Saddle River, NJ: Pearson/Prentice Hall.

Delson, E. (1973). *Fossil Colobine Monkeys of the Circum-Mediterranean Region and the Evolutionary History of the Cercopithecidae.* Ph.D. dissertation, Department of Paleontology, Columbia University, New York.

Delson, E. (1979). *Oreopithecus* is a cercopithecoid after all. *American Journal of Physical Anthropology* **50**:431-432.

Delson, E. (1986). An anthropoid enigma: historical introduction to the study of *Oreopithecus bambolii*. *Journal of Human Evolution* **15**:523-531.

Delson, E. (2000). Cercopithecinae. In *Encyclopedia of Human Evolution and Prehistory*, Delson, E., Tattersall, I., Van Couvering, J. A., & Brooks, A. S., eds. New York: Garland, pp. 166-171.

Delson, E., & Rosenberger, A. (1984). Are there any anthropoid primate living fossils? In *Living Fossils*, Eldredge, N., & Stanley, S. M., eds. New York: Springer, pp. 50–61.

Dennis, M. Y., Nuttle, X., Sudmant, P. H., *et al.* (2012). Evolution of human-specific neural *SRGAP2* genes by incomplete segmental duplication. *Cell* 149:912–922.

Desmond, A. (1997). *Huxley. From Devil's Disciple to Evolution's High Priest*. Reading, MA: Addison-Wesley.

Dewar, R. E., Radimihaly, C., Wright, H. T., Jacobs, Z., Kelly, G. O., & Berna, F. (2013). Stone tools and foraging in northern Madagascar challenge Holocene extinction models. *Proceedings of the National Academy of Sciences USA*. doi:10.1073/pnas.1306100110.

Dillon, M. E., Wang, G., & Huey, R. B. (2010). Global metabolic impacts of recent climate warming. *Nature* 467:704–706.

DiMichele, W. A., Falcon-Lang, H. J., Nelson, W. J., Elrick, S. D., & Ames, P. R. (2007). Ecological gradients within a Pennsylvanian mire forest. *Geology* 35:415–418.

Douady, C. J., Catzeflis, F., Kao, D. J., Springer, M. S., & Stanhope, M. J. (2002). Molecular evidence for the monophyly of Tenrecidae (Mammalia) and the timing of the colonization of Madagascar by Malagasy tenrecs. *Molecular and Phylogenetic Evolution* 22:357–363.

Dobson, A. (2005). Monitoring global rates of biodiversity change: challenges that arise in meeting the Convention on Biological Diversity (CBD) 2010 goals. *Philosophical Transactions of the Royal Society B* 360:229–241.

Dumont, E. R., Strait, S. G., & Friscia, A. R. (2000). Abderitid marsupials from the Miocene of Patagonia: an assessment of form, function, and evolution. *Journal of Paleontology* 74:1161–1172.

Dunbar, R. (1996). *Grooming, Gossip and the Evolution of Language*. London: Faber and Faber.

Dunsworth, H. M., Warrener, A. G., Deacon, T., Ellison, P. T., & Pontzer, H. (2012). Metabolic hypothesis for human altriciality. *Proceedings of the National Academy of Sciences USA* 109:15212–15216.

Dupont-Nivet, G., Krijgsman, W., Langereis, C. G., Abels, H. A., Dai, S., & Fang, X. (2007). Tibetan plateau aridification linked to global cooling at the Eocene-Oligocene transition. *Nature* 445:635–638.

Duval, L., Fourment, M., Nerrienet, E., *et al.* (2010). African apes as reservoirs of *Plasmodium falciparum* and the origin and diversification of the *Laverania* subgenus. *Proceedings of the National Academy of Sciences USA*. doi:10.1073/pnas.10054335107.

Eberle, J. J., & Greenwood, D. R. (2012). Life at the top of the Eocene greenhouse world—a review of the Eocene flora and vertebrate fauna from Canada's High Arctic. *Geological Society of America Bulletin* 124:3–23.

Eberle, J., Fricke, H., & Humphrey, J. (2009). Lower-latitude mammals as year-round residents in Eocene Arctic forests. *Geology* 37:499–502.

Eldredge, N., & Gould, S. J. (1972). Punctuated equilibria: an alternative to phyletic gradualism. In *Models in Paleobiology*, Schopf, T. J. M., ed. San Francisco: Freeman, Cooper, pp. 82–115.

Eldredge, N., & Cracraft, J. (1980). *Phylogenetic Patterns and the Evolutionary Process*. New York: Columbia University Press.

Elliot Smith, G. (1924). *Essays on the Evolution of Man*. Oxford: Oxford University Press.

Emmons, L. H., & Gentry, A. H. (1983). Tropical forest structure and the distribution of gliding and prehensile-tailed vertebrates. *The American Naturalist* 121:513–524.

Erikson, G. E. (1963). Brachiation in the New World monkeys and in anthropoid apes. *Symposia of the Zoological Society of London* 10:135–164.

Erickson, G. M., Makovicky, P. J., Inouye, B. D., Zhou, C.-F., & Gao, K.-Q. (2009). A life table for *Psittacosaurus lujiatunensis*: initial insights into ornithischian dinosaur population biology. *The Anatomical Record* 292:1514–1521.

Evans, A., Wilson, G. P., Fortelius, M., & Jernvall, J. (2007). High-level similarity of dentitions in carnivorans and rodents. *Nature* 445:78–81.

Ewald, P. W. (1980). Evolutionary biology and the treatment of signs and symptoms of infectious disease. *Journal of Theoretical Biology* 86:169–176.

Feibel, C. S. (2011). Anthropology: shades of the savannah. *Nature* 476:39–40.

Fischer, T. P., Burnard, P., Marty, B., et al. (2009). Upper-mantle volatile chemistry at Oldoinyo Lengai volcano and the origin of carbonatites. *Nature* 459:77–80.

Fitzgerald, J. T. (2007). *1.5 Ma Hominid Fossil Footprints Found at FwJj 14 East, Koobi Fora, Kenya*. Undergraduate Honors Thesis, Department of Anthropology, Rutgers University, New Brunswick, NJ.

Flannery, T. (2001). *The Eternal Frontier. An Ecological History of North America and Its Peoples*. New York: Atlantic Monthly Press.

Flynn, J. J. (1986). Faunal provinces and the Simpson Coefficient. In *Vertebrates, Phylogeny, and Philosophy*, Flanagan, K. M., & Lillegraven, J. A., eds. Contributions to Geology, University of Wyoming, Special Paper no. 3, pp. 317–338.

Flynn, J. J. (2009). Splendid isolation. *Natural History* 118(5):26–32.

Flynn, J. J., & Wyss, A. R. (1998). Recent advances in South American mammalian paleontology. *Trends in Ecology and Evolution* 13:449–454.

Flynn, J. J., Wyss, A. R., Charrier, R., & Swisher, C. C. (1995). An Early Miocene anthropoid skull from the Chilean Andes. *Nature* 373:603–607.

Fortelius, M. (1990). Problems with using fossil teeth to estimate body sizes of extinct mammals. In *Body Size in Mammalian Paleobiology: Estimation and Biological Implications*, Damuth, J., & MacFadden, B. J., eds. Cambridge: Cambridge University Press, pp. 207–228.

Fourie, N. H., Lee-Thorp, J. A., & Rogers Ackermann, R. (2008). Biogeochemical and craniometric investigation of dietary ecology, niche separation, and taxonomy of Plio-Pleistocene cercopithecoids from the Makapansgat Limeworks. *American Journal of Physical Anthropology* 135:121–135.

Frankel, N., Erezyilmaz, D. F., McGregory, A. P., Wang, S., Payre, F., & Stern, D. L. (2011). Morphological evolution caused by many subtle-effect substitutions in regulatory DNA. *Nature* 474:598–603.

Franzen, J. L., Gingerich, P. D., Habersetzer, J., Hurum, J. H., von Koenigswald, W., & Smith, B. H. (2009). Complete primate skeleton from the middle Eocene of Messel in Germany: morphology and paleobiology. *PLoS One* 4(5):1–27.

Franzen, J. L., Habersetzer, J., Schlosser-Sturm, E., & Franzen, E. L. (2012). Paleopathology and fate of Ida (*Darwinius masillae*, Primates, Mammalia). *Palaeobiodiversity and Palaeoenvironments*. doi:10.1007/s12549-012-0102-8.

Freedman, L. (1957). The fossil Cercopithecoidea of South Africa. *Annals of the Transvaal Museum* 23(part 2):121–262 with 52 plates.

Garber, P. A. (1992). Vertical clinging, small body size, and the evolution of feeding adaptations in the Callitrichinae. *American Journal of Physical Anthropology* 88:469–482.

Garber, P. A. (2007). Primate locomotor behavior and ecology. In *Primates in Perspective*, Campbell, C. J., Fuentes, A., Mackinnon, K. C., Panger, M., & Bearder, S. K., eds. New York: Oxford University Press, pp. 543–560.

Garcia-Castellanos, D., Estrada, F., Jiménez-Munt, I., *et al.* (2009). Catastrophic flood of the Mediterranean after the Messinian salinity crisis. *Nature* 462:778–781.

Gebo, D. L. (2004). A shrew-sized origin for primates. *Yearbook of Physical Anthropology* 47:40–62.

Gebo, D. L., Dagosto, M., Beard, K. C., & Qi, T. (2000). The smallest primates. *Journal of Human Evolution* 38:585–594.

Geschwind, D. H., & Konopka, G. (2012). Genes and human brain evolution. *Nature* 486:481–482.

Gibbons, A. (2012). Turning back the clock: slowing the pace of prehistory. *Science* 338:189–191.

Gingerich, P. D. (1973). Anatomy of the temporal bone in the Oligocene anthropoid *Apidium* and the origin of Anthropoidea. *Folia primatologica* 19:329–337.

Gingerich, P. D. (1974). Dental function in the Paleocene primate Plesiadapis. In *Prosimian Biology*, Martin, R. D., Doyle, G. A., & Walker, A. C., eds. London: Duckworth, pp. 531–541.

Gingerich, P. D. (1975a). A new genus of Adapidae (Mammalia, Primates) from the Late Eocene of southern France, and its significance for the origin of higher primates. *Contributions from the Museum of Paleontology, University of Michigan* 24:163–170.

Gingerich, P. D. (1975b). Systematic position of *Plesiadapis*. *Nature* 253:111–113.

Gingerich, P. D. (1976). *Cranial Anatomy and Evolution of Early Tertiary Plesiadapidae (Mammalia, Primates)*. Museum of Paleontology, Papers on Paleontology, no. 15. Ann Arbor: University of Michigan.

Gingerich, P. (1979). The stratophenetic approach to phylogeny reconstruction in vertebrate paleontology. In *Phylogenetic Analysis and Paleontology*, Cracraft, J., & Eldredge, N., eds. New York: Columbia University Press, pp. 41–77.

Gingerich, P. D. (1981). Early Cenozoic Omomyidae and the evolutionary history of tarsiiform primates. *Journal of Human Evolution* 10:345–374.

Gingerich, P. D., & Smith, B. H. (1985). Allometric scaling in the dentition of primates and insectivores. In *Size and Scaling in Primate Biology*, Jungers, W. L., ed. New York: Plenum Press, pp. 257–272.

Glikson, A. Y., Jablonski, D., & Westlake, S. (2010). Origin of the Mt. Ashmore structural dome, West Bonaparte Basin, Timor Sea. *Australian Journal of Earth Sciences* 57: 411–430.

Glob, P. V. (1971). *The Bog People. Iron-Age Man Preserved*. New York: Ballantine Books.

Godfray, H. C. J. (2007). Linnaeus in the information age. *Nature* 446:259–260.

Godfrey, L. R., Schwartz, G. T., Samonds, K. E., Jungers, W. L., & Catlett, K. K. (2006). The secrets of lemur teeth. *Evolutionary Anthropology* 15:142–154.

Godfrey, L. R., Jungers, W. L., & Burney, D. A. (2010). Subfossil lemurs of Madagascar. In *Cenozoic Mammals of Africa*, Werdelin, L., & Sanders, W. J., eds. Berkeley, CA: University of California Press, pp. 351–367.

Godinot, M. (2010). Paleogene prosimians. In *Cenozoic Mammals of Africa*, Werdelin, L., & Sanders, W. J., eds. Berkeley, CA: University of California Press, pp. 319–331.

Goldberg, P., & Macphail, R. I. (2006). *Practical and Theoretical Geoarchaeology*. Oxford: Blackwell Science.

Goodman, S. M., & Benstead, J. P., eds. (2003). *The Natural History of Madagascar.* Chicago, IL: University of Chicago Press.

Gordon, C. L. (2003). A first look at estimating body size in dentally conservative marsupials. *Journal of Mammalian Evolution* 10:1–21.

Gould, S. J., & Lewontin, R. C. (1979). The spandrels of San Marco and the Panglossian paradigm. A critique of the adaptationist program. *Proceedings of the Royal Society of London B* 205:581–598.

Gradstein, F. M., Ogg, J. G., Smith, A. G., et al. (2004). *A Geologic Time Scale 2004.* Cambridge: Cambridge University Press.

Grant, P. R., & Grant, B. R. (1992). Hybridization of bird species. *Science* 256:193–197.

Green, R. E., Krause, J., Briggs, A. W., et al. (2010). A draft sequence of the Neanderthal genome. *Science* 328:710–722.

Gregory, W. K. (1916). Studies on the evolution of the primates. *Bulletin of the American Museum of Natural History* 35:239–355.

Gregory, W. K. (1920). On the structure and relations of *Notharctus*, an American Eocene primate. *Memoirs of the American Museum of Natural History*, vol. 3, part 2.

Gregory, W. K. (1949). The bearing of the Australopithecinae upon the problem of man's place in nature. *American Journal of Physical Anthropology* 7:485–512.

Gregory, W. K. (1951). *Evolution Emerging. A Survey of Changing Patterns from Primeval Life to Man,* 2 vols. New York: The Macmillan Company.

Gregory, W. K., & Hellman, M. (1926). The dentition of *Dryopithecus* and the origin of man. *Anthropological Papers of the American Museum of Natural History*, vol. 28, part 1.

Groves, C. (2001). *Primate Taxonomy.* Washington DC: Smithsonian Institution Press.

Gursky-Doyen, S. (2010). Married to the mob. *Natural History* 119:20–26.

Hamon, N., Sepulchre, P., Donnadieu, Y., et al. (2012). Growth of subtropical forests in Miocene Europe: the roles of carbon dioxide and Antarctic ice sheets. *Geology* 40: 567–570.

Hamrick, M. W. (2012). The developmental origins of mosaic evolution in the primate limb skeleton. *Evolutionary Biology* 39:447–455.

Haq, B. U., Hardenbol, J., & Vail, P. R. (1987). The chronology of fluctuating sea level since the Triassic. *Science* 235:1156–1167.

Harjunmaa, E., Kallonen, A., Voutilainen, M., Hämäläinen, K., Mikkola, M. L., & Jernvall, J. (2012). On the difficulty of increasing dental complexity. *Nature* 483:324–327.

Harrison, T. (2010a). Dendropithecoidea, Proconsuloidea, and Hominoidea. In *Cenozoic Mammals of Africa*, Werdelin, L., & Sanders, W. J., eds. Berkeley, CA: University of California Press, pp. 429–469.

Harrison, T. (2010b). Later Tertiary Lorisiformes. In *Cenozoic Mammals of Africa*, Werdelin, L., & Sanders, W. J., eds. Berkeley, CA: University of California Press, pp. 333–349.

Harrison, T., Ji, X., & Zheng, L. (2008). Renewed investigations at the Late Miocene hominoid locality Leilao, Yunnan, China. *American Journal of Physical Anthropology* **135(supplement 46)**:113.

Hartenberger, J.-L. (1998). An Asian *Grande Coupure. Nature* 394:321.

Hartwig, W. C. (1995). A giant New World monkey from the Pleistocene of Brazil. *Journal of Human Evolution* 28:189–195.

Hartwig, W. C., Ed. (2002). *The Primate Fossil Record.* Cambridge: Cambridge University Press.

Hartwig, W. C., & Cartelle, C. (1996). A complete skeleton of the giant South American primate *Protopithecus*. *Nature* 381:307-310.

Hatala, K. G., Dingwall, H. L., Wunderlich, R. E., & Richmond, B. G. (2012). An experimentally-based interpretation of 1.5 million-year-old fossil hominin footprints: implications for the evolution of human foot function. *American Journal of Physical Anthropology* Supplement 54:161.

Hawks, J. (2004). How much can cladistics tell us about early hominid relationships? *American Journal of Physical Anthropology* 125:207-219.

Hayssen, V. D. (1984). Mammalian reproduction: constraints on the evolution of infanticide. In *Infanticide. Comparative and Evolutionary Perspectives*, Hausfater, G., & Hrdy, S. B., eds. New York: Aldine, pp. 105-123.

Heads, M. (2010). Evolution and biogeography of primates: a new model based on molecular phylogenetics, vicariance and plate tectonics. *Zoologica Scripta* 39:107-127.

Heard-Booth, A. N., & Kirk, E. C. (2012). The influence of maximum running speed on eye size: a test of Lueckart's law in mammals. *The Anatomical Record.* doi:10.1002/ar.22480.

Heesy, C. P. (2005). Function of the mammalian postorbital bar. *Journal of Morphology* 264:263-380.

Held, Jr., L. I. (2010). The evolutionary geometry of human anatomy: discovering our inner fly. *Evolutionary Anthropology* 19:227-235.

Hendry, A. P. (2007). Darwin in the fossils. *Nature* 451:779-780.

Henneberg, M. (1997). The problem of species in hominid evolution. *Perspectives in Human Biology* 3:21-31.

Henneberg, M., & Brush, G. (1994). Similum, a concept of flexible, synchronous classification replacing rigid species in evolutionary thinking. *Evolutionary Theory* 10:278.

Hennig, W. (1966). *Phylogenetic Systematics.* Urbana, IL: University of Illinois Press.

Henry, A. G., Ungar, P. S., Passey, B. H., *et al.* (2012). The diet of *Australopithecus sediba*. *Nature* 487:90-93.

Hershkovitz, P. (1974). A new genus of Oligocene monkey (Cebidae, Platyrrhini) with notes on postorbital closure and platyrrhine evolution. *Folia primatologica* 21:1-35.

Hershkovitz, P. (1977). *Living New World Monkeys (Platyrrhini) with an Introduction to Primates.* Chicago, IL: University of Chicago Press.

Heuvelmans, B. (1995). *On the Track of Unknown Animals*, 3rd edn. London: Kegan Paul.

Hill, T. M., Kennett, J. P., Valentine, D. L., *et al.* (2006). Climatically driven emissions of hydrocarbons from marine sediments during deglaciation. *Proceedings of the National Academy of Sciences USA.* doi:10.1073/pnas.0601304103.

Hine, C. (2008). *Systematics in Cyberscience: Computers, Change, and Continuity in Science.* Cambridge, MA: MIT Press.

Hoffstetter, R. (1971). Le peuplement mammalien de l'Amérique du Sud. Rôle des continents austraux comme centres d'origine, de diversification et de dispersion pour certains groups mammaliens. *Anais da Academia Brasileira de Ciências* 43(supplement): 125-144.

Hoffstetter, R. (1972). Relationships, origins, and history of the ceboid monkeys and caviomorph rodents: a modern reinterpretation. In *Evolutionary Biology*, vol. 6, Dobzhansky, T., Hecht, M. K., & Steere, W. C., eds. New York: Appleton-Century-Crofts, pp. 323-347.

Hoffstetter, R. (1977). Phylogénie des primates. Confrontation des résultats obtenus par les diverses voies d'approche du problème. *Bulletins et Mémoires de la Société d'Anthropologie de Paris* **4**:327–346.

Hopper, S. D. (2007). New life for systematics. *Science* **316**:1097.

Hull, D. L. (1988). *Science as a Process*. Chicago, IL: University of Chicago Press.

Hunt, G. (2007a). Evolutionary divergence in directions of high phenotypic variance in the ostracode genus *Poseidonamicus*. *Evolution* **61**:1560–1576.

Hunt, G. (2007b). The relative importance of directional change, random walks, and stasis in the evolution of fossil lineages. *Proceedings of the National Academy of Sciences USA* **104**:18404–18408.

Hürzeler, J. (1958). *Oreopithecus bambolii* Gervais. A preliminary report. *Verhandlungen der Naturforschung Gesellschaft Basel* **69(1)**:1–48.

Hürzeler, J. (1960). The significance of *Oreopithecus* in the genealogy of man. *Triangle* **4**:164–175.

Hutchinson, G. E. (1959). Homage to Santa Rosalia, or why are there so many kinds of animals? *The American Naturalist* **93**:145–159.

Huxley, J. S. (1932). *Problems of Relative Growth*. London: MacVeagh.

Huxley, T. H. (1863). *Evidence as to Man's Place in Nature*. Ann Arbor: University of Michigan Press [reprinted in 1959].

Iturralde-Vinent, M. A., & MacPhee, R. D. E. (1999). Paleogeography of the Caribbean region: implications for Cenozoic biogeography. *Bulletin of the American Museum of Natural History* **238**:1–95.

Jablonski, D. (1999). The future of the fossil record. *Science* **284**:2114–2116.

Jablonski, D., Roy, K., Valentine, J. W., Price, R. M., & Anderson, P. S. (2003). The impact of the pull of the recent on the history of marine diversity. *Science* **300**:1133–1135.

Jablonski, N. G., Ed. (1993). *Theropithecus: Rise and Fall of a Primate Genus*. Cambridge: Cambridge University Press.

Jablonski, N. G. (2002). Fossil Old World monkeys: the Late Neogene radiation. In *The Primate Fossil Record*, Hartwig, W. C., ed. Cambridge: Cambridge University Press, pp. 255–299.

Jablonski, N. G., & Frost, S. (2010). Cercopithecoidea. In *Cenozoic Mammals of Africa*, Werdelin, L., & Sanders, W. J., eds. Berkeley, CA: University of California Press, pp. 393–428.

Jacobs, G. H. (2008). Primate color vision: a comparative perspective. *Visual Neuroscience* **25**:619–633.

Jacobs, G. H., & Nathans, J. (2009). The evolution of primate color vision. *Scientific American* **April**:56–63.

Jaeger, J.-J., Beard, K. C., Chaimanee, Y., *et al.* (2010a). Late middle Eocene epoch of Libya yields earliest known radiation of African anthropoids. *Nature* **467**:1095–1098.

Jaeger, J.-J., Marivaux, L., Salem, M., *et al.* (2010b). New rodent assemblages from the Eocene Dur At-Talah escarpment (Sahara of Central Libya): systematic, biochronological, and palaeogeographical implications. *Zoological Journal of the Linnean Society* **160**:195–213.

Janečka, J. E., Miller, W., Pringle, T. H., *et al.* (2007). Molecular and genomic data identify the closest living relative of primates. *Science* **318**:792–794.

Janzen, D. H. (1967). Why mountain passes are higher in the tropics. *The American Naturalist* **101**:233–249.

Janzen, D. H., & Martin, P. H. (1983). Neotropical anachronisms: the fruits the gomphotheres ate. *Science* 215:19–27.

Jenkins, Jr., F. A., & Krause, D. W. (1983). Adaptations for climbing in North American multituberculates (Mammalia). *Science* 220:712–715.

Jenner, R. A., & Littlewood, D. T. J. (2008). Problematica old and new. *Philosophical Transactions Royal Society B.* doi:10.1098/rstb.2007.2240.

Jernvall, J., & Wright, P. C. (1998). Diversity components of impending primate extinctions. *Proceedings of the National Academy of Sciences USA* 95:11279–11283.

Ji, X., Jablonski, N. G., Su, D. F., *et al.* (2013). Juvenile hominoid cranium from the terminal Miocene of Yunnan, China. *Chinese Science Bulletin.* doi:10.1007/s11434-013-6021-x.

Jolly, A. (2004). *Lords and Lemurs: Mad Scientists, Kings with Spears, and the Survival of Diversity in Madagascar.* Boston, MA: Houghton Mifflin.

Jolly, A. (2006). A global vision. *Nature* 443:148.

Jolly, C. J. (1970). The seed-eaters: a new model of hominid differentiation based on a baboon analogy. *Man* 5:1–26.

Jones, F. C., Grabherr, M. G., Chan, Y. F., *et al.* (2012). The genomic basis of adaptive evolution in threespine sticklebacks. *Nature* 484:55–61.

Jones, N. (2010). Battle to degas deadly lakes continues. *Nature* 466:1033.

Jones, P. (1994). Biodiversity in the Gulf of Guinea. *Biodiversity and Conservation* 3:772–784.

Jungers, W. L. (1977). Hindlimb and pelvic adaptations to vertical climbing and clinging in *Megaladapis*, a giant subfossil prosimian from Madagascar. *Yearbook of Physical Anthropology* 20:508–524.

Jungers, W. L. (1978). The functional significance of skeletal allometry in *Megaladapis* in comparison to living prosimians. *American Journal of Physical Anthropology* 19:303–314.

Junker, J., Blake, S., Boesch, C., *et al.* (2012). Recent decline in suitable environmental conditions for African great apes. *Diversity and Distributions* 18:1077–1091.

Kangas, A. T., Evans, A. R., Thesleff, L., & Jernvall, J. (2004). Nonindependence of mammalian dental characters. *Nature* 432:211–214.

Kaplan, M. (2012). Primates were always tree-dwellers. *Nature.* doi:10.1038/nature.2012.11423.

Karanth, K. P., Delefosse, T., Rakotosumimanana, B., Parsons, T. J., & Yoder, A. D. (2005). Ancient DNA from giant extinct lemurs confirms single origin of Malagasy primates. *Proceedings of the National Academy of Sciences USA* 102:5090–5095.

Kavanagh, K. D., Evans, A. R., & Jernvall, J. (2007). Predicting evolutionary patterns of mammalian teeth from development. *Nature* 449:427–432.

Kay, R. F. (1975). The functional adaptations of primate molar teeth. *American Journal of Physical Anthropology* 43:195–216.

Kay, R. F. (1980). Platyrrhine origins. A reappraisal of the dental evidence. In *Evolutionary Biology of the New World Monkeys and Continental Drift*, Ciochon, R. L., & Chiarelli, A. B., eds. New York: Plenum Press, pp. 159–188.

Kay, R. F. (2012). Evidence of an Asian origin for stem anthropoids. *Proceedings of the National Academy of Sciences USA.* doi:10.1073/pnas.1207933109.

Kay, R. F., Thorington, Jr., R. W., & Houde, P. (1990). Eocene plesiadapiform shows affinities with flying lemurs and primates. *Nature* 345:342–344.

Kay, R. F., Williams, B A., & Anaya, F. (2001). The adaptations of Branisella boliviana, the earliest South American monkey. In *Reconstructing Behavior in the Primate Fossil Record*, Plavcan, J. M., van Schaik, C., Kay, R. F., & Jungers, W. L., eds. New York: Kluwer/Plenum, pp. 339-370.

Kay, R. F., Campbell, V. M., Rossie, J. B., Colbert, M. W., & Rowe, T. B. (2004). Olfactory fossa of *Tremacebus harringtoni* (Platyrrhini, Early Miocene, Sacanana, Argentina): implications for activity pattern. *The Anatomical Record A* 281:1157-1172.

Kay, R. F., Simons, E. L., & Ross, J. L. (2008). The basicranial anatomy of African Eocene/Oligocene anthropoids. Are there any clues for platyrrhine origins? In *Elwyn Simons: A Search for Origins*, Fleagle, J. G., & Gilbert, C. C., eds. New York: Springer, pp. 125-158.

Kay, R. F., Perry, J. M. G., Malinzak, M., et al. (2012). Paleobiology of Santacrucian primates. In *Early Miocene Paleobiology in Patagonia: High-Latitude Paleocommunities of the Santa Cruz Formation*, Vizcaino, S. F., Kay, R. F., & Bargo, M. S., eds. New York: Cambridge University Press, pp. 306-330.

Kermack, D. M., & Kermack, K. A. (1984). *The Evolution of Mammalian Characters*. Washington DC: Kapitan Szabo.

King, S. J., Arrigo-Nelson, S., Pochron, S. T., et al. (2005). Dental senescence in a long-lived primate links infant survival to rainfall. *Proceedings of the National Academy of Sciences USA* 102:16579-16583.

Kirk, E. C. (2004). Effects of activity pattern on eye and orbit morphology in primates. *American Journal of Physical Anthropology* 123(supplement 38):126.

Kirk, E. C. (2006). Effects of activity pattern on eye size and orbital aperture size in primates. *Journal of Human Evolution* 51:159-170.

Kirk, E. C., & Simons, E. L. (2001). Diets of fossil primates from the Fayum Depression of Egypt: a quantitative analysis of molar shearing. *Journal of Human Evolution* 40:203-229.

Kist, R., Watson, M., Wang, X., et al. (2005). Reduction of *Pax9* gene dosage in an allelic series of mouse mutants causes hypodontia and oligodontia. *Human Molecular Genetics* 14:3605-3617.

Kivell T. L., & Schmitt D. (2009). Independent evolution of knuckle-walking in African apes shows that humans did not evolve from a knuckle-walking ancestor. *Proceedings of the National Academy of Sciences USA* 106(34):14241-14246.

Klepinger, L. L. (2006). *Fundamentals of Forensic Anthropology*. Hoboken, NJ: John Wiley & Sons.

Knapp, S., Polaszek, A., & Watson, M. (2007). Spreading the word. *Nature* 446:261-262.

Knott, C. D., & Thompson, M. E. (2012). C-peptide and the cost of reproduction in Bornean orangutans. *American Journal of Physical Anthropology* Supplement 54:184.

Köhler, M., & Moyà-Solà, S. (1997). Ape-like or hominid-like? The positional behavior of *Oreopithecus bambolii* reconsidered. *Proceedings of the National Academy of Sciences USA* 94:11747-11750.

Köhler, M., & Moyà-Solà, S. (1999). A finding of Oligocene primates on the European continent. *Proceedings of the National Academy of Sciences USA* 96: 14664-14667.

Köhler, M., Marín-Moratalla, N., Jordana, X., & Aanes, R. (2012). Seasonal bone growth and physiology in endotherms shed light on dinosaur physiology. *Nature* 487: 358-361.

Kolbe, J. J., Leal, M., Schoener, T. W., Spiller, D. A., & Losos, J. B. (2012). Founder effects persist despite adaptive differentiation: a field experiment with lizards. *Science* **335**:1086–1089.

Kordos, L., & Begun, D. R. (2002). Rudabánya: a Late Miocene subtropical swamp deposit with evidence of the origin of African apes and humans. *Evolutionary Anthropology* **11**:45–57.

Krause, D. W. (1986). Competitive exclusion and taxonomic displacement in the fossil record: the case of rodents and multituberculates in North America. In *Vertebrates, Phylogeny, and Philosophy*, Flanagan, K. M., & Lillegraven, J. A., eds. Contributions to Geology, University of Wyoming, Special Paper no. 3, pp. 95–117.

Krause, D. W. (1991). Were paromomyids gliders? Maybe, maybe not. *Journal of Human Evolution* **21**:177–188.

Krause, D. W. (2010). Biogeography: washed up in Madagascar. *Nature* **463**:613–614.

Kull, C. A. (2004). *Isle of Fire. The Political Ecology of Landscape Burning in Madagascar.* Chicago, IL: University of Chicago Press.

Ladevèze, S., de Muizon, C., Beck, R. M. D., Germain, D., & Cespedes-Paz, R. (2011). Earliest evidence of mammalian social behavior in the Basal Tertiary of Bolivia. *Nature* **474**:83–86.

Laidler, K. (2005). *Female Caligula. Ranavalona, the Mad Queen of Madagascar.* Hoboken, NJ: John Wiley & Sons.

Langergraber, K. E., Prüfer, K., Rowney, C., *et al.* (2012). Generation times in wild chimpanzees and gorillas suggest earlier divergence times in great ape and human evolution. *Proceedings of the National Academy of Sciences USA* **109**:15716–15721.

Laporte, L. F. (2000). *George Gaylord Simpson. Paleontologist and Evolutionist.* New York: Columbia University Press.

Lawlor, T. E. (1986). Comparative biogeography of mammals on islands. *Biological Journal of the Linnean Society* **28**:99–125.

Le Gros Clark, W. E. (1964). *The Fossil Evidence for Human Evolution*, rev. edn. Chicago, IL: University of Chicago Press.

Le Gros Clark, W. E. (1971). *The Antecedents of Man*, 3rd edn. Chicago, IL: Quadrangle Books.

Lee, A. K., & Cockburn, A. (1985). *Evolutionary Ecology of Marsupials.* Cambridge: Cambridge University Press.

Lee-Thorp, J., & Sponheimer, M. (2006). Contributions of biogeochemistry to understanding hominin dietary ecology. *Yearbook of Physical Anthropology* **49**:131–148.

Lee-Thorp, J. A., & Sponheimer, M. (2013). Hominin ecology from hard tissue biogeochemistry. In *Early Hominin Paleoecology*, Sponheimer, M., Lee-Thorp, J. A., Reed, K. E., & Ungar, P. S., eds. Boulder, CO: University Press of Colorado, pp. 281–324.

Leendertz, F. H., Lankester, P., Guislain, P., *et al.* (2006). Anthrax in Western and Central African great apes. *American Journal of Primatology* **68**:928–933.

Lei, R., Engberg, S. E., Andriantompohavana, R., *et al.* (2008). *Nocturnal Lemur Diversity at Masoala National Park.* Museum of Texas Tech University, Special Publication no. 53.

Lewin, R. (1987). *Bones of Contention: Controversies in the Search for Human Origins.* New York: Simon and Schuster.

Lewitus, E., Hof, P. R., & Sherwood, C. C. (2012). Phylogenetic comparison of neuron and glial densities in the primary visual cortex and hippocampus of carnivores and primates. *Evolution* **66**:2551–2563.

Lewton, K. L. (2012). Evolvability of the primate pelvic girdle. *Evolutionary Biology* 39: 126–139.

Li, W. H., & Tanimura, M. (1987). The molecular clock runs more slowly in man than in apes and monkeys. *Nature* 326:93–96.

Locke, D. P., Hillier, L. W., Warren, W. C., et al. (2011). Comparative and demographic analysis of orang-utan genomes. *Nature* 469:529–533.

Lockley, M. (1999). *The Eternal Trail. A Tracker Looks at Evolution*. Reading, MA: Perseus Books.

Lockley, M. (2007). A tale of two ichnologies: the different goals and potentials of invertebrate and vertebrate (tetrapod) ichnotaxonomy and how they relate to ichnofacies analysis. *Ichnos* 14:39–57.

Loope, D. B., Dingus, L., Swisher III, C. C., & Minjin, C. (1998). Life and death in a Cretaceous dune field, Nemegt Basin, Mongolia. *Geology* 26:27–30.

Lucas, P. W. (2004). *Dental Functional Morphology: How Teeth Work*. Cambridge: Cambridge University Press.

Luo, Z.-X., Yuan, C.-X., Meng, Q.-J., & Ji, Q. (2011). A Jurassic eutherian mammal and divergence of marsupials and placentals. *Nature* 476:442–445.

Mack, J. (1986). *Madagascar. Island of the Ancestors*. London: British Museum Publications.

MacLeod, N., Benfield, M., & Culverhouse, P. (2010). Time to automate identification. *Nature* 467:154–155.

MacPhee, R. D. E., & Fleagle, J. G. (1991). Postcranial remains of *Xenothrix mcgregori* (Primates, Xenotrichae) and other late Quaternary mammals from Long Mile Cave, Jamaica. *Bulletin of the American Museum of Natural History* 206:287–321.

MacPhee, R. D. E., & Horovitz, I. (2004). New craniodental remains of the Quaternary Jamaican monkey *Xenothrix mcgregori* (Xenotrichini, Callicebinae, Pitheciidae) with a reconsideration of the *Aotus* hypothesis. *American Museum Novitates* 3434:1–51.

MacPhee, R. D. E., & Iturralde-Vinent, M. A. (1995). Earliest monkey from Greater Antilles. *Journal of Human Evolution* 28:197–200.

MacPhee, R. D. E., & Jacobs, L. L. (1986). Nycticeboides simpsoni and the morphology, adaptations, and relationships of Miocene Siwalik Lorisidae. In *Vertebrates, Phylogeny, and Philosophy*, Flanagan, K. M., & Lillegraven, J. A., eds. Contributions to Geology, University of Wyoming. Special Paper no. 3, pp. 131–161.

Maiolano, S., Boyer, D. M., Bloch, J. I., Gilbert, C. C., & Groenke, J. (2012). Evidence for a grooming claw in a North American Adapiform primate: implications for anthropoid origins. *PLoS ONE*. doi:10.1371/journal.pone.0029135.

Malukiewicz, J., Grativol, A. D., Ruiz-Miranda, C. R., & Stone, A. C. (2012). Almost Carioca: hybridization between introduced populations of *Callithrix jacchus* and *C. penicillata* in Rio de Janeiro State, Brazil. *American Journal of Physical Anthropology* **Supplement 54**:202.

Marmi, J., Casanovas-Vilar, I., Robles, J. M., Moyà-Solà, S., & Alba, D. H. (2012). The paleoenvironment of *Hispanopithecus laietanus* as revealed by paleobotanical evidence from the Late Miocene of Can Llobateres 1 (Catalonia, Spain). *Journal of Human Evolution* 62:412–423.

Marques-Bonet, T., Kidd, J. M., Ventura, M., et al. (2009). A burst of segmental duplications in the genome of the African great ape ancestor. *Nature* 457:877–881.

Marris, E. (2007a). The species and the specious. *Nature* 446:250–253.

Marris, E. (2007b). What to let go. *Nature* 450:152–155.

Ni, X., Wang, Y., Hu, Y., & Li, C. (2004). A euprimate skull from the early Eocene of China. *Nature* **427**:65–68.

Ni, X., Gebo, D. L., Dagosto, M., *et al.* (2013). The oldest known primate skeleton and early haplorhine evolution. *Nature* **498**:60–64.

Niedźwiedzki, G., Szrek, P., Narkiewicz, K., Narkiewicz, M., & Ahlberg, P. E. (2010). Tetrapod trackways from the early Middle Devonian period of Poland. *Nature* **463**:43–48.

Nilsson, M. A., Churakov, G., Sommer, M., *et al.* (2010). Tracking marsupial evolution using archaic genomic retroposon insertions. *PloS Biology* **8(7)**:e1000436.

Nordt, L., Atchley, S., & Dworkin, S. (2003). Terrestrial evidence for two greenhouse events in the latest Cretaceous. *GSA Today* **13**:4–9.

Norell, M., & Ellison, M. (2005). *Unearthing the Dragon. The Great Feathered Dinosaur Discovery*. New York: Pi Press.

Norell, M. A., Clark, J. M., Chiappe, L. M., & Dashzeveg, D. (1995). A nesting dinosaur. *Nature* **378**:774–776.

Novacek, M. (2007). *Terra. Our 100-Million-Year-Old Ecosystem—and the Threats That Now Put It at Risk*. New York: Farrar, Straus and Giroux.

Novacek, M. J., Wyss, A. B., & McKenna, M. J. (1988). The major groups of eutherian mammals. In *The Phylogeny and Classification of the Tetrapods,* vol. 2: *Mammals,* Benton, M. J., ed. Oxford: Oxford University Press, pp. 31–71.

Novacek, M. J. (1993). Patterns of diversity in the mammalian skull. In *The Skull,* vol. 2: *Patterns of Structural and Systematic Diversity,* Hankin, J., & Hall, B. K., eds., Chicago, IL: University of Chicago Press, pp. 438–545.

Nowak, R. M. (1999). *Walker's Mammals of the World*, 2 vols., 6th edn. Baltimore, MD: Johns Hopkins University Press.

O'Leary, M. A., Bloch, J. I., Flynn, J. J., *et al.* (2013). The placental mammal ancestor and the post-K-Pg radiation of placentals. *Science* **339**:662–667.

Olsen, E. C., & Miller, R. L. (1958). *Morphological Integration*. Chicago, IL: University of Chicago Press.

Osborn, H. F. (1908). New fossil mammals from the Fayûm Oligocene, Egypt. *American Museum of Natural History Bulletin* **26**:415–424.

Osman Hill, W. C. (1972). *Evolutionary Biology of the Primates*. London: Academic Press.

Pagani, M., Pedentchouk, N., Huber, M., *et al.* (2006). Arctic hydrology during global warming at the Paleocene/Eocene thermal maximum. *Nature* **442**:671–675.

Pälike, H., Norris, R. D., Herrle, J. O., *et al.* (2006). The heartbeat of the Oligocene climate system. *Science* **314**:1894–1898.

Pancost, R. D., Steart, D. S., Handley, L., *et al.* (2007). Increased terrestrial methane cycling at the Palaeocene-Eocene thermal maximum. *Nature* **449**:332–335.

Pakskewitz, S. (2011). Species complexes: confusion in identifying the true vector of malaria and other parasites. The Biomedical and Life Sciences Collection. London: Henry Stewart Talks, Ltd. (http://hstalks.com/bio).

Paterson, H. E. H. (1981). The continuing search for the unknown and unknowable: a critique of contemporary ideas on speciation. *South African Journal of Science* **77**: 113–119.

Paterson, H. E. H. (1982). Perspective on speciation by reinforcement. *South African Journal of Science* **78**:53–57.

Paterson, H. E. H. (1986). Environment and species. *South African Journal of Science* **82**:62–65.

Patterson, B. (2013). (Re)naming the lions of Africa. *In the Field* **Summer 2013**:9.

Pearson, P. N., Foster, G. L., & Wade, B. S. (2009). Atmospheric carbon dioxide through the Eocene-Oligocene climate transition. *Nature* **461**:1110–1113.

Pennisi, E. (2012). The great guppy experiment. *Science* **337**:904–908.

Perelman, P., Johnson, W. E., Roos, C., et al. (2011). A molecular phylogeny of living primates. *PLoS Genetics* **7(3)**:e1001342.

Perry, G. H., Dominy, N., Claw, K. G., et al. (2007). Diet and the evolution of human amylase gene copy number variation. *Nature Genetics* **39**:1256–1260.

Peters, S. E. (2005). Geologic constraints on the macroevolutionary history of marine animals. *Proceedings of the National Academy of Sciences USA* **102**: 12326–12331.

Peterson, K. J., Cotton, J. A., Gehling, J. G., & Pisani, D. (2008). The Ediacaran emergence of bilaterians: congruence between the genetic and geological fossil records. *Philosophical Transactions of the Royal Society B*. doi:10.1098/rstb.2007.2233.

Pilbeam, D. R. (1969). Tertiary Pongidae of East Africa: evolutionary relationships and taxonomy. *Bulletin Peabody Museum of Natural History Yale University*, no. 31.

Pilbeam, D. (1982). New hominoid skull material from the Miocene of Pakistan. *Nature* **295**:232–234.

Pontzer, H., Raichlen, D. A., Shumaker, R. W., Ocobock, C., & Wich, S. A. (2010). Metabolic adaptation for low energy throughput in orangutans. *Proceedings of the National Academy of Sciences USA* **107**:14048–14052.

Powe, C. E., Knott, C. D., & Conklin-Brittain, N. (2010). Infant sex predicts breast milk energy content. *American Journal of Human Biology* **22**:50–54.

Pozzi, L., Hodgson, J. A., Burrell, A. S., & Disotell, T. R. (2011). The stem catarrhine *Saadanius* does not inform the timing of the origin of crown catarrhines. *Journal of Human Evolution* **61**:209–210.

Prüfer, K., Munch, K., Hellmann, I., et al. (2012). The bonobo genome compared with the chimpanzee and human genomes. *Nature* **486**:527–531.

Pruvost, M., Schwarz, R., Bessa Correia, V., et al. (2007). Freshly excavated fossil bones are best for amplification of ancient DNA. *Proceedings of the National Academy of Sciences USA*. doi:10.1073/pnas.0610257104.

Quental, T. B., & Marshall, C. R. (2013). How the Red Queen drives terrestrial mammals to extinction. *Science*. doi:10.1126/science.1239431.

Radinsky, L. (1979). *The Fossil Record of Primate Brian Evolution*. Forty-Ninth James Arthur Lecture on the Evolution of the Human Brain. New York: American Museum of Natural History.

Raia, P., Carotenuto, F., Passaro, F., Fulgione, D., & Fortelius, M. (2012). Ecological specialization in fossil animals explains Cope's rule. *The American Naturalist* **179**: 328–337.

Rainger, R. (1989). What's the use: William King Gregory and the functional morphology of fossil vertebrates. *Journal of the History of Biology* **22**:103–139.

Rak, Y. (1983). *The Australopithecine Face*. New York: Academic Press.

Ramaswamy, S. (2004). *The Lost Land of Lemuria. Forbidden Geographies, Catastrophic Histories*. Berkeley, CA: University of California Press.

Ramdarshan, A., Merceron, G., & Marivaux, L. (2012). Spatial and temporal ecological diversity amongst Eocene primates of France: evidence from teeth. *American Journal of Physical Anthropology* **147**:201–216.

Ramo, M. E., & Huybers, P. (2008). Unlocking the mysteries of the ice ages. *Nature* **451**:284–285.

Rasmussen, D. T. (1990). Primate origins: lessons from a neotropical marsupial. *American Journal of Primatology* **22**:263–277.

Reed, K. (2011). Weighing in on African mammals. *Evolutionary Anthropology* **20**:76–77.

Reed, W. M. (1930). *The Earth for Sam. The Story of Mountains, Rivers, Dinosaurs and Men*. New York: Harcourt, Brace.

Retallack, G. J. (1991). *Miocene Paleosols and Ape Habitats of Pakistan and Kenya*. Oxford: Oxford University Press.

Reznick, D. N., Shaw, F. H., Rodd, F. H., & Shaw, R. G. (1997). Evaluation of the rate of evolution in natural populations of guppies (*Poecilia reticulata*). *Science* **275**: 1934–1937.

Rhesus Macaque Genome Sequencing and Analysis Consortium. (2007). Evolutionary and biomedical insights from the rhesus macaque genome. *Science* **316**:222–234.

Rich, T. H., Hopson, J. A., Musser, A. M., Flannery, T. F., & Vickers-Rich, P. (2005). Independent origins of middle ear bones in monotremes and therians. *Science* **307**: 910–914.

Richard, A. F., Goldstein, S. J., & Dewar, R. E. (1989). Weed macaques: the evolutionary implications of macaque feeding ecology. *International Journal of Primatology* **10**: 569–594.

Ridley, M. (2010). *The Rational Optimist. How Prosperity Evolves*. New York: HarperCollins.

Robbins, M. M., Gray, M., Fawcett, K. A., *et al*. (2011). Extreme conservation leads to recovery of the Virunga mountain gorilla. *PLoS One* **6(6)**:e19788. doi:10.10.1371/journal.pone.0019788.

Robinson, J. T. (1972). *Early Hominid Posture and Locomotion*. Chicago, IL: University of Chicago Press.

Rook, L., Bondioli, L., Köhler, M., Moyà-Solà, S., & Macchiavelli, R. (1999). *Oreopithecus* was a bipedal ape after all: evidence from the iliac cancellous architecture. *Proceedings of the National Academy of Sciences USA* **96**:8795–8799.

Rook, L., Mottura, A., & Gentili, S. (2001). Fossil *Macaca* remains from RDB quarry (Villafranca d'Asti, Italy): new data and overview. *Journal of Human Evolution* **40**: 187–202.

Rosenberger, A. L. (2012a). Hitting on a fossil skull in the subaquatic underworld. *AnthroQuest* **25**:1 & 11.

Rosenberger, A. L. (2012b). New World monkey nightmares: science, art, use, and abuse (?) in platyrrhine taxonomic nomenclature. *American Journal of Primatology* **74**: 692–695.

Rosenberger, A., & Laitman, J. T., eds. (2011). *Evolutionary and Functional Morphology of New World Monkeys*. Special issue. *The Anatomical Record* **294(12)**:1951–2221.

Ross, C. F., & Kirk, E. C. (2007). Evolution of eye size and shape in primates. *Journal of Human Evolution* **52**:294–313.

Rowe, T. B., Marrini, T. E., & Luo, Z.-X. (2011). Fossil evidence on origin of the mammalian brain. *Science* **332**:955–957.

Rudd, J. (1960). *Taboo. A Study of Malagasy Customs and Beliefs*. Oslo: Oslo University Press.

Ruff, C. B. (2002). Long bone articular and diaphyseal structure in Old World monkeys and apes. I. Locomotor effects. *American Journal of Physical Anthropology* **119**:305–342.

Ruff, C. B., Holt, B., & Trinkaus, E. (2006). Who's afraid of the big bad Wolff?: "Wolff's Law" and bone functional adaptation. *American Journal of Physical Anthropology* **129**: 484–498.

Ruse, M. (1999). *Mystery of Mysteries. Is Evolution a Social Construction?* Cambridge, MA: Harvard University Press.

Russo, G. A., & Shapiro, L. J. (2013). Reevaluation of the lumbosacral region of *Oreopithecus bambolii*. *Journal of Human Evolution.* http://dx.doi.org/10.1016/j.jhevol.2013.05.004.

Ryan, T. M., & Shaw, C. N. (2012). Unique sets of trabecular bone features characterize locomotor behavior in human and non-human anthropoid primates. *PLoS ONE* **7(7)**: e41037. doi:10.1371/journal.pone.0041037.

Ryan, T. M., Silcox, M. T., Walker, A., *et al.* (2012). Evolution of locomotion in Anthropoidea: the semicircular canal evidence. *Proceedings of the Royal Society B.* doi:10.1098./rspb.2012.0939.

Sallam, H. M., Seiffert, E. R., Simons, E. L., & Brindley, C. (2010). A large-bodied anomaluroid rodent from the earliest late Eocene of Egypt: phylogenetic and biogeographic implications. *Journal of Vertebrate Paleontology* **30**:1579–1593.

Samuel, M. D., & Garton, E. O. (1985). Home range: a weighted normal estimate and tests of underlying assumptions. *Journal of Wildlife Management* **49**:513–519.

Santana, S. E., Grosse, I. R., & Dumont, E. R. (2012). Dietary hardness, loading behavior, and the evolution of skull form in bats. *Evolution* **66**:2587–2598.

Scally, A., & Durbin, R. (2012). Revising the human mutation rate: implications for understanding human evolution. *Nature Reviews Genetics* **13**:745–753.

Scally, A., Dutheil, J. Y., Hillier, L. W., *et al.* (2012). Insights into hominid evolution from the gorilla genome sequence. *Nature* **483**:169–175.

Scannella, J., & Horner, J. R. (2010). *Torosaurus* Marsh, 1891, is *Triceratops* Marsh. 1889 (Ceratopsidae: Chasmosaurinae): synonymy through ontogeny. *Journal of Vertebrate Paleontology* **30**:1157–1168.

Schaal, S., & Ziegler, W., eds. (1992). *Messel. An Insight into the History of Life and of the Earth.* Oxford: Clarendon Press.

Schillaci, M. A., Froehlich, J. W., Supriatna, J., & Jones-Engel, L. (2005). The effects of hybridization on growth allometry and craniofacial form in Sulawesi macaques. *Journal of Human Evolution* **49**:335–369.

Schluter, D. (2000a). Ecological character displacement in adaptive radiation. *The American Naturalist* **156(supplement)**:S4–S16.

Schluter, D. (2000b). *The Ecology of Adaptive Radiation.* New York: Oxford University Press.

Schubert, B. A., Jahren, A. H., Eberle, J. J., Sternberg, L. S. L., & Eberth, D. A. (2012). A summertime rainy season in the Arctic forests of the Eocene. *Geology* **40**:523–526.

Schultz, A. H. (1926). Studies on the variability of platyrrhine monkeys. *Journal of Mammalogy* **7**:286–304.

Schultz, A. H. (1936a). Characters common to higher primates and characters specific for man. *Quarterly Review of Biology* **11**:259–283.

Schultz, A. H. (1936b). Characters common to higher primates and characters specific for man (continued). *Quarterly Review of Biology* **11**:425–455.

Schultz, A. H. (1950). The physical distinctions of man. *Proceedings of the American Philosophical Society* **94**:428–449.

Schweitzer, M. H., Wittmeyer, J. L., Horner, J. R., & Toporski, J. K. (2005). Soft-tissue vessels and cellular preservation in *Tyrannosaurus rex*. *Science* **307**:1952–1955.

Sclater, W. L., & Sclater, P. L. (1899). *The Geography of Mammals*. London: Kegan Paul, Trench, Trübner.

Scott, J. E., Hogue, A. S., & Ravosa, M. J. (2012). The adaptive significance of mandibular symphyseal fusion in mammals. *Journal of Evolutionary Biology* **25**:661–673.

Scott, J. J., Renaut, R. W., & Owen, R. B. (2010). Taphonomic controls on animal tracks at saline, alkaline Lake Bogoria, Kenya Rift Valley: impact of salt efflorescence and clay mineralogy. *Journal of Sedimentary Research* **80**:639–665.

Scott, W. B. (1937). *A History of Land Mammals of the Western Hemisphere*, rev. edn. New York: Macmillan.

Scott-Ram, N. R. (1990). *Transformed Cladistics, Taxonomy, and Evolution*. Cambridge: Cambridge University Press.

Secord, R., Bloch, J. L., Chester, S. G. B., et al. (2012). Evolution of the earliest horses driven by climate change in the Paleocene-Eocene Thermal Maximum. *Science* **335**:959–962.

Seehausen, O. (2004). Hybridization and adaptive radiation. *Trends in Ecology and Evolution* **19**:198–207.

Seiffert, E. R., Simons, E. L., Boyer, D. M., Perry, J. M. G., Ryan, T. M., & Sallam, H. M. (2010a). A fossil primate of uncertain affinities from the earliest Late Eocene of Egypt. *Proceedings of the National Academy of Sciences USA* **107**:9712–9717.

Seiffert, E. R., Simons, E. L., Fleagle, J. G., & Godinot, M. (2010b). Paleogene anthropoids. In *Cenozoic Mammals of Africa*, Werdelin, L., & Sanders, W. J., eds. Berkeley, CA: University of California Press, pp. 369–391.

Shapiro, M. D., Bell, M. A., & Kingsley, D. M. (2006). Parallel genetic origins of pelvic reduction in vertebrates. *Proceedings of the National Academy of Sciences USA*. doi:10.1073/pnas.0604706103.

Shoval, O., Sheftel, H., Shinar, G., et al. (2012). Evolutionary trade-offs, Pareto optimality, and the geometry of phenotype space. *Science* **336**:1157–1160.

Shuker, K. P. N. (2012). *The Encyclopedia of New and Rediscovered Animals*. Landisville, PA: Coachwhip.

Silcox, M. T. (2008). The biogeographic origins of primates and euprimates: East, West, North, or South of Eden. In *Mammalian Evolutionary Morphology. A Tribute to Frederick S. Szalay*, Sargis, E. J., & Dagosto, M., eds. Dordrecht: Springer, pp. 199–232.

Simons, E. L. (1960). *Apidium* and *Oreopithecus*. *Nature* **186**:824–826.

Simons, E. L. (1961). The phyletic position of *Ramapithecus*. *Postilla*, no. 57. Peabody Museum Yale University.

Simons, E. L. (1967). The earliest apes. *Scientific American* **217(6)**:28–35.

Simons. E. L. (1969). Late Miocene hominid from Fort Ternan, Kenya. *Nature* **221**:448–451.

Simons, E. L. (1977). *Ramapithecus*. *Scientific American* **236(5)**:28–35.

Simons, E. L. (1992). Diversity in the early Tertiary anthropoidean radiation in Africa. *Proceedings of the National Academy of Sciences USA* **89**:10743–10747.

Simons, E. (2008). Eocene and Oligocene mammals of the Fayum, Egypt. In *Elwyn Simons: A Search for Origins*, Fleagle, J. G., & Gilbert, C. C., eds. New York: Springer, pp. 87–105.

Simons, E. L., & Chopra, S. R. K. (1969). *Gigantopithecus* (Pongidae, Hominoidea). A new species from North India. *Postilla*, no. 138. Peabody Museum Yale University.

Simons, E. L., & Kay, R. F. (1983). *Qatrania*, a new basal anthropoid primate from the Fayum, Oligocene of Egypt. *Nature* **304**:624–626.

Simons, E. L., & Pilbeam, D. R. (1965). A preliminary revision of the Dryopithecinae (Pongidae, Anthropoidea). *Folia primatologica* **3**:81–152.

Simons, E. L., Seiffert, E. R., Ryan, T. M., & Attia, Y. (2007). A remarkable female cranium of the early Oligocene anthropoid *Aegyptopithecus zeuxis* (Catarrhini, Propliopithecidae). *Proceedings of the National Academy of Sciences USA* **104**:8731–8736.

Simpson, G. G. (1937). The Fort Union of the Crazy Mountain field, Montana, and its mammalian faunas. *U.S. National Museum Bulletin* **169**:1–287.

Simpson, G. G. (1940a). Mammals and land bridges. *Journal of the Washington Academy of Sciences* **30**:137–163.

Simpson, G. G. (1940b). Studies on earliest primates. *Bulletin of the American Museum of Natural History* **77**:185–212.

Simpson, G. G. (1943). Mammals and the nature of continents. *American Journal of Science* **241**:1–31.

Simpson, G. G. (1944). *Tempo and Mode in Evolution.* New York: Columbia University Press.

Simpson, G. G. (1978). Early mammals in South America: fact, controversy, and mystery. *Proceedings of the American Philosophical Society* **122**:318–328.

Simpson, G. G. (1980). *Splendid Isolation. The Curious History of South American Mammals.* New Haven, CT: Yale University Press.

Simpson, G. G. (1984). *Discoverers of the Lost World. An Account of Some of Those Who Brought Back to Life South American Mammals Long Buried in the Abyss of Time.* New Haven, CT: Yale University Press.

Slack, N. G. (2011). *G. Evelyn Hutchinson and the Invention of Modern Ecology.* New Haven, CT: Yale University Press.

Sluijs, A, Brinkhuis, H., Schouten, S., et al. (2007). Environmental precursors to rapid light carbon injection at the Palaeocene/Eocene boundary. *Nature* **450**:1218–1221.

Smith, A. (2003). Making the best of a patchy fossil record. *Science* **301**:321–322.

Smith, F. A. (2012). Some like it hot. *Science* **335**:924–925.

Smith, T., Rose, K. D., & Gingerich, P. D. (2006). Rapid Asia-Europe-North America geographic dispersal of earliest Eocene primate *Teilhardina* during the Paleocene-Eocene Thermal Maximum. *Proceedings of the National Academy of Sciences USA* **103**:11223–11227.

Smith, T. D., Rossie, J. B., & Bhatnagar, K. P. (2007). Evolution of the nose and nasal skeleton in primates. *Evolutionary Anthropology* **16**:132–146.

Smith, T. D., Deleon, V. B., & Rosenberger, A. L. (2013). At birth, tarsiers lack a postorbital bar or septum. *The Anatomical Record* **296**:365–377.

Sodikoff, G. (2007). Animal taboos and environmental expiations in Madagascar. Paper presented in the Department of Human Ecology, February 7, 2007, Cook College, Rutgers University, New Brunswick, NJ.

Sokal, R. R. (1974). Classification: purposes, principles, progress, prospects. *Science* **185**:1115–1123.

Sokal, R. R. (1983a). A phylogenetic analysis of the caminalcules. I. The data base. *Systematic Zoology* **32**:159–184.

Sokal, R. R. (1983b). A phylogenetic analysis of the caminalcules. II. Estimating the true cladogram. *Systematic Zoology* **32**:185–201.

Sokal, R. R. (1983c). A phylogenetic analysis of the caminalcules. III. Fossils and classification. *Systematic Zoology* **32**:248–258.

Sokal, R. R. (1983d). A phylogenetic analysis of the caminalcules. IV. Congruence and character stability. *Systematic Zoology* 32:259–275.

Sokal, R. R., & Sneath, P. H. A. (1963). *Principles of Numerical Taxonomy.* San Francisco, CA: Freeman.

Soligo, C. (2007). Invading Europe: did climate or geography trigger Early Eocene primate dispersals? *Folia primatologica* 78:297–313.

Sponheimer, M., Passey, B. H., de Ruiter, D. J., Guatelli-Steinberg, D., Cerling, T. E., & Lee-Thorp, J. A. (2006). Isotopic evidence for dietary variability in the early hominin *Paranthropus robustus. Science* 314:980–982.

Sponheimer, M., Alemseged, Z., Cerling, T. E., *et al.* (2013). Isotopic evidence of early hominin diets. *Proceedings of the National Academy of Sciences USA* 110:10513–10518.

Stehli, F. G., & Webb, S. D., eds. (1985). *The Great American Biotic Interchange.* New York: Plenum Press.

Stehlin, H. G. (1909). Remarques sur les faunules de Mammifères des couches éocènes et oligocènes du Bassin de Paris. *Bulletin de la Société Géologique de France* 9:488–520.

Steiper, M. E., & Seiffert, E. R. (2012). Evidence for a convergent slowdown in primate molecular rates and its implications for the timing of early primate evolution. *Proceedings of the National Academy of Sciences USA* 109:6006–6011.

Stevens, G. C. (1989). The latitudinal gradient in geographical range: how so many species coexist in the tropics. *The American Naturalist* 133:240–256.

Stevens, N. J., Seiffert, E. R., O'Connor, P. M., *et al.* (2013). Palaeontological evidence for an Oligocene divergence between Old World monkeys and apes. *Nature* 497:611–614.

Storey, M., Duncan, R. A., & Swisher III, C. C. (2007). Paleocene-Eocene Thermal Maximum and the opening of the Northeast Atlantic. *Science* 316:587–589.

Straus, Jr., W. L. (1963). The classification of *Oreopithecus.* In *Classification and Human Evolution*, Washburn, S. L., ed. Chicago, IL: Aldine Publishing Company, pp. 146–177.

Susman, R. L. (2004). *Oreopithecus bambolii*: an unlikely case of hominid like grip capability in a Miocene ape. *Journal of Human Evolution* 46:105–117.

Sussman, R. W. (1991). Primate origins and the evolution of angiosperms. *American Journal of Primatology* 23:209–223.

Sussman, R. W., & Raven, P. H. (1978). Pollination by lemurs and marsupials: an archaic coevolutionary system. *Science* 200:731–736.

Sussman, R. W., Rasmussen, D. T., & Raven, P. H. (2013). Rethinking primate origins again. *American Journal of Primatology* 75:95–106.

Sutter, N. B., Bustamante, C. D., Chase, K., *et al.* (2007). A single *IGF1* allele is a major determinant of small size in dogs. *Science* 316:112–115.

Sutton, M. D. (2008). Tomographic techniques for the study of exceptionally preserved fossils. *Proceedings of the Royal Society B.* doi:10.1098/rspb.2008.0263.

Szalay, F. S. (1968). The beginnings of primates. *Evolution* 22:19–36.

Szalay, F. S. (1976). Systematics of the Omomyidae (Tarsiiformes, Primates) taxonomy, phylogeny, and adaptations. *Bulletin of the American Museum of Natural History*, vol. 156, Article 3.

Szalay, F. S., & Delson, E. (1979). *Evolutionary History of the Primates.* New York: Academic Press.

Takai, M., Anaya, F., Shigehara, N., & Setoguchi, T. (2000). New fossil materials of the earliest New World monkey, *Branisella boliviana*, and the problem of platyrrhine origins. *American Journal of Physical Anthropology* 111:263–281.

Tang, H. Y., Smith-Caldas, M. S. B., Driscoll, M. V., Salhadar, S., & Shingleton, A. W. (2011). FOXO regulates organ-specific phenotypic plasticity in *Drosophila*. *PLoS Genetics* **7**. doi:10.1371/journal.pgen.1002373.

Tarver, J. E., Donoghue, P. C. J., & Benton, M. J. (2010). Is evolutionary history repeatedly rewritten in light of new fossil discoveries? *Proceedings of the Royal Society B*. doi:10.1098/rspb.2010.0663.

Tattersall, I. (1969a). Ecology of North Indian *Ramapithecus*. *Nature* **221**:451–452.

Tattersall, I. (1969b). More on the ecology of North Indian *Ramapithecus*. *Nature* **224**: 821–822.

Tattersall, I. (1973). Cranial anatomy of the Archaeolemurinae (Lemuroidea, Primates). *Anthropological Papers of the American Museum of Natural History*, vol 52, part 1.

Tattersall, I. (1975). *The Evolutionary Significance of Ramapithecus*. Minneapolis, MN: Burgess.

Tattersall, I. (1986). Species recognition in human paleontology. *Journal of Human Evolution* **15**:165–175.

Tattersall, I. (2008). Vicariance vs. dispersal in the origin of the Malagasy mammal fauna. In *Elwyn Simons: A Search for Origins*, Fleagle, J. G., & Gilbert, C. C., eds. New York: Springer, pp. 397–408.

Tavaré, S., Marshall, C. R., Will, O., Soligo, C., & Martin, R. D. (2002). Using the fossil record to estimate the age of the last common ancestor of extant primates. *Nature* **416**:726–729.

Taylor, L. (2008). Old lemurs: preliminary data on behavior and reproduction from the Duke University Primate Center. In *Elwyn Simons: A Search for Origins*, Fleagle, J. G., & Gilbert, C. C., eds. New York: Springer, pp. 319–334.

Telford, M. J., & Littlewood, D. T. J. (2008). The evolution of the animals: introduction to a Linnean tercentenary celebration. *Philosophical Transactions of the Royal Society B*. doi:10.1098/rstb.2007.2231.

The *Heliconius* Genome Consortium. (2012). Butterfly genome reveals promiscuous exchange of mimicry adaptations among species. *Nature* **487**:94–98.

Thewissen, J. G. M., Cooper, L. N., & Behringer, R. R. (2012). Developmental biology enriches paleontology. *Journal of Vertebrate Paleontology* **32**:1223–1234.

Tilden, C. (2008). Low fetal energy deposition rates in lemurs: another energy conservation strategy. In *Elwyn Simons: A Search for Origins*, Fleagle, J. G., & Gilbert, C. C., eds. New York: Springer, pp. 311–318.

Tobias, J. A., Seddon, N., Spottiswoode, C. N., Pilgrim, J. D., Fishpool, L. D. C., & Collar, N. J. (2010). Quantitative criteria for species delimitation. *Ibis* **152**:724–746.

Tomiya, S. (2009). Differential patterns of taxonomic and morphological succession of carnivorous and non-carnivorous mammals from the Uintan to the Duchesnean in Southern California. Abstracts 9th North American Paleontological Convention, p. 61.

Towal, R. B., Quist, B. W., Gopal, V., Solomon, J. H., & Hartmann, M. J. Z. (2011). The morphology of the rat vibrissal array: a model for quantifying spatio-temporal patterns of whisker-object contact. *PLoS Computational Biology* **7(4)**:e1001120.

Tripati, A. K., Roberts, C. D., & Eagle, R. A. (2009). Coupling of CO_2 and ice sheet stability over major climate transitions of the last 20 million years. *Science* **326**: 1394–1397.

Trivedi, B. (2010). The primate connection. *Nature* **466**:55.

Tucker, A. (2009). Ethiopia's exotic monkeys. *Smithsonian* **40(9)**:72–77.

Turner, A., & Antón, M. (2004). *Evolving Eden. An Illustrated Guide to the Evolution of the African Large-Mammal Fauna.* New York: Columbia University Press.

Tuttle, R. H. (1974). Darwin's apes, dental apes, and the descent of man: normal science in evolutionary anthropology. *Current Anthropology* 15:389-426.

Tuttle, R. H. (1994). Up from electromyography. Primate energetics and the evolution of human bipedalism. In *Integrative Paths to the Past*, R. S. Corruccini, & R. L. Ciochon, eds. Englewood Cliffs, NJ: Prentice-Hall, pp. 269-284.

Tuttle, R. H., Buxhoeveden, D. P., & Cortright, G. W. (1979). Anthropology on the move: progress in experimental studies of nonhuman primate positional behavior. *Yearbook of Physical Anthropology* 22:187-214.

Tyson, P. (2000). *The Eighth Continent: Life, Death, and Discovery in the Lost World of Madagascar.* New York: William Morrow and Company.

Ungar, P. S., & Sponheimer, M. (2011). The diets of early hominins. *Science* 334:190-193.

Utescher, T., Bruch, A. A., Micheels, A., Mosbrugger, V., & Popova, S. (2011). Cenozoic climate gradients in Eurasia—a palaeo-perspective on future climate change? *Palaeogeography, Palaeoclimatology, Palaeoecology* 304:351-358.

Vail, P. R., & Hardenbol, J. (1979). Sea level changes during the Tertiary. *Oceanus* 22:71-79.

Vajda, V., Raine, J. L., & Hollis, C. J. (2001). Indications of global deforestation at the Cretaceous-Tertiary boundary by New Zealand fern spike. *Science* 294:1700-1702.

Van Bocxlaer, B., Van Damme, D., & Feibel, C. S. (2008). Gradual versus punctuated equilibrium evolution in the Turkana Basin molluscs: evolutionary events or biological invasions? *Evolution* 62:511-520.

Van Couvering, J. A., Aubry, M.-P., Berggren, W. A., et al. (2009). What, if anything, is Quaternary? *Episodes* 32(2):1-2.

Van Valen, L. (1965). Treeshrews, primates and fossils. *Evolution* 19:137-151.

Van Valen, L. (1973a). A new evolutionary law. *Evolutionary Theory* 1:1-30.

Van Valen, L. (1973b). Are categories in different phyla comparable? *Taxon* 22:333-373.

Van Valen, L. (1973c). Festschrift. A review of Evolutionary Biology, vol. 6. *Science* 180:488.

Van Valen, L. (1978). The beginning of the Age of Mammals. *Evolutionary Theory* 4:45-80.

Van Valen, L., & Sloan, R. E. (1965). The earliest primates. *Science* 150:743-745.

Venditti, C., Meade, A., & Pagel, M. (2011). Multiple routes to mammalian diversity. *Nature* 479:393-396.

Vizcaíno, S. F., Kay, R. F., & Bargo, M. S., eds. (2012). *Early Miocene Paleobiology in Patagonia: High-Latitude Paleocommunities of the Santa Cruz Formation.* New York: Cambridge University Press.

Vogel, C. (1966). Morphologische Studien am Gesichtsschädel catarrhiner Primaten. *Bibliotheca Primatologica* 4:1-226.

Vogel, C. (1975). Remarks on the reconstruction of the dental arcade of Ramapithecus. In *Paleoanthropology. Morphology and Paleoecology*, Tuttle, R. H., ed. The Hague: Mouton, pp. 87-98.

Vogel, E. R., Knott, C. D., Crowley, B. E., Blakely, M. D., Larsen, M. D., & Dominy, N. J. (2012). Bornean orangutans on the brink of protein bankruptcy. *Biology Letters* 8: 333-336.

von Bitter, P. H., Purnell, M. A., Tetreault, D. K., & Stott, C. A. (2007). Eramosa Lagerstätte—exceptionally preserved soft-bodied biotas with shallow-marine shelly and bioturbating organisms (Silurian, Ontario, Canada). *Geology* 35:879-882.

Vrba, E. S. (1992). Mammals as a key to evolutionary theory. *Journal of Mammalogy* **73**: 1–28.

Vrba, E. S. (1995). On the connections between paleoclimate and evolution. In *Paleoclimate and Evolution with Emphasis on Human Origins*, Vrba, E. S., Denton, G. H., Partridge, T. C., & Burckle, L. H., eds. New Haven, CT: Yale University Press, pp. 24–45.

Wagner, C. E., Harmon, L. J., & Seehausen, O. (2012). Ecological opportunity and sexual selection together predict adaptive radiation. *Nature* **487**:366–369.

Walker, A., & Shipman, P. (2005). *The Ape in the Tree. An Intellectual and Natural History of Proconsul.* Cambridge, MA: Belknap Press.

Walker, P., & Murray, P. (1975). An assessment of masticatory efficiency in a series of anthropoid primates with special reference to the Colobinae and Cercopithecinae. In *Primate Functional Morphology and Evolution*, Tuttle, R. H., ed. The Hague: Mouton Press, pp. 135–150.

Wallace, A. R. (1876). *The Geographical Distribution of Animals*, 2 vols. London: MacMillan.

Ward, C. V. (2013). Early hominin posture and locomotion. In *Early Hominin Paleoecology*, Sponheimer, M., Lee-Thorp, J. A., Reed, K. E., & Ungar, P. S., eds. Boulder, CO: University Press of Colorado, pp. 163–201.

Warren, W. C., Hillier, L. W., Marshall Graves, J. A., *et al.* (2008). Genome analysis of the platypus reveals unique signatures of evolution. *Nature* **453**:175–183.

Webb, S. D., Hulbert, Jr., R. C., & Lambert, W. D. (1995). Climatic implications of large-herbivore distributions in the Miocene of North America. In *Paleoclimate and Evolution with Emphasis on Human Origins*, Vrba, E. S., Denton, G. H., Partridge, T. C., & Burckle, L. H., eds. New Haven, CT: Yale University Press, pp. 91–108.

Weidenreich, F. (1946). *Apes, Giants, and Man.* Chicago, IL: University of Chicago Press.

Wells, N. A. (2003). Some hypotheses on the Mesozoic and Cenozoic paleoenvironmental history of Madagascar. In *The Natural History of Madagascar*, Goodman, S. M., & Benstead, J. P., eds. Chicago, IL: University of Chicago Press, pp. 16–34.

Westgate, J. (2009). Paleoecology of a Gulf Coast Uintan-Age tropical rain forest/mangrove swamp community from Laredo, Texas. Abstracts of the 9th North American Paleontological Convention, p. 55.

Whitfield, J. (2007). We are family. *Nature* **446**:247–249.

Wildman, D. E., Chen, C., Erez, O., Grossman, L. J., Goodman, M., & Romero, R. (2006). Evolution of the mammalian placenta revealed by phylogenetic analysis. *Proceedings of the National Academy of Sciences USA* **103**:3203–3208.

Wilf, P., & Johnson, K. R. (2004). Land plant extinction at the end of the Cretaceous: a quantitative analysis of the North Dakota megafloral record. *Paleobiology* **30**:347–368.

Wilf, P., Labandeira, C. C., Johnson, K. K., & Ellis, B. (2006). Decoupled plant and insect diversity after the end-Cretaceous extinction. *Science* **313**:1112–1115.

Williams, G. C., & Nesse, R. M. (1991). The dawn of Darwinian medicine. *Quarterly Review of Biology* **66**:1–22.

Williams, S. A. (2010). Morphological integration and the evolution of knuckle-walking. *Journal of Human Evolution* **58**:432–440.

Williamson, P. G. (1981). Paleontological documentation of speciation in Cenozoic mollusks from Turkana Basin. *Nature* **293**:437–443.

Wilme, L., Goodman, S. M., & Ganzhorn, J. U. (2006). Biogeographic evolution of Madagascar's microendemic biota. *Science* **312**:1063–1065.

Winker, K. (2010). Is it a species? *Ibis* **152**:679–682.

Witmer, L. M., & Ridgely, R. C. (2009). New insights into the brain, braincase, and ear region of tyrannosaurs (Dinosauria, Theropoda) with implications for sensory organization and behavior. *The Anatomical Record* **292**:1266–1296.

Wolfe, J. A. (1978). A paleobotanical interpretation of Tertiary climates in the Northern Hemisphere. *American Scientist* **66**:694–703.

Wolfe, J. A. (1993). A method of obtaining climate parameters from leaf assemblages. *US Geological Survey Bulletin*, no. 2040.

Wolfe, N. (2011). *The Viral Storm. The Dawn of a New Pandemic Age.* New York: Times Books.

Wolfe, N. D., Dunavan, C. P., & Diamond, J. (2007). Origins of major human infectious diseases. *Nature* **447**:279–283.

Wood, B. A., & Collard, M. (1999). The human genus. *Science* **284**:65–71.

Wood Jones, F. (1916). *Arboreal Man.* New York: Longmans Green.

Wynn, J. G., Alemseged, Z., Bobe, R., et al. (2006). Geological and paleontological context of a Pliocene juvenile hominin at Dikika, Ethiopia. *Nature* **443**:332–336.

Yoder, A., & Yang, Z. (2004). Divergence dates for Malagasy lemurs estimated from multiple gene loci: geological and evolutionary context. *Molecular Ecology* **13**:757–773.

Yoon, C. K. (2009). *Naming Nature. The Clash Between Instinct and Science.* New York: W.W. Norton.

Young, N. M., & Devlin, M. J. (2012). Finding our inner animal: understanding human evolutionary variation via experimental model systems. *American Journal of Physical Anthropology* Supplement **54**:45–46.

Zachos, J. C., Shackleton, N. J., Revenaugh, J. S., Pälike, H., & Flower, B. P. (2001a). Climate response to orbital forcing across the Oligocene-Miocene boundary. *Science* **292**: 274–278.

Zachos, J., Pagani, M., Sloan, L., Thomas, E., & Billups, K. (2001b). Trends, rhythms, and aberrations in global climate 65 Ma to present. *Science* **292**:686–693.

Zachos, J. C., Dickens, G. R., & Zeebe, R. E. (2008). An early Cenozoic perspective on greenhouse warming and carbon-cycle dynamics. *Nature* **451**:279–283.

Zalmout, I. S., Sanders, W. J., MacLatchy, L. M., et al. (2010). New Oligocene primate from Saudi Arabia and the divergence of apes and Old World monkeys. *Nature* **466**:360–364.

Zanazzi, A., Kohn, M. J., MacFadden, B. J., & Terry, Jr., D. O. (2007). Large temperature drop across the Eocene-Oligocene transition in central North America. *Nature* **445**: 639–642.

Zhou, C. F., Wu, S., Martin, T., & Luo, Z.-X. (2013). A Jurassic mammaliaform and the earliest mammalian evolutionary adaptations. *Nature* **500**:163–167.

INDEX

2 standard deviations, 21, 25
2 standard deviations rule, 25

adapid, 177
adapiform, 179
Adapis, 144
adapoid, 149–150
adapoid (adapiform), 142, 144–145, 147, 150–151, 179
adaptation (fitness), 22, 95–96, 98–99, 111, 121, 152–154, 207, 248
adaptive radiation (radiation), 19, 39, 95–96, 98–99, 120–121, 126, 140, 152, 157–159, 164, 178, 187, 189, 191, 194, 197, 207, 214, 216, 237, 248–249, 253, 259
adaptive zone, 95–96, 106, 122, 182, *See* niche
Aegyptopithecus zeuxis, 127, 189–193, 202, 245, 249
Afradapis, 193
Afrasia djijidae, 189
Afropithecus turkanensis, 227
Afrotarsius, 187, 189
Allocebus trichotis, 173
allometric equations, 69
allometry (allometric), 71, 81, 192
allometry, definition of, 68
allopatric speciation, 19, 103, 258
Alouatta, 203–204
Alouatta palliata, 204
alpha taxonomy, 32
Altanius orlovi, 151
AMY1 gene, 102
anagenesis, 94
Anapithecus, 230
anthropoids, convergent evolution in, 185, 202, 211
anthropoids, earliest, 142, 175, 179, 187
anthropoids, origin of, 149–150, 174, 179, 181, 183, 193
Antillothrix bernensis, 201, 212
Aotus, 31, 183, 203, 211
Aotus trivirgatus, 209
Apidium, 127, 190, 202, 234
Apidium phiomense, 190
Archaeoindris, 164–166

Archaeoindris fontoynontii, 166
Archaeolemur, 167–168
archaeolemurids, 152, 167–168
Archicebus achilles, 142
Arctocebus, 153
Arctodontomys, 135
Ardipithecus, 246
Ardipithecus ramidus, 78
Arsinoea, 189
Ateles, 203, 205
Australopithecus afarensis, 44, 240–241
Australopithecus africanus, 41, 77, 85
Australopithecus [Paranthropus] boisei, 78, 167, 241, 252–253
Australopithecus robustus, 77, 85, 252
Australopithecus sediba, 41, 52, 78
Avahi mooreorum, 31

Beached Viking Funeral Ships, 60
Bergmann's rule, 73
beta taxonomy, 32
bilophodont teeth (bilophodonty), 222, 247–248
binomial nomenclature, 16
biogeography, 90
biological species concept, 20–21, 23, 203
Biretia, 187, 189–190, 193
Biretia megalopsis, 190
body size (body weight, body mass), 67–68, 70–74, 76, 81, 87–88, 93, 99, 107, 109–110, 112, 114–115, 117–118, 127, 135, 142–143, 150, 164–165, 177–178, 182, 189, 191, 200, 203, 207–208, 211, 213, 214, 219, 237, 245, 251
body size, diet and. *See* Kay's Threshold
body size, genomics and, 74
bone biology, 82, 89, *See* Wolff's Law
bone strain, 82
Brachyteles, 203, 205
brain evolution, 4, 72, 94, 112, 151, 191
brain size (endocranial volume), 5, 41, 67, 72, 88, 94, 107, 109, 137, 147, 167–168, 191–192, 211, 215
Branisella boliviana, 198–199

Index

Cacajao, 203
Caipora bambuiorum, 213
Callicebus, 203, 212
Callicebus moloch, 209
Callimico, 203
Callimico goeldi, 190
Callithrix, 152, 203–204
Callithrix jacchus, 21, 115, 204, 206
Callithrix penicillata, 21, 204
callitrichids, derived state of, 203, 206
Caluromys, 72, 123
caminalcules, 29–30
Cantius, 135, 145–146
carbon isotopes, 63, 76, 79, 241, *See* stable isotopes
Carpolestes simpsoni, 114, 124, 127
catarrhines, morphological similarity of, 194
Catopithecus browni, 190
Cebuella, 152, 203
Cebus, 88, 201, 203, 207
Cebus apella, 247
Cebus capucinus, 88
cercopithecoids, conservative dental and skeletal morphology of, 248
Cercopithecus, 96, 249
Cheirogaleus, 161, 169, 172, 200
Cheirogaleus medius, 169
Chilecebus carrascoensis, 198, 206–207
Chiromyoides major, 127
Chororapithecus abyssinicus, 228
cladistics, 18, 25, 27–29, 144
Climate Pulse model, 90
coefficient of variation (CV), 94
collagen, 20, 37, 77, 79–80, 98
colobines, sacculated stomachs of, 247
colugos, 10, 103, 116, 123, 140
community structure, 39, 62–64, 76, 91–92, 130, 172, 187, 193, 257, *See* ecosystem structure
comparative genomics, 7, 17, 74, 97, 101, 194, 229, 262
competition, 7, 38, 65, 73, 79, 90–91, 114, 128, 130, 169, 191, 208, 225, 252, 260
Contemporary Evolution, 22–23, 55, 94–95, *See* evolution, rates of
continental drift. *See* plate tectonics
convergent evolution, 10, 17, 28–29, 97–98, 107–108, 110, 121, 123–124, 126, 132–133, 158, 166, 169, 181, 195, 228
coordinated stasis, 65
Cope's rule, 73

criterion of diagnosability, 31–32
crown catarrhines, 121, 191, 232
crown hominoids, postcranial anatomy of, 217, 223
crown species, 1
cryptic species, 31, 163
Cuvier, Georges (Baron Cuvier), 66, 74, 144

Darwinius masillae, 16, 115, 148–150, 181
Daubentonia madagascariensis, 108, 132, 173, 191
Daubentonia robusta, 164
Dead Clade Walking, 57
Deep Time, 10, 106, 111
Dendropithecus macinnesi, 227
Denisova Cave, 17
dental formula, 2, 109, 124, 144, 212
dental morphology, genomics and, 80
Diablomomys dalquesti, 178
diagenesis, 53, 76
diagonal sequence diagonal-couplets gait, 110, 115, 123, *See* primates, locomotion in
diet and dentition, 74, 76–77, 79–80, 112, 123, 250, *See* bilophodont teeth
Digital Automated Identification System (DAISY), 18
digitigrady (digitigrade), 5, 250
disparity, 173, 254, 259
divergence time, molecular data and, 56
DNA, ancient, 10, 17, 20–21, 37, 93, 98, 159, 215
Dolichocebus gaimanensis, 205, 208
Dryomomys szalayi, 115, 126–127
Dryopithecus, 73, 221, 223
Dryopithecus fontani, 220
Dryopithecus laietanus, 223

Ebola, 259, 262
ecological species concept, 20, 22
ecosystem structure, 64, 106, 129, 163, 169, 173, 233, 260–261, *See* community structure
ectodysplasin, 80
Ekgmowechashala philotau, 150, 178, 183
electromyography (EMG), 82–83, 183
Elvis taxa, 57
enamel, 2, 37, 42, 76, 79–80, 85–86, 89, 108, 126, 131, 137, 176, 187, 225, 228, 240–241, 252, *See* stable isotopes
Encyclopedia of Life, 18
endocasts, 41–42, 147, 151, 211
endocranial volume. *See* brain size

Eocene radiation, brain evolution in, 147, 151
Eocene radiation, dental evolution in, 142, 146
Eocene radiation, diet in, 142, 149, 151
Eocene radiation, locomotion in, 142–143, 179
Eocene/Oligocene boundary, 174, 176–177, 193, 197
Eosimias, 117, 142
Erythrocebus patas, 88
euprimates, 10, 103, 105, 110, 114–115, 117, 124, 126–127, 129, 131, 136, 140–143, 147, 259
Europolemur, 149
evolution, rates of, 55, 92, 102, 120–121, 145, 214, 250, 258, *See* Contemporary Evolution
evolutionary development, 80, 98–99
evolutionary taxonomy, 28
extinction, 7, 12, 17–18, 20, 54–55, 57, 59, 61, 65–66, 74, 91, 94, 105, 111, 120, 129, 150, 165, 167, 170, 172–174, 177, 191, 214, 216, 232, 237, 248, 252, 254–255, 257, 259, 262, *See* mass extinction
eye size, temporal patterning of behavior and, 86–88, 142, 179, 181, 198, 209, 211

FAD (First Appearance Datum), 55–57, 91
Fayum, 53, 75, 152, 186–187, 189–191, 193–194, 198, 202, 234
First Appearance Datum. *See* FAD
fossilization, process of, 25, 37–38
FOXO transcription factor, 99
FOXP2 gene, 102
fractal geometry, 81
fruit bats, 10, 106–107

Galago, 109, 144
gamma taxonomy, 32
gene trees, 17, 215
genetic analysis and placental relationships, 10
geoarchaeology, 53
Gigantopithecus, 221, 233
Gigantopithecus blacki, 221, 237–239
Gigantopithecus giganteus, 237
global ice ages, causes of, 61–62, 255
Gorilla, 219, 225
Gorilla beringei, 214
Gorilla gorilla, 88, 214
Gregory, William King, 143–144, 150, 181, 221, 234

Hadropithecus stenognathus, 167–168
Hapalemur aureus, 173

Hapalemur simus, 173
hemochorial placenta, 181
heterodont, 2, 5
Hispanopithecus laietanus, 223–224, 230, 245
home range, 60, 74, 84
hominoids, subtropical forests and, 216, 230
Homo, 47, 84, 102, 181, 215, 218, 225, 237
Homo erectus, 45, 47–48, 100, 237, 239, 252–253
Homo habilis, 48
Homo sapiens, 237
Homo sapiens sapiens, 238
homoplasy, 17, 29–30, 97–98, 150, 181, 185, 196, 217
Homunculus, 208, 210–211
Homunculus patagonicus, 206, 209, 211
Hox genes, 99, 101
HoxPax9 gene, 80
HoxC6 gene, 100
HoxD13 gene, 101
HoxPrx1 gene, 100
Hutchinsonian ratios, 225, 227
Huxley, Thomas Henry, 8–9
hybridization, 19, 21, 23, 96, 215, 250
Hylobates, 70, 219, 258

icehouse earth, 174
ichnofossils. *See* trace fossils
IGF1 gene, 74
Ignacius, 130, 140–141
Ignacius clarkforkensis, 124, 127
index fossils, 12
Indri indri, 173
International Code of Zoological Nomenclature (ICZN), 15–16, 26, 237
International Commission on Stratigraphy (ICS), 12–13
island arcs, 60, 184, 195, 200–202, 211, *See* volcanic islands
isometric (isometry), 81, 107
isometry, definition of, 68
Isthmus of Panama, 61, 195

K/Ar (potassium/argon) dating, 44
Kamoyapithecus, 187, 194, 214
Karanisia arenula, 189, 193
Karanisia clarki, 152
Kay's Threshold, 74, 189, *See* body size, diet and
Kenyapithecus, 223, 229
Kenyapithecus wickeri, 229
Khoratpithecus chiangmuanensis, 228–229

Khoratpithecus piriyai, 228
Komba, 152

La Grande Coupure, 174, 177–178
lactation, 7, 68, 169
LAD (Last Appearance Datum), 55–57, 238
Lagerstätten, 33, 49
Lagonimico conclucatus, 206
Lagothrix, 201, 203, 205
land bridges, 35, 59–60, 138, 155, 258
Last Appearance Datum. *See* LAD
Last Glacial Maximum (LGM), 258
Lazarus taxa, 57
leaf edges (margins) and paleoclimate, 62
Lemur, 109
lemurs, energetic stress in, 169
Leontopithecus, 203
Leontopithecus rosalia, 206
Lepilemur scottorum, 31
Leptadapis magnus, 151
Leuckart's law, 88
Linnaean taxonomy (Linnaean hierarchy), 15–16, 18, 28, 34–35
Linnaeus, Carolus, 15–18
locomotion, kinematics and, 83
locomotion, semicircular canals and, 83, 190, 192–193, 205, 211, 225, 245, 249
Lophocebus, 222
Lufengpithecus, 231, 233
Lufengpithecus lufengensis, 231, 249

Macaca, 186, 199, 218, 222, 249–251, 258
Macaca fuscata, 7, 225, 251
Macaca mulatta, 101, 225
Macaca robusta, 251
macaques as living fossils, 250
macroevolution, xvi, 100
Madagascar, geological history of, 157
Madagascar, human colonization of, 170
Mammalia, Class, 1, 32
mammal-like reptiles, 1, 123
mammals, definition of, 1–5, 6
mandibular symphysis, 3, 142, 183
Mandrillus sphinx, 69, 250
Marmosa murina, 128
mass extinction, 57, 92, 105, *See* extinction
Megaconus, 1
Megaladapidae, Family, 166–167
Megaladapis, 163, 165, 167–168, 189
Megaladapis edwardsi, 163, 192
Mesopithecus, 232, 249–250

Mesopithecus pentelicus, 247, 250
Messel site (Messel fauna), 49, 51, 132, 148, 150
Messinian Salinity Crisis (Messinian Crisis), 255, 257
metabolic rate, 3, 67, 89, 92, 118, 166, 169, 262
Mico, 203
Microcebus, 117, 128, 143–144
microevolution, 55
Micropithecus clarki, 12
middle ear, evolution of, 1, 6, 10, 108
Milanković cycles, 61
Miocene hominoids, postcranial anatomy of, 217, 221, 223, 225, 228–229, 235–236
Mioeuoticus, 152
molecular clock, 93, 97, 110–111, 223, 229
monito del monte, 196–197, 208
monkey-lemurs, adaptations of, 167–169
monotreme genome, 7
morphological integration, 81, 110, 112, 245
morphospace, 95
morphospecies, 20, 39
multituberculates, 7, 117, 126, 130–132, 136, 197

Nakalipithecus nakayamai, 232, 236
natural selection (selection pressure), 14, 28, 55, 64, 76, 81, 93–95, 99, 167, 182–183, 220, 241
Neanderthal, 17, 21–22, 79, 102, 215
Necrolemur, 148
Necrolemur antiquus, 151
niche separation (niche differentiation), 38, 79, 225, 227
niche structure, 10, 225, 227, 254
niche, definition of, 67
niche, major determinants of, 67
nitrogen isotopes, 77, 79, 87, *See* stable isotopes
Noah's Ark, 60, 103, 212
node density effect, 29
Nosmips aenigmaticus, 187
Notharctus, 109, 143, 145, 150
Notharctus osborni, 145
Notharctus tenebrosus, 150
Nsungwepithecus gunnelli, 194, 247–248
numerical taxonomy, 28–29
Nycticeboides simpsoni, 152–153
Nycticebus, 153

olfactory lobes (olfactory bulbs), 5, 41, 88, 107, 147, 191, 211

olfactory receptor genes, 5
omomyid, 142, 144–145, 147–151, 177, 179, 181–182, 184, 198, 201
orbit size, 87, 108, 148, 179, 181, 191, 198, 211, *See* eye size
Oreopithecus, 236
Oreopithecus bambolii, 233–235, 245
Orrorin tugenensis, 223
OTU (operational taxonomic unit), 14
Ouranopithecus, 243
Ourayia uintensis, 178, 182
oxygen isotope ratios, 62, 77, 105, 137, 174, 176, *See* stable isotopes
oxygen isotope stages (OIS), 257

Palaeopropithecidae, Family, 166
Palaeopropithecus, 168
Paleocene/Eocene Thermal Maximum (PETM), 73, 133, 135–136
paleosols, 63, 105
Pan, 219, 225
Pan paniscus, 19
Papio, 167, 220, 250
parallel evolution, 59, 98, 214, *See* convergent evolution
Paralouatta varonai, 201, 212
parapithecids (parapithecoids), 187, 189–191, 202
parapithecids, adaptive zone of, 191
Parapithecus grangeri, 88, 190–191
petrosal bulla, 108, 127
phenetic species concept, 20, 24
phenetic taxonomy, 29, *See* numerical taxonomy
phenotypic plasticity, 81, 99
PhyloCode, 18
phylogenetic species concept, 23–24, 28, 215
phytoliths, 76, 78, 240
Pierolapithecus catalaunicus, 217, 223
Pithecia, 183, 203
Pitx1 gene, 98
plagiaulacoid premolar, 126, 130, 133
plate tectonics, 12, 59, 161, 202
platyrrhine diversity, 202
platyrrhines in the Greater Antilles, 201, 211–212
platyrrhines, competition with ancient marsupials, 197, 208
platyrrhines, exudate eating in, 203, 206
platyrrhines, origin of, 201–202
platyrrhines, prehensile tails in, 203–204

plesiadapiform primates, 10, 103, 105, 114, 122, 131, 150, *See* plesiadapoids
Plesiadapis, 115, 124, 126–127, 129
Plesiadapis cookei, 127
Plesiadapis dubius, 127
Plesiadapis fodinatus, 125
Plesiadapis tricuspidens, 124, 126
plesiadapoids, 117, 123–124, 126–127, 129–133, 147, 259
plesiadapoids, adaptive zone of, 127, 129–132
plesiadapoids, dentition of, 129, *See* plagiaulacoid premolar
Plesiopithecus teras, 191
pliopithecid hominoids, survival of, 248
Pliopithecus, 221
Pongo, 219
Pongo abelii, 214
Pongo pygmaeus, 214, 262
postorbital bar, 108, 124, 142, 181
postorbital septum, 108, 181–182, 211
Presbytis, 249, 258
Presbytis entellus, 250
primate communities, 92, 254
primate conservation, 9–10, 31–32, 129, 172, 260
primate niche structure, 10, 225, 227
primates as disease reservoirs, 261–263
primates, definition of, 9, 106–109
primates, dentition of, 108
primates, generalized traits of, 10, 106
primates, locomotion in, 109–111, *See* diagonal sequence diagonal-couplets gait
Primates, Order, 7, 9–11, 113, 116, 150, 155
primates, origin of, 106, 110–114, 116–117, 120, 123, 152
primates, relative digit length in, 106, 115, 127–128
primates, shrew-like body size of earliest, 117–119
primates, traditional ideas about, 7, 9
primates, visual specializations in, 107–108, 185
Problematica, 17
Proconsul, 53, 73, 194, 221, 223–225
Proconsul africanus. See *Proconsul heseloni*
Proconsul heseloni, 224–225, 245
Proconsul major, 225, 227
Proconsul nyanzae, 225
Progalago, 152
Prohylobates, 221

Propithecus tattersalli, 173
Propliopithecus chirobates, 229
Proteopithecus sylviae, 190, 198
Protopithecus brasiliensis, 213
Pseudoloris godinoti, 178
Pseudoloris parvulus, 151
Pseudotetonius, 146
pull of the recent, 54
punctuated equilibria, 55, 100
Purgatorius, 108, 111, 123

Ramapithecus, 221, 223
Ramapithecus brevirostris, 222
Rangwapithecus gordoni, 227
Rapaport's rule, 65, 92
recognition species concept, 20, 23–24
Red Queen's hypothesis, 91–92
refuge areas (refugia), 20, 178, 258
retino-tectal system, 107
Rhinopithecus, 166
Rhinopithecus roxellanae, 166
Rooneyia viejaensis, 42, 150–151, 183, 201
Rudapithecus hungaricus, 223, 230, 232, 245
Rukwapithecus fleaglei, 194, 214, 227
Rusinga Island, ashfalls at, 50

Saadanius hijazensis, 187, 193–194, 245, 249
Saguinus, 186, 203
Saguinus oedipus, 206
Saharagalago, 152, 193
Saimiri, 142, 203
Sapajus, 207
sea-floor spreading, 59–60, 133, 199
selection pressure. *See* natural selection
sexual dimorphism, 67–68, 73–74, 193, 211, 221, 228
Shoshonius, 142, 148
Shoshonius cooperi, 148, 179
Sima de los Huesos, 20
similum, 21
Simpson Coefficient, 90, 140
SIV (simian immunodeficiency virus), 261
Sivapithecus, 73, 221, 223, 228, 231
Sivapithecus indicus, 228, 235
Sivapithecus parvada, 223, 233
Sivapithecus sivalensis, 221
SLOSS (Single Large Or Several Small), 260
sloth-lemurs, adaptations of, 166
Smilodectes, 145
Smilodectes gracilis, 147

speciation, 19, 22–23, 28, 32, 54–55, 61, 91, 96, 98, 120–121, 157, 216, 249
species thresholds, 30, *See* 2 standard deviations rule
species, definition of, 15, 18–19
SRGAP2 gene, 102, 215
SRGAP2C gene, 102, 215
stable isotopes, 37, 76, 78–79
stem catarrhines (basal catarrhines), 121, 191, 193, 206, 224, 234, 249
stratigraphy, 12
stratophenetics, 28, 181
strontium isotopes, 42, 85–86, 89
subfossil lemurs, body size of, 164
subfossil lemurs, dental development in, 168
subfossil lemurs, niches of, 166, 172
subfossil lemurs, possible survivorship of, 164–166
subfossil lemurs, state of preservation of, 163, 167
sweepstakes dispersal, 60, 161, 174, 184, 198–201
Symphalangus syndactylus, 70
synapsid, 2–3, 6

Talahpithecus, 189
taphonomy (taphonomic processes), 36, 51–53, 57, 73, 76, 127
tarsiers, adaptations of, 179–181
tarsiers, antiquity of, 152, 179
tarsiers, phyletic position of, 179–180
tarsiers, postorbital septum of, 181
tarsiiforms, Eocene, 144, *See* omomyid
Tarsius, 108, 144, 151, 179–180
Tarsius syricta, 182
taxonomic inflation, 27, 31, 162, 203, 221, 223
taxonomy, definition of, 14
Teilhardina, 142
Teilhardina asiatica, 87, 142–143
Tetonius, 142, 146, 148
Tetonius homunculus, 147–148
therian, 3–4, 7, 111, 123, 130, 197
Theropithecus, 167, 189, 250–252
Theropithecus brumpti, 71, 192, 252
Theropithecus darti, 252
Theropithecus gelada, 71, 259
Theropithecus oswaldi, 252–253
time-averaging, 94, 187, 258
tooth-scraper (tooth-comb, dental comb), 108–109, 150, 152, 167

torpor (hibernation), 60, 118, 140, 169, 200, 208
trace fossils (ichnofossils), 36, 43, 49
trackways, 36, 43–45, 57
transformed cladistics, 29
tree-shrews, 10, 103, 108
Tremacebus harringtoni, 206, 208, 211–212
tribosphenic molar, 3
Triceratops, 26
triconodont, 7
Turkanapithecus kalakolensis, 226–227
type specimen, 16, 18, 25, 48, 142, 148–149, 151, 194, 211–212, 214, 226, 237

Van Valen, Leigh, 91, 98
variability, relation of taxonomy to, 17, 21, 25, 94

vibrissae, 1, 4–5, 211
Victoriapithecus macinnesi, 194, 245, 247–249
volcanic islands, 59–60, 202, *See* island arcs
vomeronasal organ, 107

Wadilemur, 152
weaning, 79
weed macaques, 251
Wolff's Law, 82

Xanthorhysis, 179
Xenothrix mcgregori, 201, 212

Y-5 pattern (dryopithecine pattern), 221–222, 234
Yuanmoupithecus, 229